Yucheng Wu
Tungsten in Nuclear Fusion Applications

Also of interest

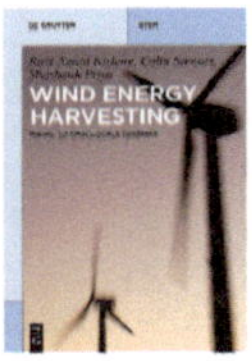

Wind Energy Harvesting.
Micro-to-Small Scale Turbines
Ravi Kishore, Shashank Priya, Colin Stewart, 2023
ISBN 978-1-61451-565-4
e-ISBN (PDF) 978-1-61451-417-6

Chemical Energy Storage
Robert Schlögl (Ed.), 2022
ISBN 978-3-11-060843-4
e-ISBN (PDF) 978-3-11-060845-8

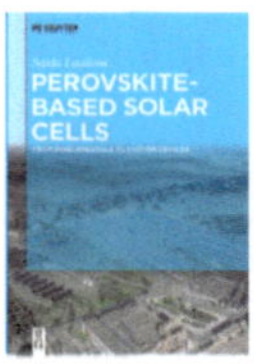

Perovskite-Based Solar Cells.
From Fundamentals to Tandem Devices
Saida Laalioui, 2022
ISBN 978-3-11-076060-6
e-ISBN (PDF) 978-3-11-076061-3

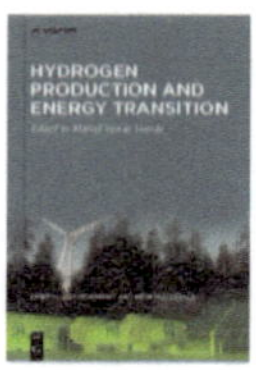

Hydrogen Production and Energy Transition
Volume 1
Marcel Van de Voorde (Ed.), 2021
ISBN 978-3-11-059622-9
e-ISBN (PDF) 978-3-11-059625-0

Hydrogen Storage for Sustainability
Volume 2
Marcel Van de Voorde (Ed.), 2021
ISBN 978-3-11-059623-6
e-ISBN (PDF) 978-3-11-059628-1

Yucheng Wu

Tungsten in Nuclear Fusion Applications

Plasma-Facing Materials

DE GRUYTER

Author
Yucheng Wu (吴玉程)
School of Materials Science and Engineering
Hefei University of Technology
230009 Hefei, Republic of China
ycwu@hfut.edu.cn

Translation Team
Translator: Tang Jun (唐军)
Other translators:
Fang Yuanyuan (方媛媛)   Li Kangxi (李康熙)

Edit Team
Editor: Yao Shuo (姚硕)
Other editors:
Li Fanyu (李芳宇)   Song Yuan (宋 愿)
Yang Xinqi (杨昕琦)   Zhao Qianqian (赵倩倩)

ISBN 978-3-11-155689-5
ISBN 978-3-11-155698-7 (PDF)
ISBN 978-3-11-155715-1 (EPUB)
DOI https://doi.org/10.1515/9783111556987

Library of Congress Control Number: 2025943777

Bibliographic information published by the Deutsche Nationalbibliothek
The Deutsche Nationalbibliothek lists this publication in the Deutsche Nationalbibliografie;
detailed bibliographic data are available on the internet at http://dnb.dnb.de.

De Gruyter and Walter de Gruyter GmbH are part of De Gruyter Brill.
www.degruyterbrill.com

Questions about General Product Safety Regulation:
productsafety@degruyterbrill.com

Cover illustration: Kiryl Pro motion/iStock/Getty Images Plus
Printing: CPI books GmbH, Leck

MIX
Papier | Fördert gute Waldnutzung
FSC® C083411

Preface

Human beings are now undergoing a gradual transition from fossil fuels to renewable energy, aiming to achieve low emission or zero emission through the generation and utilization of clean, efficient, and green energy. Overexploitation and utilization of energy have brought about a series of environmental problems, thus countries around the world have put forward plans and timelines for carbon peaks and carbon neutrality. The energy produced by nuclear fusion reaction is the ultimate and ideal energy that is clean, safe, and inexhaustible. The fuel deuterium is so abundant in seawater to be used for tens of billions of years, and the development and utilization of nuclear fusion energy is a challenge for human beings. In 1986, the International Thermonuclear Experiment Reactor (ITER) was proposed to be jointly constructed, and China officially joined the ITER program in 2003. The technical goal of ITER is to verify the feasibility of fusion energy and to provide engineering and physical support for future demonstration power plants (DEMO), and plasma-facing materials (PFMs) turn out to be both vital and challenging in the development and engineering.

The author describes the evolution of fusion energy with introduction to the progress and goals of the ITER program and illustration of the important role and service conditions of PFMs. Meanwhile, the composition and organization of tungsten matrix materials and their performance characteristics are expounded, and processing methods and the toughening path of brittle materials are discussed. Furthermore, the performance and damage mechanism of tungsten matrix materials to resist thermal load and irradiation damage are examined, and refinement for divertor material selection, module manufacturing, and thermal load testing are put forward.

The author's research has been funded by the National Key Research and Development Program of China, the National Natural Science Foundation of China, and the Key International Cooperation Project of the National Natural Science Foundation of China. And the author would like to express sincere gratitude for the constant help and support from Professor Li Jiangang (academician of China), and researchers Chen Junling and Luo Guangnan from the Institute of Plasma Physics, Chinese Academy of Sciences. Professor Luo Laima, Professor Liu Jiaqin, Professor Zan Xiang, Associate Researcher Zhu Xiaoyong, and Doctor Ding Xiaoyu, and Tan Xiaoyue from Hefei University of Technology, as key members of the author's research team, have also made significant contributions to this work. Thanks also go to Professor Tang Jun, Associate Professor Li Kangxi, and Fang Yuanyuan for their translation work.

Fusion energy, as a representative of future clean energy, is of a promising development prospect. The author will further his study on PFMs for fusion reactions, thus promoting the development of fusion energy research and application.

Wu Yucheng
Hefei, China
2025-11-06

 | https://doi.org/10.1515/9783111556987-202

Contents

Preface —— V

Author's biography —— XI

Contributors to this work —— XIII

Chapter 1
Fusion energy and plasma-facing materials —— 1

1.1 Introduction —— **1**
1.2 Fundamentals of fusion —— **1**
1.3 Types of nuclear fusion —— **3**
1.4 Development of fusion energy —— **4**
1.4.1 Development of international fusion reactor device and fusion energy research —— **8**
1.4.2 Development of fusion energy in China —— **9**
1.5 Plasma-facing materials (PFMs) —— **10**
1.5.1 Plasma-material interactions —— **10**
1.5.2 Damage forms of PFM materials —— **11**
1.5.3 Selection of PFMs —— **13**
1.6 Summary —— **21**

Chapter 2
Tungsten-based plasma-facing materials —— 24

2.1 Introduction —— **24**
2.2 Tungsten-based materials for fusion —— **25**
2.2.1 Application of tungsten in International fusion devices —— **25**
2.2.2 Application of tungsten in China's EAST —— **25**
2.2.3 Applications and challenges of tungsten materials for future fusion reactors —— **28**
2.3 Novel tungsten-matrix materials —— **30**
2.3.1 Alloyed tungsten —— **30**
2.3.2 Oxide dispersion-strengthened tungsten (ODS-W) —— **53**
2.3.3 Tungsten-carbide composites (CDS-W) —— **66**
2.3.4 Ultrafine-grained/nanocrystalline tungsten —— **84**
2.4 Summary —— **87**

Chapter 3
Preparation and processing of tungsten matrix materials for fusion reactors — 97

3.1 Introduction — **97**
3.2 Preparation of tungsten composite powders with different nanostructures — **97**
3.2.1 Top-down technology — **99**
3.2.2 Bottom-up technology — **100**
3.3 Sintering densification of tungsten alloy — **103**
3.3.1 Conventional sintering — **104**
3.3.2 Special sintering technologies — **105**
3.3.3 Powder activation pretreatment — **108**
3.4 Forming and processing of tungsten materials — **110**
3.4.1 Rolling processing — **111**
3.4.2 Rotary forging — **113**
3.4.3 Hydrostatic extrusion — **113**
3.4.4 Equal-channel angular pressing — **114**
3.4.5 High-pressure torsion — **115**
3.4.6 Multidirectional forging — **115**
3.5 Summary — **117**

Chapter 4
Strengthening and toughening of tungsten-matrix materials for fusion reactors — 122

4.1 Introduction — **122**
4.2 Brittleness of tungsten — **122**
4.2.1 Low-temperature brittleness — **122**
4.2.2 Recrystallization brittleness — **123**
4.2.3 Neutron irradiation effects — **124**
4.3 The intrinsic cause of tungsten brittleness — **125**
4.3.1 Chemical bonds — **125**
4.3.2 Dislocation mobility — **125**
4.3.3 Internal defects in actual materials — **126**
4.3.4 Grain boundary characteristics — **127**
4.3.5 Cleavage fracture — **128**
4.4 Ways to solve tungsten's brittleness — **128**
4.4.1 Addition of alloying elements — **128**
4.4.2 Second phase doping — **129**
4.4.3 Severe plastic deformation — **130**
4.5 Summary — **131**

Chapter 5
Radiation damage of tungsten-matrix materials in fusion reactors — 134
5.1 Introduction — **134**
5.2 Irradiation damage behavior of tungsten materials — **135**
5.2.1 Damage and retention of tungsten materials irradiated with hydrogen isotope — **135**
5.2.2 Damage behavior of tungsten materials under helium ion irradiation — **144**
5.2.3 Damage behavior of tungsten materials under neutron irradiation — **151**
5.2.4 Damage behavior of tungsten materials under heavy ion irradiation — **157**
5.3 Summary — **165**

Chapter 6
Transient thermal shock behavior and damage of tungsten-matrix materials for fusion reactors — 171
6.1 Introduction — **171**
6.2 Transient thermal load experiments on tungsten and tungsten alloys — **172**
6.2.1 Transient thermal load experimental simulation device — **172**
6.2.2 Transient thermal load damage behavior of commercial tungsten — **174**
6.2.3 Transient thermal load damage behavior of ultrafine-grained tungsten — **190**
6.2.4 Transient thermal load damage behavior of tungsten alloy — **192**
6.2.5 Transient thermal load damage behavior of dispersion-strengthened tungsten composites — **198**
6.3 Summary — **204**

Chapter 7
Manufacture and evaluation of tungsten-copper modules for fusion reactor divertor — 209
7.1 Introduction — **209**
7.2 Divertor for fusion reactor — **209**
7.2.1 Design of the divertor — **209**
7.2.2 Selection of materials for divertor components — **211**
7.2.3 Connection between the tungsten divertor and heat sink — **215**
7.2.4 ITER full-tungsten divertor certification test — **221**
7.3 High heat load testing of mono-block tungsten-copper module at home and abroad — **222**

7.3.1 Japan W/Cu-mono-block test evaluation — **222**
7.3.2 Evaluation of W/Cu-mono-block test in Europe — **226**
7.3.3 China W/Cu-mono-block test evaluation — **230**
7.4 Summary — **237**

Index — 243

Author's biography

Prof. Yucheng Wu now is a distinguished professor of materials science and engineering at Hefei University of Technology (HFUT). He received his PhD in condensed physics in 2000 from the Institute of Solid State Physics, Chinese Academy of Sciences, and worked as senior visiting scholar in the Institute of Theoretical and Applied Physics at the University of Stuttgart in 2001–2002. Currently, he is also the honorary professor of the University of St Andrews since 2013, guest professor of RMIT University since 2012, and visiting professor of the York University (England), Hanyang University (Korea), and Nanyang University of Technology, as well as guest professor of the Institute of Solid State Physics, Chinese Academy of Sciences since 2002. His research interests mainly include the fusion materials, plasma-facing tungsten matrix composites, energy-related materials, functional nanomaterials, and their processing technology. In recent years, he presided over more than 50 scientific research projects, including National Major Basic Research ITER Project (China), International Cooperation Project of Ministry of Science and Technology of China, National Natural Science Foundation of China, and Major Science and Technology Research Projects of the Ministry of Education of China. He has published more than 380 peer-reviewed scientific papers in international journals such as *Nature Communications*, *Science Advances*, *Advanced Materials*, *Advanced Functional Materials*, *ACS Nano*, *Nuclear Fusion*, and *Journal of Nuclear Materials* with total citation of over 15,000, and he has also filed over 60 China invention patent applications. He has been ranked as "2% top scientists in the world" released by Stanford University for three times and received "Zhou Zhihong Scientific Achievements Award" granted by China Society of Materials Heat Treatment, and became a fellow of Chinese Society of Micro-Nanotechnology.

He worked as the vice president (2008–2017) and executive vice president at Hefei University of Technology (HUT) (2020–2023), and the president of Taiyuan University of Technology (TYUT) (2017–2020). Currently, he serves as the president of Hefei Institute of Technology (HFIT) since 2024. As an eminent researcher, he takes charge of National-Local Joint Engineering Research Center of Nonferrous Metals and Processing Technology since 2010, China International S&T Cooperation Base for Advanced Energy and Environmental Materials since 2017, Base of Introducing Talents of Discipline to Universities for Advanced Clean Energy Materials and Technology (111 Project, 2018–present). He works with some academic organizations, such as the vice chairman of Chinese Heat Treatment Society since 2016 and Functional Materials Science Society since 2000.

 | https://doi.org/10.1515/9783111556987-204

Contributors to this work

Prof. Luo Naima

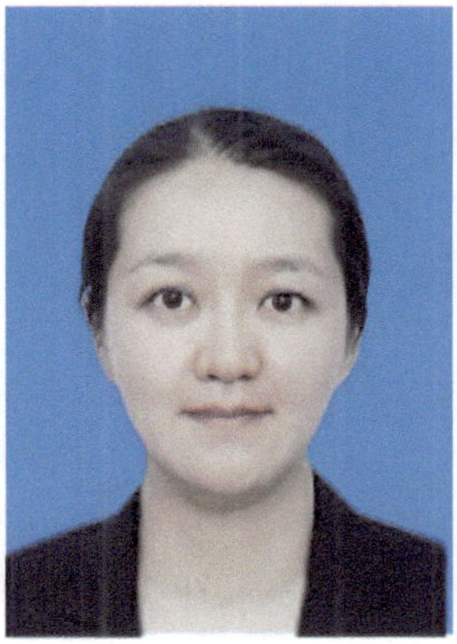

Prof. Liu Jiaqin

Prof. Zan Xiang

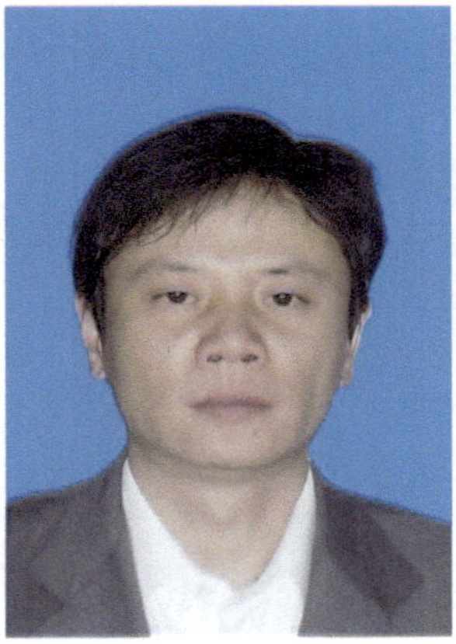

A.P. Zhu Xiaoyong

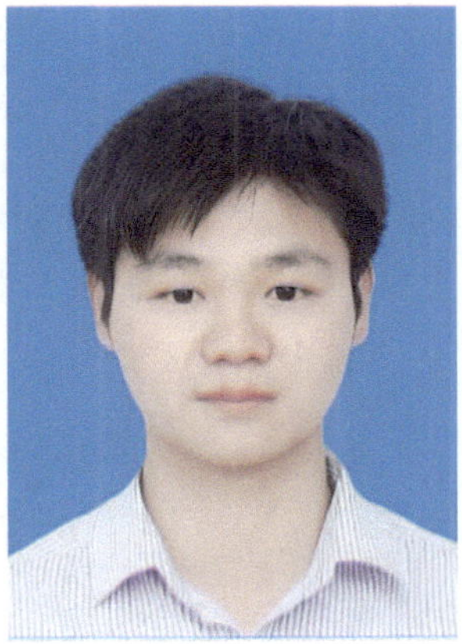

A.P. Tan Xiaoyue

A.P. Ding Xiaoyu

 | https://doi.org/10.1515/9783111556987-205

Prof. Tang Jun

A.P. Fang Yuanyuan

A.P. Li Kangxi

Chapter 1
Fusion energy and plasma-facing materials

1.1 Introduction

The fusion energy, generated by the fusion reaction of light atomic nuclei, is regarded as an ideal ultimate clean energy source and a significant potential solution to human energy challenges. It brings almost no environmental problems, such as radioactive pollution and emissions, and is safe, clean, and inexhaustible. There are approximately 4.6×10^{21} L of seawater on the Earth, which contains a large amount of deuterium, the fuel of the fusion reaction. It is estimated that the deuterium contained in 1 L of seawater can generate about 1.1×10^{10} J of energy during the fusion reaction, and the total energy of the deuterium can reach 5×10^{31} J. The stock of deuterium is extremely rich and enables to sustain human usage for tens of billions of years.

Under human control, the nuclei of deuterium, an isotope of hydrogen, fuse at temperatures of hundreds of millions of degrees Celsius, resulting in thermonuclear reactions that release enormous amounts of energy. The energy produced by nuclear fusion is higher than that of nuclear fission, and its generation can be controlled and programmed according to human needs. Therefore, fusion energy can serve as a source, output, and application of clean energy.

In recent years, the fusion reactor technology has made great progress, but there are still many complex technical and engineering challenges that remain to be addressed. Currently, the technology is still in the stage of engineering feasibility demonstration and validation. A fusion device contains tens of thousands of components, which must withstand high thermal loads and high-energy irradiation. Hence, the selection and application of plasma-facing materials (PFMs) are a key issue to be addressed.

1.2 Fundamentals of fusion

Fusion reactions, also known as thermonuclear reactions, need to be carried out at high temperatures. To achieve a fusion reaction, the generated plasma must reach sufficiently high temperatures, and the higher the heating temperature, the greater the probability of the fusion reaction occurring. For fusion energy to be useful, the energy released must be greater than the energy needed to sustain the fusion reaction, satisfying the conditions of the Lawson criterion (plasma temperature × density × energy confinement time). From the point of view of obtaining energy, the following fusion reactions are particularly important:

 | https://doi.org/10.1515/9783111556987-001

$$D + D \rightarrow {}^{3}He\ (0.8\ MeV) + n\ (2.5\ MeV) \tag{1.1}$$

$$D + D \rightarrow {}^{3}T\ (1.0\ MeV) + p\ (3.0\ MeV) \tag{1.2}$$

$$D + T \rightarrow {}^{4}He\ (3.5\ MeV) + n\ (14.1\ MeV) \tag{1.3}$$

$$D + {}^{3}He \rightarrow {}^{4}He\ (3.5\ MeV) + n\ (14.1\ MeV) \tag{1.4}$$

Among them, D, T, He, p, and n represent deuterium, tritium, helium, protons, and neutrons, respectively. Among these reactions, reaction (1.3) has a faster reaction rate and requires lower conditions than other reactions. Relatively speaking, it is easier to achieve and emits more energy. Compared with the D-D reaction, the temperature of the D-T reaction can be reduced by one order of magnitude. For fusion reactions to occur, the plasma temperature must reach 2×10^{8} °C, the particle density must reach 10^{20} m^{-3}, and the energy confinement time must be more than 1 s.

Under laboratory conditions, the deuterium-tritium reaction is shown in Figure 1.1. Due to the very small reaction cross section between protons, the deuterium-tritium reaction is the easiest to achieve energy output. Firstly, it is necessary to obtain a high temperature of approximately 100,000 °C, heat the deuterium-tritium mixture as fuel to the plasma state of a completely ionized substance (composed of electrons and ions), and then allow the two nuclei to come together despite strong electrostatic repulsion. When they overcome the repulsive force and approach each other to approximately three trillionths of a millimeter, the nuclear force pulls the two together, turning into a new nucleus and releasing a huge amount of energy. The process conditions of the fusion reaction are related to temperature. The nucleus must run at a very fast speed, breaking through the repulsion between the positively charged nuclei. To ensure that the nucleus achieves this operating state, high-temperature conditions are needed until the temperature reaches hundreds of millions of degrees Celsius, so that the nuclei of deuterium and tritium will collide at great speed to produce a deuterium-tritium nuclear fusion reaction.

According to the principle of fusion reaction, an annular vacuum chamber is designed where gaseous hydrogen fuel (D + T) is transformed into a high-temperature plasma through external auxiliary heating. In this state, hydrogen atoms collide and fuse, releasing enormous energy manifested as the kinetic energy of helium (He) particles and neutrons. After a period of operation, the internal reactor can sustain itself without the need for additional external heating. The temperature generated by nuclear fusion is enough for the nucleus to continue to fuse. A small portion of the generated energy remains within the reactor to support the chain reaction, while the larger portion of the energy could be extracted through a heat exchange system to the outside of the reactor. This allows the output energy to exceed the input energy, enabling integration into the power grid.

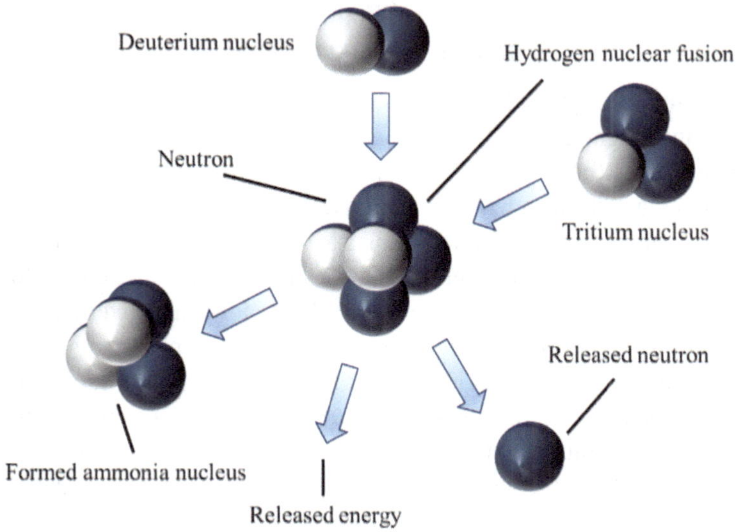

Figure 1.1: Schematic diagram of the deuterium-tritium fusion reaction.

1.3 Types of nuclear fusion

There are currently two main types of controlled nuclear fusion: inertial confinement using high-energy lasers, and magnetic confinement using the tokamak device:

(1) Inertial confinement fusion [1] (ICF)
Using the inertia of matter to achieve confinement is called inertial confinement, which can be realized through experiments where a high-power energy pulse is delivered in an extremely short time to a small number of fusion fuel pellets. These pellets are typically hollow spheres about 1 mm in diameter, filled with a few milligrams of a deuterium and tritium gas or solid mixture. The rapid energy input causes the pellet to be intensely heated, turning the fuel into high-temperature plasma as particles are propelled at high speeds in all directions. Since the inertial flight of a certain range of particles must take a certain amount of time, and because the heating time is very short, the thermonuclear fusion reaction is completed before these particles fly out of the reaction area. Therefore, the key to achieving fusion in this manner is the need for highly directional energy beams and extremely high power.

A few milligrams of deuterium and tritium are packed into a small sphere with a diameter of about a few millimeters, and then a high-energy laser beam or particle beam is uniformly directed at the sphere from the outside, which causes the deuterium-tritium gas inside the capsule to be compressed inward, increasing the pressure and consequently raising the temperature sharply. When the temperature reaches the level required for the nuclear reaction, the gas in the sphere undergoes a nuclear fusion reaction. This process is known as laser inertial confinement.

(2) Magnetic confinement fusion [2] (MCF)
In a closed and encircling magnetic field line space formed by a strong magnetic field, magnetic confinement forces the plasma to run along the magnetic field lines, and various heating methods are used to heat the deuterium-tritium plasma to a level where fusion reactions would occur, resulting in nuclear fusion reactions in a quasi-steady state or even a steady state. Tokamak, which uses a strong magnetic field to confine high-temperature plasma, is the most promising device for controlled thermonuclear fusion reactions. Since the 1980s, international research has been mainly focused on tokamak devices, such as the world's largest fusion reactor, Joint European Torus (JET) [1], the Tokamak Fusion Test Reactor (TFTR), and the DIII-D National Fusion Facility (DIII-D) [2] in the United States, and Japan's Jaeri Tokamak 60 (JT-60) [3]. China's controllable nuclear fusion devices started late, such as the China Circulator No. 2A (HL-2A) [4] and the world's first fully superconducting tokamak device, EAST (Experimental Advanced Superconducting Tokamak) [5]. However, after more than half a century of arduous exploration, remarkable progress has also been made.

1.4 Development of fusion energy

In 1986, at the initiative of the United States and the Soviet Union, under the framework of the International Atomic Energy Agency (IAEA), the United States, Europe, Japan, and the Soviet Union initiated the joint construction of the International Thermonuclear Experimental Reactor (ITER). The goal is to combine the poloidal magnetic field generated by the strong current with the toroidal magnetic field to confine the high-temperature plasma, so that the plasma can be separated from the wall of the vacuum chamber, finally realizing the magnetic confinement of the high-temperature plasma [6]. The goal is to imitate the process of energy production by the Sun, polymerize hydrogen isotopes into helium, generate hundreds of millions of degrees of heat like the Sun, and continuously release an inexhaustible source of thermonuclear fusion energy, which is why it is also vividly called the "artificial sun."

In the early days of the ITER program, there were many setbacks and adjustments. From 1992 to 1998, it was determined to be the ITER Engineering Design Agency (EDA) and included some technical pre-research stages, with an estimated construction investment of nearly USD 10 billion. Due to the large budget investment estimate, the United States announced its withdrawal from ITER in July 1998. In 1998, after receiving the engineering design report, the ITER participants began to consider revising the original plan, and the joint working group of Europe, Japan, and Russia completed the new EDA, main components, and technology research and development of the ITER-Fusion Energy Advanced Tokamak (ITER-FEAT) after 3 years of hard work. The aim was to reduce construction costs as much as possible while meeting the main objectives [7]. In 2003, the United States announced its return to the ITER program, and in February of the same year, China was officially approved as a participating country, in June, South Korea joined the program, and in December 2005, India also joined the ITER organization. The increase in ITER participants shows the willingness of the international community to actively cooperate in the development of fusion energy. On May 24, 2006, China, the European Union, the United States, South Korea, Japan, Russia, and India completed the negotiation on all planned legal documents, and in Brussels, Belgium, the parties initialed the "Agreement on the Establishment of the ITER International Fusion Energy Organization for the Joint Implementation of the ITER Project" and the "Agreement on the Privileges and Immunities of the ITER International Fusion Energy Organization for the Joint Implementation of the ITER Project," marking a new stage in the implementation of the ITER program [8–10].

In terms of participants and investment scale, the ITER program is another major project involving multilateral international scientific and technological cooperation after the International Space Station, indicating that thermonuclear fusion technology has entered the stage of engineering feasibility validation from the basic research stage [11–13]. China not only dispatched a large team of management and technical personnel to the ITER headquarters in France, but also undertook many important engineering projects and successfully completed several procurement packages on time, such as the PF conductor package, which features a complex manufacturing process and unprecedented technologies and equipment challenges, involving welding processes, non-destructive testing techniques, conductor shaping, and winding techniques, among others. This highlights China's first-class quality assurance and makes a notable contribution to the successful construction and engineering preparation of the ITER facility.

Figure 1.2 shows the schematic diagram of the ITER structural device [14]. After the completion of the engineering design, construction began in Cadarache, in the south of France, with the objectives of the proposed installation: based on the nuclear reactor level, capable of having the integrated technology to verify the main engine of the installation; inspecting the stable operational ability of the device; and ensuring that the output power obtained from the fusion reaction is at least 10 times the input power ($Q \geq 10$, where the energy gain factor is the fusion power gain factor), so as to provide a technology and device demonstration for the reliable operation of 500,000 kW of fusion reaction power. JET has performed relatively well so far, producing 70% of its input in 1997, and no fusion reactor has yet been able to produce more energy than it consumes. On August 31, 2020, the lifting of the lower barrel of the Dewar of the ITER tokamak device was successfully completed, which was the first major component installation since the launching ceremony of the ITER program.

The technical goal of ITER is to verify the feasibility of nuclear fusion to generate energy, as well as to provide the physical research and engineering foundation support for future fusion demonstration power plants (DEMO). Table 1.1 lists the material selection of the vacuum chamber and its components in the ITER device [15], with different functional requirements. For steady-state fusion reactors, the service requirements are that the neutron wall thermal load is generally 3–5 MW/m^2; the load factor is less than 80%; the average displacement per atom (DPA) is more than 200; and the operation time is up to 30 years. In view of such a service environment, the blanket scheme to choose, including first wall materials, multiplied neutrons, degraded neutrons, and coolants, must be considered comprehensively. The upgrading of traditional materials, the development of new materials, and breakthroughs in manufacturing technology must be verified, ranging from basic research to engineering applications. It will take a long time to truly realize the application of fusion energy.

DEMO is a fusion reactor model used to demonstrate future fusion energy power generation, and is the only way to realize the commercial application of fusion energy, laying a scientific, theoretical, and engineering application foundation for the construction and operation of fusion reactor power stations [16]. Compared to the conditions of the ITER device, the DEMO device will have stronger neutron irradiation, and the materials applied will need to adapt to the requirements of low neutron activation. In addition, a cooling system with higher thermal cycling efficiency must be developed [17]. Table 1.2 lists a comparison between the operating conditions of the ITER experimental reactor and the DEMO demonstration reactor [18].

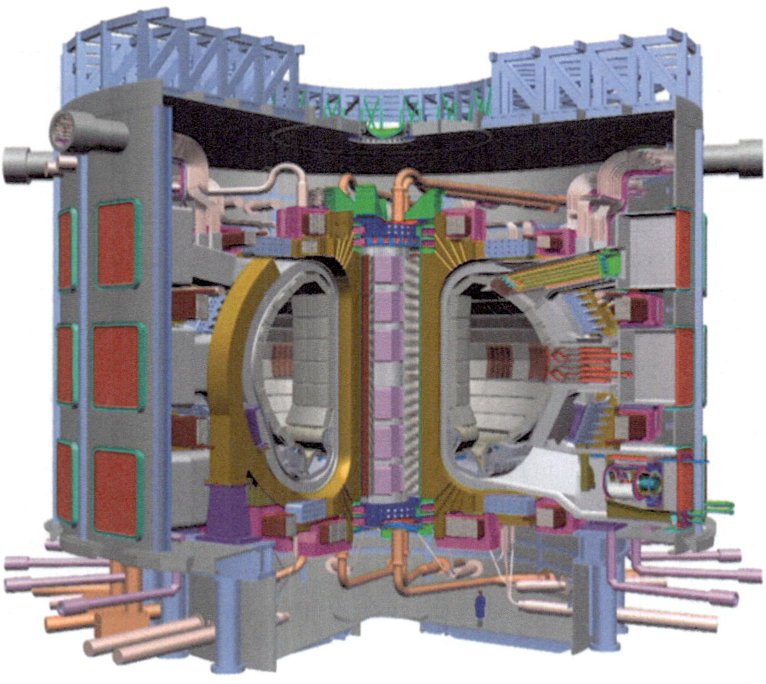

Figure 1.2: Schematic diagram of the ITER system [14].

Table 1.1: Material selection for ITER's vacuum chamber and its components [15].

Material	Brand	Component
Beryllium	S-65C VHP(DShG-200)	Armor tiles for the first wall and limiter
Carbon fiber composites	SEP NB 31, NIC 01 (CX2002U, SEP NS31)	Armor tiles for divertor target plate
Tungsten	Pure sintered W	Armor tiles for divertor
	316L(N)-IG	Shielding module, vacuum chamber, blanket, cooling pipe, and material behind the divertor body; thin tube wall of the first wall, and cooling tube in the vacuum chamber
Austenitic steel	AISI 660	Fixing parts for window welding (fixing wedges and bolts)
	SS30467	Boride steel is used in the walls of shielding plates
Carbon steel	SS 430	Magnetic steel is embedded in a hidden place within a longitudinal field coil to reduce longitudinal field waviness
Copper and copper alloys	CuCrZr-IG	Heat-absorbing materials in plasma-facing materials components
	CuAl25-IG	Heat-absorbing materials in plasma-facing materials components
Niobium alloys	Inconel 718	Bolt material in movable braced, blanket cooling water pipeline
Titanium alloy	Ti-6Al-4V	Movable supporting material to support the blanket module
Ceramics	Al_2O_3, $MgAl_2O_4$	Electrical insulating material in the support blanket module

Table 1.2: Comparison of the operating conditions of the ITER experimental reactor and the DEMO reactor [18].

ITER	DEMO
Operational flexibility (experimental setup)	Realization of fusion energy conversion
No fusion energy conversion (≤100 °C water-cooled)	(>300 °C water-cooled/ >400 °C helium-cooled)
Pulse discharge (~400 s), low duty ratio (~3%)	Quasi-steady-state operation (several hours of discharge) with a high duty ratio (>50%)
Steady-state heat flow 10 MW/m^2 (divertor)	Steady-state heat flow ∼10 MW/m^2 (divertor)
Transient heat flux loading with plasma disruption/VDS/ELMs (20 MW/m^2, <10 s)	Very limited plasma disruption/ELMs events
Low neutron irradiation dose (FW: ~1 dpa)	High neutron wall loading
Mature material systems (W, Be, CuCrZr, 316L)	Low-activation materials (W alloy, RAFM, ODS-FS, etc.)

1.4.1 Development of international fusion reactor device and fusion energy research

After years of theoretical and experimental exploration, researchers have found that there is still a significant gap between the experimentally obtained plasma energy confinement time and theoretical predictions. Owing to the complexity of nuclear fusion and the challenges in the research process, international exchange and cooperation have been conducted to help achieve control over magnetic confinement thermonuclear fusion and face the fundamental engineering challenges. Over the years, a lot of large-scale tokamak devices have been constructed or refurbished worldwide and significant progress has been made in both theoretical and experimental research:

(1) European JET device
Built in 1982 by the European Atomic Energy Commission at the Karam Laboratory in the United Kingdom, the JET is one of the largest tokamak installations in the world. In the experiment, the JET device satisfies three conditions for critical plasma: the temperature reaches more than 100,000,000 °C, the neutron number density of 1 cm^3 is 25 trillion, and the magnetic field confinement time is 0.6 s. Since then, many tokamak devices have been built in Europe, such as ASDEX-U in Germany, FTU in Italy, and MAST in the United Kingdom. ASDEX-U was the first device in the world to obtain the High Confinement Mode (H-Mode) of plasma, with the earliest divertor, vertical elongation of the plasma cross section, and triangular deformation. H-Mode is regarded as an important constraint mode for tokamak and is also the standard mode of operation for ITER. In 1991, the JET device was the first in the world to obtain a fusion power output.

(2) TFTR and DIII-D installations in the United States
The TFTR in the United States is built on a large scale in the Plasma Physics Laboratory of Princeton University. The device gained fusion power output immediately after JET, and then shut down. The tokamak device built by the General Atomic (GA) in the United States is smaller and more flexible, allowing for more experimental research. DIII-D in San Diego was the first tokamak device in the world to use D-shaped cross-sectional constraints without obtaining fusion power output.

(3) JT-60 device in Japan
JT-60 was the world's first device to use a horizontal divertor and was built by the Japan Atomic Energy Research Institute (JAERI) (now renamed the Japan Atomic Energy Agency (JAEA)). Later, the device was changed to a vertical divertor and upgraded to JT-60U to obtain the high confinement mode of the plasma. JT-60U uses electromagnetic waves to heat plasma through a "high-frequency heating" device, breaking through the difficulty of continuous operation in a short time. The fusion power output was obtained by using deuterium as fuel, and the fusion power output gain was 1.3 when converted to deuterium-tritium fusion. In addi-

tion, JAERI built a JFT series of small tokamak devices. Japan's National Institute of Fusion Science (NIFS) is another well-known fusion research institute that is at the forefront of the world, particularly with its spiral devices (stellarators).

(4) Tokamak device in Russia
Russia (the former Soviet Union) has a world-class level of expertise in magnetic confinement device construction, having built the world's first tokamak device, the first modern improved confinement Tokamak device T-3, the first large superconducting Tokamak device T-15, and the first superconducting Tokamak device T-7 (in the 1990s, the Institute of Plasma Physics of the Chinese Academy of Sciences introduced and transformed it into HT-7).

1.4.2 Development of fusion energy in China

China's controlled nuclear fusion energy research began in the mid-1950s of the last century, and has been on the way of stable and gradual development in difficulty for a long time, with the establishment of professional research institutions, such as the Southwestern Institute of Physics (SWIP) of the China National Nuclear Corporation and the Institute of Plasma Physics at the Chinese Academy of Sciences (ASIPP). The University of Science and Technology of China, Huazhong University of Science and Technology, Dalian University of Technology, Tsinghua University, and other colleges and universities have set up nuclear fusion and plasma physics majors to train professional talent.

In 1965, the first professional fusion research center, the Institute of Nuclear Industry 585 (SWIP), was established, and 22 small and medium-sized fusion test devices were built successively, among which the representative tokamak devices are China Circulator No. 1 (HL-1, 1984), China Circulator No. 1 (HL-1M, 1995), and China Circulator No. 2A (HL-2A, 2002). In particular, the successful completion of the HL-1 device indicates that China's controlled nuclear fusion has entered the stage of large-scale physical experimental research. HL-1 has been upgraded to HL-1M, and various parameters such as high density, strong auxiliary heating, high current drive, and higher parameters under pellet injection conditions have been significantly improved, and plasma confinement physics experiments have entered the advanced ranks of similar devices worldwide. In December 2002, the HL-2A main engine was installed and operated. In May 2007, the plasma temperature successfully reached 55,000,000 °C, and the plasma current was 433 kA, which took a new step on the road of hundreds of millions of nuclear fusion ignition temperatures.

Founded in September 1978, the Institute of Plasma Physics of the Chinese Academy of Sciences (ASIPP) has entered the international advanced ranks in high-temperature plasma physics experiments and nuclear fusion engineering technology research after more than 40 years of development, forming extensive international exchanges and cooperation, and is also one of the important supporting units of the China working team of the ITER Program. In 2006, China independently designed and successfully built the first fully superconducting tokamak device, HT-7U (later renamed the EAST device), which has the characteristics of a large-scale noncircular cross section, full superconductivity, and an advanced divertor, and has become the only magnetic confinement nuclear fusion device in the world capable of carrying out high-power and long-pulse operation and is directly related to ITER technology and operation, which marks China as the fourth country in the world with a superconducting tokamak device. Consequently, the international fusion community has referred to it as "an outstanding achievement and an important milestone in the development of fusion energy in the world." The scientific goal of EAST is to build a large-scale fully superconducting tokamak device with a noncircular cross section and its experimental system, to develop and establish the various technologies required for steady-state operation in advanced tokamak fusion reactors, and to carry out experimental research on the fundamental physical problems of steady-state, safe, and efficient operation.

In the context of ITER construction, China attaches great importance to engineering construction and the development of supporting domestic facilities. In 2007 and 2008, China successively launched scientific research projects to support domestic ITER, and carried out relevant physical theory discussions, engineering simulations, and other research. The selection of materials and service performance were gradually included in the research theme. In 2018, Hefei University of Technology undertook a significant number of national key R&D projects and talent projects in the research of PFMs and carried out extensive international cooperation with the Forschungszentrum Jülich Research Center in Germany, Kyoto University in Japan, and the Japan Institute of Nuclear Fusion. It has been approved as the national higher education discipline innovation and intelligence project of "New Materials and Technology for Advanced Energy," which is supported by the Ministry of Education of China and the Foreign Affairs Bureau. The author is a leading scholar in overall management.

1.5 Plasma-facing materials (PFMs)

1.5.1 Plasma-material interactions

The confined plasma interacts with the surrounding space and materials, and the plasma-material interaction (PMI) causes inducements: the spiral magnetic field, composed of the toroidal magnetic field and the poloidal magnetic field in the fusion device, can effectively confine the plasma, but there is no boundary for the plasma. Owing to the incompleteness of the magnetic field on the plasma confinement, some charged particles collide, transport anomalously, and diffuse and drift in the direction perpendicular to the magnetic surface, thus contacting and acting on the surface of the component. In addition, neutral particles, neutrons, and photons are not confined by the magnetic field and act directly on the vacuum wall [19]. Originally, a limiter or divertor was proposed to isolate the main plasma from the source of impurities by making the plasma interact with the material through a special magnetic field structure, in a divertor located away from the main plasma [20].

The divertor has now become one of the key components in ITER devices, serving as collectors and exchangers for heat and particle flow [21]. It endures quite severe heat loads and irradiation. The function of the limiter or divertor is to confine the plasma within a certain volume, so there is a plasma boundary – the last closed flux surface (LCFS). The LCFS is the core plasma area inward, and the scraped-off layer is outward, including the vacuum chamber wall, the limiter and the divertor target plate outward, and the part that is in direct contact with the plasma is collectively called the first wall, and the material of the first wall surface is the PFM. Figure 1.3 shows the schematic diagram of the plasma-wall interaction. PMI can be divided into two aspects. One is the particle flow and energy flow bombarding the wall of the device. The mechanisms of impurities are: physical sputtering, desorption, evaporation, chemical sputtering, arc igniting, surface blistering, and hydrogen precipitation at grain boundaries, etc. Impurities enter the main confinement area, which adversely affects plasma confinement and its quality. The second aspect is the particle flow and energy flow bombarding the wall, which is mainly manifested as sputtering corrosion, thermal corrosion, and irradiation damage of PFM [22], resulting in damage to the first wall material.

1.5.2 Damage forms of PFM materials

During the interaction between the plasma and the first wall, the first wall bears a high thermal load and high particle flux to produce impurities, and at the same time, it causes a series of material damage. The main manifestations are as follows:

(1) Sputtering corrosion and hot corrosion
PFM sputtering corrosion and hot corrosion include local ablation, melting, cracking, and thermal fatigue, as well as surface blistering and surface ejection caused by charged particles, neutral particles, neutrons, and photon bombardment. Sputtering corrosion will lead to a gradual thinning of PFM, which will largely limit the service capacity and service life of the plasma-facing component (PFC). Once evaporation and melting occur (metal PFM will lose the melting layer under the action of gravity and electromagnetic force), it will further affect the normal application of PFC and shorten the service life as well.

(2) Irradiation damage
Irradiation damage can be divided into two types: surface damage and body damage: (1) Surface damage is mainly caused by charged particles, neutral atoms, and photon

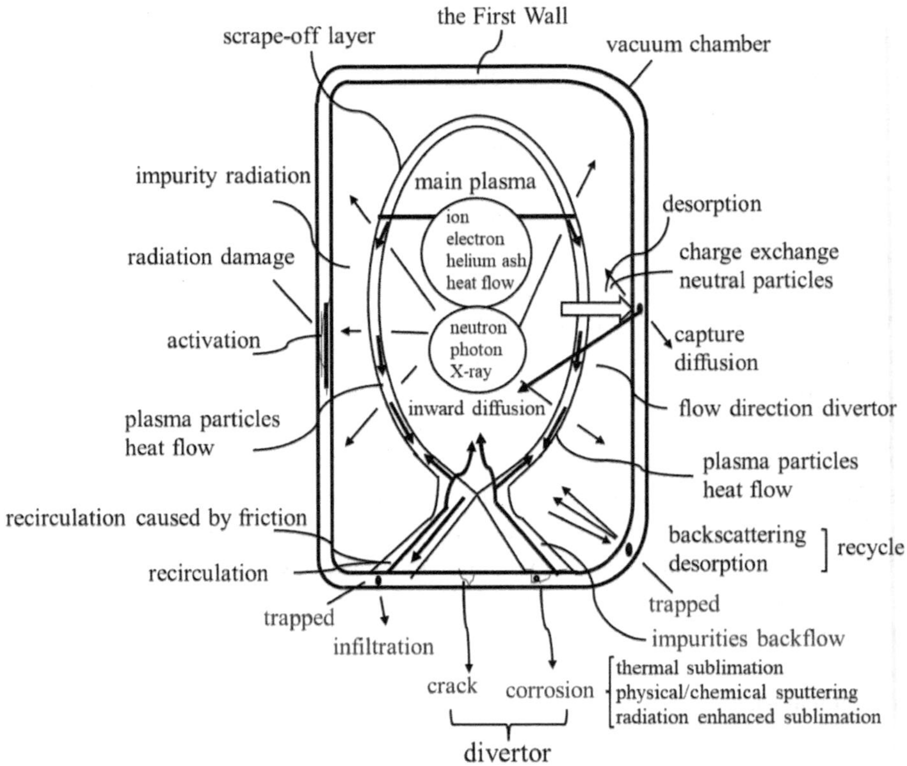

Figure 1.3: Schematic diagram of plasma-wall interaction.

bombardment, and the main forms are surface sputtering, surface blistering, and evaporation. Damage caused would result in fusion devices of materials, as well as impurity pollution to the plasma. (2) Body damage is mainly caused by neutrons, being the main factor in damaging the material itself and shortening its service life. The two basic physical processes caused by neutrons in materials are atomic displacement and nuclear reaction, and the main forms of damage include material activation, swelling, and changes in mechanical and physical properties. This type of damage adversely affects the overall material, leading to performance degradation and a shortened lifespan of the PFC.

The neutron irradiation effect is divided into the displacement damage effect and the impurity atom effect caused by nuclear transmutation. There is no essential difference in the displacement damage generated by 14 MeV neutrons compared to the off-position damage generated by 1–2 MeV neutrons in a fission reactor. When the neutron energy increases from 1 MeV to 14 MeV, the damage ratio is almost constant for light atoms. However, for heavy atoms, the displacement damage will increase sharply several times over. Although there is almost no difference in the spatial distribution of irradiation defects generated by neutrons of these two energies, there is a tendency to concentrate the irradiation defects and form large aggregates in high-energy neutron irradiation [23].

1.5.3 Selection of PFMs

1.5.3.1 Performance requirements for PFMs

PFMs in fusion reactor devices are the first barrier to be directly irradiated by high-temperature plasma, which protect the vacuum chamber wall and various internal components. In the process of stable operation, in addition to the high-energy and high-throughput beam current particle irradiation, the high-temperature plasma in the core carries a huge amount of energy, which burdens the energy on the surface in the form of thermal radiation and particle impact, generating a thermal load of about tens of kilowatts to hundreds of kilowatts.

During the operation of thermonuclear fusion devices, PFMs face extremely harsh service environments, so extremely high requirements are put forward for PFMs, structural materials, and cooling technologies [24, 25]. Typical service conditions of PFMs are shown in Table 1.3 [11, 26]. The main functions of PFMs are: (1) effectively transfer heat radiated to the surface of the material; (2) effectively control the impurities entering the plasma; (3) protect other parts from damage caused by plasma bombardment during abnormal stop operation. Table 1.4 [11] lists the properties required for some PFMs, ensuring the operation of the device:

Table 1.3: Typical operating parameters of PFMs [11].

Parameter	Value or range
First wall surface heat flux	0.1–1.0 MW/m^2
Divertor surface heat flux	3–30 MW/m^2
Neutron damage rate	10–50 dpa/year
Total neutron current	50–100 dpa
The ionic energy of the particle bombardment on the surface of the divertor	1–200 ev
The energy of the particle bombardment on the surface of the divertor	10^{22}–10^{23}/sm^2

(1) High melting point, good thermal conductivity, and thermal shock resistance: During the normal operation of a fusion reactor, PFMs are exposed to direct scouring from high-temperature plasma and are subjected to a huge thermal load (about 5 MW/m^2 in normal operation and more than 20 MW/m^2 in unstable conditions). PFMs should have good thermal conductivity, a high melting point, and thermal shock resistance to ensure reliable operation and prevent melting.

(2) Low sputtering yield to ensure plasma quality: Control the impurity amount generated by physical sputtering, chemical sputtering, and irradiation-enhanced sublimation

Table 1.4: Significant performance of PFMs [11].

Thermophysical properties	Mechanical properties	Surface properties	Radiation effects
Thermal conductivity	Strength and toughness	Physical sputtering	Swelling
Thermal expansion	Hydrogen embrittlement	Chemical sputtering	Changes in mechanical properties after irradiation
Coefficient melting point	Thermal shock resistance	Absorption and degassing of hydrogen	Changes in thermophysical properties after irradiation
	Thermal fatigue		Activation after irradiation

to avoid power loss, which, in turn, affects the combustion of plasma. Impurities with a high atomic number (*Z*), such as W, Mo and other metals, have a strong tendency to gather toward the central confinement region, resulting in the heating power of the central confinement region being lower than its radiation power, which becomes the main source of power loss and forms a hollow distribution of electron temperature. For fusion reactors, the concentration of heavy impurities should not exceed ten thousandths; otherwise, the plasma cannot sustain a self-sustaining burn. The impurities with a low atomic number (*Z*) are mainly oxygen, nitrogen, carbon, and boron, which are difficult to control and can cause similar results when the amount is too large.

(3) Low gaseous absorption and discharge of hydrogen (deuterium and tritium): Once PFMs retain large amounts of hydrogen, the hydrogen will enter the plasma during the plasma discharge, causing a hydrogen cycle that gradually intensifies. This situation is very unfavorable for the fusion reaction and needs to be controlled and avoided as much as possible. Therefore, the retention sensitivity and magnitude of the material itself to hydrogen (deuterium and tritium) must also be considered when selecting materials.

(4) Low radioactivity: The radioactive energy of the radioactive material can be reduced to a safe level as soon as possible, so that the radioactivity is low. Nuclear fusion is different from nuclear fission because there are no fission products, making it a clean reaction or clean energy source. However, PFMs are also radiated while being damaged by radiation due to the high-energy neutrons produced by the D-T reaction. Therefore, PFMs must have low radioactivity [27]. Table 1.5 [28] lists some of the restricted radioactive elements and those recommended for PFMs.

(5) PFMs need to be easy to process, mold, and install to meet the requirements of engineering design.

Table 1.5: Low radioactive elements and restricted elements [28].

Low radioactive elements	C, Cr, W, V, Ta, Ti, Mn, Si, B
Restricted elements	Nb, Mo, Ni, Cu

1.5.3.2 Properties comparison of candidate PFMs

It is difficult for existing fusion materials to meet the demanding requirements of high temperature, high pressure, and strong neutron irradiation of fusion reactors in the future, and how to improve the performance of existing materials or/and develop new high-performance materials is a task facing the research and development of fusion materials. The construction and operation of different types of tokamaks, as well as the gradual implementation of the ITER program, are useful for the development, evaluation, and verification of relevant fusion materials during the experimental operation of the device [29, 30].

The research results of material application indicate that the candidate PFCs in different high heat load regions of the tokamak are beryllium (Be), carbon (C), and wolfram (W), and the advantages and disadvantages of the three materials can be compared in Table 1.6 [22]. The basic properties of several hot spot candidate materials at 600 °C are shown in Table 1.7 [31]. Table 1.8 [32] shows the PFMs being studied in the ITER EDA R&D project.

Table 1.6: Properties of wolfram, beryllium, and carbon as the first wall materials [22].

	Wolfram (W)	Beryllium (Be)	Carbon (CFC)
Advantages	1) High melting point (3,430 °C) 2) Low physical sputtering rate (high sputtering emission threshold) 3) High thermal conductivity 4) Low tritium retention 5) Low swelling 6) No chemical corrosion (H incidence) 7) Possibility of in situ plasma repairment 8) Mature joining process	1) Good compatibility with plasma (low *Z*, elimination of breaking density limit capacity, less escape) 2) High thermal conductivity 3) Possibility of in-situ plasma repairment 4) Strong oxygen absorption capacity 5) Low activity 6) No chemical sputtering 7) Mature joining process	1) High resistance to shock and thermal fatigue 2) Sublimation at high temperatures without melting 3) Low *Z* 4) Vast experience 5) High thermal conductivity

Table 1.6 (continued)

	Wolfram (W)	Beryllium (Be)	Carbon (CFC)
Disadvantages	1) High *Z* (in plasma, low permissible presence concentration) 2) The incidence of oxygen impurities causes poor chemical corrosion 3) Low-temperature brittleness 4) High-temperature recrystallization 5) As the temperature increases, the decline of the strength is significant	1) Low melting point (1,250 °C) 2) High physical sputtering yield, poor corrosion resistance, and short lifespan 3) Poor oxidation resistance above 800 °C 4) BeO is highly toxic, and special security measures are required 5) Low resistance to neutron irradiation (swelling, high He production, and brittleness)	1) Large tritium storage 2) Radiation sublimation increases at high temperatures 3) Low physical sputtering threshold 4) High chemical corrosion reduces material life 5) High tritium retention capacity 6) Dust generation 7) Requires special wall-clearing treatment
Application area (ITER)	Divertor target plate, top cover (baffle and dome)	First wall (first wall)	Divertor high-flux area (strike point)

Each material has its own characteristics and limitations, which will add new inspiration for the performance improvement of traditional materials and the research and development of new materials in the future, gradually adapting to the material requirements of fusion reactor components:

(1) Beryllium

Beryllium has many advantages, such as a low atomic number, high thermal conductivity, good adaptability to plasma, high specific strength, high elastic modulus, low plasma pollution, and can be used as an oxygen absorber, with a small neutron absorption cross section and a large scattering cross section, and has been selected as PFMs in ITER [33–35]. The successful application of beryllium in JET has attracted much attention as low-Z PFMs. Table 1.8 lists several commonly used Be materials, among which S-65C has low BeO content and low metal impurities, and has good ductility at high temperature. DShG-200 has high thermal stress resistance; TR-30 has high irradiation resistance, and TShG-56 has moderate performance.

However, the disadvantages of beryllium as PFMs are obvious, including its low melting temperature (1,284 °C), high vapor pressure, and high physical sputtering yield. Under the condition of neutron irradiation, the structure of beryllium crystal will change and its properties will deteriorate, such as the reduction of thermal conductivity, especially under the conditions of low irradiation temperature (70 °C) and high irradiation dose (32 dpa), which will lead to a sharp decrease in its thermal

Table 1.7: Basic properties of several PFMs (600 °C) [31].

Material	Atomic number	Melting point (°C)	Density (g/cm^3)	Thermal conductivity (W/(m · K))	Coefficient of thermal expansion (10^{-5}/K)	Modulus of elasticity (GPa)	Operating temperature rate (°C)	Self-sputteringtritium retention rate	
								1,000 °C	(%)
Graphite	6		1.8~2.1	90~300	4.5	8.2~28.0	RT ~ 2,000	>1	>1 (after irradiation)
CFC	6		1.8	100~400	1.5	11.3	RT ~ 2,000	>1	>1 (after irradiation)
Be	4	1,284	1.85	96	18.4	200	RT ~ 1,000	<1	<1
W	74	3,400	19.25	176	4.5	370	RT ~ 1,000	<1 (>100 ev)	<1

Table 1.8: PFMs being studied in the ITER EDA R&D project [32].

Materials	EU	JA	RF	US
Be	S-65-C	S-65-C	DShG-200	S-65-C
			TShG-56	Plasma sprayed
			TR-30	
			PVD	
			S-65-C	
Carbon based	SEP NS31 (CFC + Si)	MFC-1		
	SEP N31			
	Dunlop P120	CX-2002U		
	SEP NS11 (CFC + Si)	NIC-01		
	SEP N112	B, Si-doped CFC		
W	W-1%La_2O_3	CVD W	Sintered W (cw. recrystallized)	Sintered (rolled, stress relieved, recrystallized)
	W-5Re	Plasma sprayed		
	Plasma sprayed	W/Cu graded	W-Y (melted)	
			Monocrystal	Plasma sprayed
			CVD W	W-Li, W-Al
			W-Re-Mo-H_fC	W/Cu graded

conductivity (from 200 W/m · K to 35 W/m · K) [36]. These shortcomings will also limit its use, not to mention that its thermal shock resistance is also a matter of concern. Qian Ronghui [37] used high-power scanning electron beam heating to simulate the high thermal load in the fusion device, and carried out thermal shock experiments on beryllium materials, and the beryllium samples began to melt significantly when the input energy exceeded 2 MJ/m^2, and the depth of the ablation crater formed within the thermal shock energy density of 9.62 MJ/m^2 did not exceed 150 μm, and the material had a trace weight loss due to sublimation and splashing. Beryllium is banned in many commercial fields because of its toxicity [38].

(2) Carbon-based materials and composite materials.
In ITER's early design, carbon was chosen as the vertical target material for the divertor. Carbon-based materials have the advantages of low atomic number and high thermal conductivity, maintaining certain strength at high temperatures, high thermal shock resistance, good compatibility with plasma, and high ability to withstand abnormal events (including plasma disruptions and edge zone modes) in tokamak devices. Plasma rupture and slow transient processes in ITER will bring great thermal load energy to armor and materials. Carbon fiber composites (CFCs) have a significant advantage as PFMs due to their high thermal conductivity (300 W/m · K at 20 °C and 145 W/m · K at 800 °C). Currently, CFCs can only be used in areas in direct contact with the plasma, such as the vertical target of the divertor and collection plates, because CFCs will not melt under high-power operating conditions (slow transients and disruptions), exhibiting a high ablation lifetime and excellent thermodynamic properties at high heat flux densities. Commonly used carbon-based materials are shown in Table 1.8, and rich experimental data have been accumulated for the application of CFC in major devices [39–41].

However, there is no carbon material compensating for the two major defects: 1) Poor sputtering resistance and high chemical corrosion rate. Carbon materials have strong chemical sputtering around 800 K, and exhibit irradiation-enhanced sublimation above 1,200 K. The rapid increase of carbon corrosion will cause carbon impurities to overflow in the experimental device, resulting in the deterioration of plasma quality. 2) The porosity is high (about 19%) with high adsorption to deuterium and tritium and a serious impact on the plasma produced by deuterium-tritium burning. In addition, according to the current design of the divertor of ITER devices, neutron radiation mainly affects the thermal conductivity of CFCs [42]. Therefore, more data are needed to evaluate and predict the thermal response of irradiated CFCs.

(3) Tungsten and tungsten-matrix materials
Tungsten has been considered and explored as a PFM for divertors. Tungsten metal has a high melting point (3,430 °C), high thermal conductivity, and a very small adsorption capacity for deuterium and tritium, only equivalent to one-tenth of the graphite. It does not react with hydrogen and has the advantages of low radioactivity, strong anti-sputtering ability, and high resistance to plasma erosion. Based on the performance characteristics of tungsten and the results of existing studies, tungsten is the most promising candidate for PFMs [43–46].

However, tungsten has poor thermal shock resistance and a low tolerance for impurities (2 to 3 orders of magnitude smaller than carbon impurities) due to physical sputtering and irradiation effects, which easily cause physical sputtering and pollute the plasma. When the ion energy is greater than 100 eV, the self-sputtering yield of W-W will be greater than 1, so tungsten could only be used in fusion systems with energy below this level. In recent years, with the in-depth understanding of the plasma physics of the tokamak edge, coupled with the improvement of the plasma confinement level and the development of the divertor configuration, the temperature of the plasma edge could be reduced below the critical temperature of physical sputtering of tungsten, and the sputtering yield of tungsten has been greatly dropped. Therefore, tungsten has become the focus of research on first-wall materials and components.

Figure 1.4(a) is a schematic diagram of the sputtering rate of three candidate first-wall materials by D^+ [47], which shows that the sputtering rate decreases rapidly with the increase of the atomic number of the plasma material under the same particle energy bombardment. Considering the PFMs from this perspective, choosing a tungsten material with high *Z* is very beneficial for impurity generation control. As shown in Figure 1.4(b), the sputtering thresholds for deuterium, helium, and carbon to tungsten are 220, 110, and 80 eV, respectively. Therefore, under the conditions of pure deuterium plasma and cold edge plasma, a small amount of sputtering mainly comes from impurity ions such as oxygen, carbon, and tungsten self-sputtering, and tungsten sputtering can be ignored, which is the main reason why tungsten materials are selected as PFMs by devices such as ITER and JET [48–50].

As a future engineering application, the main bottlenecks of tungsten materials are their low-temperature brittleness, recrystallization brittleness, and irradiation brittleness. After continuous research and experimental verification, it is found that the brittleness of tungsten-based materials can be reduced, and the comprehensive properties can be reliably guaranteed through advanced material preparation and processing technologies such as alloying, dispersion strengthening, and compounding.

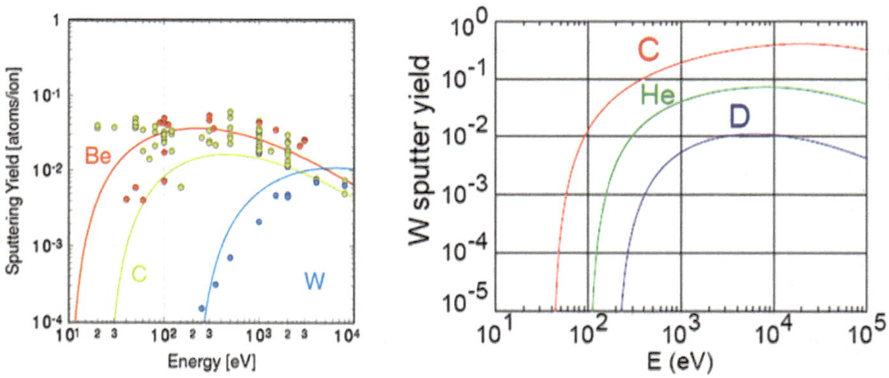

Figure 1.4: Sputtering rate (a) of D^+ for Be, C, and W PFM, and sputtering rate (b) for W of different particles [47].

1.6 Summary

The comprehensive properties of high-beam-resistant low-energy plasma (including D/T/He particles), steady-state heat flow (~10 MW/m^2), transient thermal shock (~GW/m^2), and high-energy 14 MeV fusion neutron irradiation for PFMs are fundamental for the first barrier of the high-temperature plasma periphery and the steady-state operation of plasma with high parameters.

As PFMs, beryllium, carbon-based materials, and tungsten-based materials have different advantages and disadvantages. The strength of beryllium is of high thermal conductivity, good mechanical properties, and strong oxygen absorption capacity. The weakness is low melting point, high sputtering rate, and toxicity. The strength of carbon-based materials is good thermodynamic properties, low *Z*, non-melting, good thermal shock resistance, and thermal fatigue performance. The weakness is poor anti-sputtering ability and high hydrogen isotope retention. Tungsten-based materials are regarded as the most likely PFMs to be fully used in future tokamaks due to their high melting point, low sputtering rate, non-reactivity with hydrogen, and low tritium retention. Nevertheless, their weaknesses include low-temperature brittleness, recrystallization embrittlement, and neutron radiation embrittlement.

Therefore, research should be carried out on the design, preparation, and processing of tungsten-based PFMs to find and develop new PFMs for future nuclear fusion devices.

References

[1] J. Jacquinot, M. Keilhacker, P.-H. Rebut. Chapter 2: Mission and highlights of the JET joint undertaking: 1978–1999 [J]. Fusion Science and Technology, 2008, 53(4): 866–890.

[2] J.L. Luxon, T.C. Simonen, R.D. Stambaugh, DIII-D Team. Overview of the DIII-D fusion science program [J]. Fusion Science and Technology, 2005, 48(2): 807–827.

[3] L. Jiangang. Current status and development of tokamak research [J]. Physics, 2016, 45(2): 88–97.

[4] J. Cheng, L.W. Yan, K.J. Zhao, J.Q. Dong, W.Y. Hong, J. Qian, Q.W. Yang, X.T. Ding, X.R. Duan, Y. Liu. Density fluctuation of geodesic acoustic mode on the HL-2A tokamak [J]. Nuclear Fusion, 2009, 49(8): 085030.

[5] J. Li, H.Y. Guo, B.N. Wan, X.Z. Gong, Y.F. Liang, G.S. Xu, K.F. Gan, J.S. Hu, H.Q. Wang, L. Wang, L. Zeng, Y.P. Zhao, P. Denner, G.L. Jackson, A. Loarte, R. Maingi, J.E. Menard, M. Rack, X.L. Zou. A long-pulse high-confinement plasma regime in the Experimental Advanced Superconducting Tokamak [J]. Nature Physics, 2013, 9(12): 817–821.

[6] W. Burkart. Status report on fusion research [J]. Nuclear Fusion, 2005, 45(10): A1–A28.

[7] R. Aymar. The ITER reduced cost design [J]. Fusion Engineering and Design, 2000, 49–50: 13–25.

[8] G. Shuangquan. Joining technology of tungsten and heat sink materials for plasma materials [D]. Ph.D., Xi'an: Xi'an Jiaotong University, 2011.

[9] R.W. Conn, D.E. Baldwin, R.J. Briggs, J.D. Callen, R.J. Goldston, R.D. Hazeltine, R.E. Siemon, T.S. Taylor. Panel report to FESAC: Review of the 1996 ITER detailed design report [J]. Journal of Fusion Energy, 1999, 18(4): 213–299.

[10] W. Davis, A. Wallander, I. Yonekawa, The Io-ct Control System Division. Current status of ITER I&C system as integration begins [J]. Fusion Engineering and Design, 2016, 112: 788–795.

[11] Z. Zhangjian, Z. Zhihong, S. Weiping, G. Changchun. Research progress on plasma-oriented materials in fusion reactors [J]. Materials Reports, 2005, 19(12): 5–8.

[12] W. Xi. Research status and prospect of magnetic confinement nuclear fusion [J]. Modern Physics Knowledge, 1999, 11(5): 17–19.

[13] Q. Lijian. 50 years of nuclear fusion research [J]. Nuclear Science and Engineering, 2001, 21(1): 29–38.

[14] M. Shimada, D.J. Campbell, V. Mukhovatov, M. Fujiwara, N. Kirneva, K. Lackner, M. Nagami, V.D. Pustovitov, N. Uckan, J. Wesley, N. Asakura, A.E. Costley, A.J.H. Donne', E.J. Doyle, A. Fasoli11, C. Gormezano, Y. Gribov, O. Gruber, T.C. Hender, W. Houlberg, S. Ide, Y. Kamada, A. Leonard, B. Lipschultz, A. Loarte, K. Miyamoto, V. Mukhovatov, T.H. Osborne, A. Polevoi, A.C.C. Sips. Overview and summary [J]. Nuclear Fusion, 2007, 47(6): S1–S17.

[15] Q. Lijian. Fusion energy and its application [M]. Beijing: Science Press, 2008.

[16] F. Kaiming. Review of ITER experimental cladding plan [J]. Nuclear Fusion and Plasma Physics, 2006, 26(3): 161–169.

[17] G. Federici, W. Biel, M.R. Gilbert, R. Kemp, N. Taylor, R. Wenninger. European DEMO design strategy and consequences for materials [J]. Nuclear Fusion, 2017, 57: 092002.

[18] L. Feng, L. Guangnan, L. Qiang, W. Wanjing. Research on the application of tungsten in nuclear fusion reactor [J]. China Tungsten, 2017, 32(2): 41–48.

[19] D. Reiter. Edge plasma physics overview [J]. Transaction Fusion Technology, 1996, 19: 267–270.

[20] H. Jiakun. Fusion reactor materials [M]. Chemical Industry Press, 2007.

[21] Y. Hao. Study on copper-based brazing metal for tungsten/CuCrZr alloy bonding [D]. Master's Degree, Dalian: Dalian University of Technology, 2015.

[22] Y. Chen, Y. Wu, Cheng. Preparation and properties of tungsten matrix composites with face plasma [M]. Hefei University of Technology Press, 2008.

[23] L. Wenyi. Introduction to nuclear materials [M]. Beijing: Chemical Industry Press, 2007.

[24] G.M. McCracken, P.E. Stott. Plasma-surface interactions in tokamaks [J]. Nuclear Fusion, 1979, 19(7): 889–981.

[25] W.B. Gauster, W.R. Spears. Requirements and selection criteria for plasma-facing materials and components in the ITER EDA design [J]. Nuclear Fusion, 1994, 5: 7–18.

[26] T. Hino. Japanese universities' activities for PFC development and PMI studies[J]. Fusion Engineering and Design, 1998, 39–40: 439–444.

[27] W. Jingrong. Nuclear fusion and international experimental thermonuclear fusion reactor [J]. Rare Metals Letters, 2002, 10: 1–5.

[28] L. Yunkai, J. Kangjun. Plasma facing materials for fusion reactors [J]. Materials Reports, 1999, 13(3): 3–5.

[29] G. Kalinin, V. Barabash, A. Cardella, J. Dietz, K. Ioki, R. Matera, R.T. Santoro, R. Tivey, The ITER Home Teams. Assessment and selection of materials for ITER in-vessel components [J]. Journal of Nuclear Materials, 2000, 283–287: 10–19.

[30] K. Ioki, V. Barabash, A. Cardella, F. Elio, Y. Gohar, G. Janeschitz, G. Johnson, G. Kalinin, D. Lousteau, M. Onozuka, R. Parker, G. Sannazzaro, R. Tivey. Design and material selection for ITER first wall/ blanket, divertor and vacuum vessel [J]. Journal of Nuclear Materials, 1998, 258–263: 74–84.

[31] X. Zengyu. Current status and prospect of fusion materials research [J]. Atomic Energy Science and Technology, 2003, 37: 105–110.

[32] V. Barabash, M. Akiba, I. Mazul, M. Ulrickson, G. Vieider. Selection, development and characterization of plasma facing materials for ITER [J]. Journal of Nuclear Materials, 1996, 233–237: 718–723.

[33] R.W. Conn, R.P. Doerner, J. Won. Beryllium as the plasma-facing material in fusion energy systems-experiments, evaluation, and comparison with alternative materials [J]. Fusion Engineering and Design, 1997, 37(4): 481–513.
[34] B. Patel, W. Parsons. Operational beryllium handling experience at JET [J]. Fusion Engineering and Design, 2003, 69(1–4): 689–694.
[35] E.B. Deksnis, A.T. Peacock, H. Altmann, C. Ibbot, H.D. Falter. Beryllium plasma-facing components: JET experience [J]. Fusion Engineering and Design, 1997, 37(4): 515–530.
[36] V. Barabash, G. Federice, J. Linke, C.H. Wu. Material/plasma surface interaction issues following neutron damage [J]. Journal of Nuclear Materials, 2003, 313–316: 42–51.
[37] Q. Ronghui. Thermal shock resistance of beryllium in the face of plasma [J]. Nuclear Materials and Engineering, 1999, 19(1): 57–60.
[38] Z. Xiaofeng, L. Weiliang, G. Shuangquan, J. Linwen. Research progress on plasma-oriented materials in fusion reactors [J]. Science and Technology Innovation Herald, 2010, 3: 118–119.
[39] S. Jinren, Z. Gengtai, L. Lang. Research trends of graphite materials for fusion devices [J]. New Carbon Materials, 1996, 11(3): 27.
[40] W. Mingxu, Z. Nianman, W. Zhiwen, L. Bo, L. Li, G. Changchun. Study on the interaction between plasma and graphite and its coating [J]. Nuclear Fusion and Plasma Physics, 2000, 20(1): 31.
[41] W. Mingxu, X. Zengyu, C. Jiming, W. Jihong, Y. Longwen, B. Gangyi. Study on graphite components of the first wall of HL-2A device [J]. Nuclear Fusion and Plasma Physics, 2004, 24(1): 24.
[42] V. Barabash, M. Akiba, J.P. Bonal, G. Federici, R. Matera, K. Nakamura, H.D. Pacher, M. Rödig, G. Vieider, C.H. Wu. Carbon fiber composites application in ITER plasma facing components [J]. Journal of Nuclear Materials, 1998, 258–263: 149–159.
[43] N. Noda, V. Philipps, R. Neu. A review of recent experiments on W and high Z materials as plasma-facing components in magnetic fusion devices [J]. Journal of Nuclear Materials, 1997, 241–243: 227–243.
[44] J.W. Davis, V.R. Barabash, A. Makhankov, L. Plöchl, K.T. Slattery. Assessment of tungsten for use in the ITER plasma facing components [J]. Journal of Nuclear Materials, 1998, 258–263: 308–312.
[45] I. Smid, M. Akiba, G. Vieider, L. Plöchl. Development of tungsten armor and bonding to copper for plasma-interactive components [J]. Journal of Nuclear Materials, 1998, 258–263: 160–172.
[46] M. Roedig, W. Kuehnlein, J. Linke, M. Merola, E. Rigal, B. Schedler, E. Visca. Investigation of tungsten alloys as plasma facing materials for the ITER divertor [J]. Fusion Engineering and Design, 2002, 61–62: 135–140.
[47] M. Kaufmann. Tungsten as First wall material in fusion devices, 24th Symposium on Fusion Technolog (SOFT). Warsaw, Poland, September 11–15, 2006.
[48] R. Andreani, E. Diegele, R. Laesser, B. van der Schaaf. The European intergrated materials and technology programme in fusion [J]. Journal of Nuclear Materials, 2004, 329: 20–30.
[49] H. Maier, T. Hirai, M. Rubel, R. Neu, P. Mertens, H. Greuner, C. Hopf, G.F. Matthews, O. Neubauer, G. Piazza, E. Gauthier, J. Likonen, R. Mitteau, G. Maddaluno, B. Riccardi, V. Philipps, C. Ruset, C.P. Lungu, I. Uytdenhouwen, Jet EFDA Contributors. Tungsten and beryllium armour development for the JET ITER-like wall Project [J]. Nuclear Fusion, 2007, 47: 222–227.
[50] T. Hirai, H. Maier, M. Rubel, P. Mertens, R. Neu, E. Gauthier, J. Likonen, C. Lungu, G. Maddaluno, G.F. Matthews, R. Mitteau, O. Neubauer, G. Piazza, V. Philipps, B. Riccardi, C. Ruset, I. Uytdenhouwen, Jet EFDA Contributors. R&D on full tungsten divertor and beryllium wall for JET ITER-like wall project [J]. Fusion Engineering and Design, 2007, 82: 1839–1845.

Chapter 2
Tungsten-based plasma-facing materials

2.1 Introduction

In the magnetic confinement fusion device, the first wall material facing the plasma is subjected to a harsh service condition characterized by multifield coupling with multiple superimposed loads: (1) high thermal field: the PFMs directly surround the high-temperature plasma, facing a steady-state thermal load of 5–20 MW/m^2 and a transient thermal shock of ~GW/m^2 [1, 2], which will lead to recrystallization and creep cracking of the material; (2) strong irradiation field: PFMs are exposed to a variety of strong beam particles such as a large flux (1×10^{20}–1×10^{24} m^{-2} s^{-1}) of H^+, He^+, and high-energy 14 MeV neutrons, which will lead to increased lattice defects, blistering, severe embrittlement, and even permanent damage of the material [3]; (3) high stress field: the temperature gradient inside the material caused by high heat flow will induce the thermal stress field, resulting in fatigue and cracking of the material. In high thermal fields, PFMs are required to have high thermal conductivity, temperature strength, thermal shock resistance, and good connection performance with heat sink materials. Additionally, under irradiation by various strong particle streams, they should exhibit properties such as strong resistance to D/T/He plasma etching, low fuel particle retention, low neutron irradiation embrittlement, and activity. Therefore, PFMs should possess good thermal shock resistance, radiation resistance, comprehensive mechanical properties, and high temperature stability.

Tungsten (W) is characterized by a high melting point and thermal conductivity, low sputtering rate, fuel retention, and neutron activation, making it the most promising candidate material for plasma-facing first walls in nuclear fusion devices and components [4–7]. However, PFMs that fully meet these requirements do not currently exist. Pure tungsten materials present several challenges that need to be addressed including difficult machining, high brittleness, high ductile-brittle transition temperature (DBTT), low recrystallization temperature, susceptibility to cracking under high heat loads, and irradiation-induced embrittlement. These factors severely affect the fabrication, processing, and overall performance of the material in applications.

At present, there are two frequently used methods to improve the performance of tungsten-based materials. One is to prepare new tungsten-based materials through the composition and structural design of tungsten-based materials including alloying and dispersion strengthening. The other approach involves using thermal processing techniques such as severe plastic deformation (SPD), repetitive rolling, and rotary forging to refine the grain structure of tungsten materials, control the recrystallization process, and hence enhance their performance [8, 9].

 | https://doi.org/10.1515/9783111556987-002

2.2 Tungsten-based materials for fusion

2.2.1 Application of tungsten in International fusion devices

In the largest ITER under construction, tungsten is mainly used in the upper part of the vertical target plate of the divertor, that is, the Baffle region [10], as shown in Figure 2.1. The ITER organization has clearly indicated a goal to transition from beryllium/carbon/tungsten to beryllium/tungsten, ultimately achieving the goal of all-tungsten applications [11, 12]. There is consensus that all-tungsten will be used in the design of future ITER reactors, including DEMO fusion reactors [13, 14]. The ITER divertor utilizes technologically advanced water-cooling methods, with the primary W/Cu cooling components being either mono-block or flat-type. Copper alloys are used as heat sink materials owing to their high thermal conductivity and moderate neutron irradiation activation. For both types of ITER W/Cu components, the acceptance criteria for the high heat load test are specified as enduring 5,000 cycles at 10 MW/m^2, 300 cycles at 20 MW/m^2, and 1,000 cycles at 3 MW/m^2 plus 1,000 cycles at 5 MW/m^2.

Since 1999, ASDEX has begun to use the physical vapor deposition method to upgrade its graphite walls to graphite-based tungsten-coated walls (with a deposition thickness of 4 μm), making it the only fusion device with walls fully coated in tungsten (Figure 2.2a) [15]. The vacuum plasma spraying (VPS) process was used to prepare a tungsten coating with a thickness of 200 μm, but the VPS tungsten tile fell off in large blocks under the heat flow load of 10 MW/m^2, and finally the divertor was replaced by a combined magnetron sputtering and ion implantation technology (CMSII) to prepare a tungsten coating with a thickness of 10 μm [16]. In 2011, JET also modified the ITER wall material [17], with bulk tungsten in the impact point area and a tungsten layer of 10–20-μm thick on the CFC for the rest of the divertor using CMSII technology (Figure 2.2b). At the end of 2012, the Tore Supra device in France was renamed WEST (tungsten environment in steady state tokamak), changing its original ring limiter position to a D-shaped divertor position [18, 19], and the first plasma discharge was achieved at the end of 2016 after modification (Figure 2.2c). The WEST tungsten divertor with similarity to ITER in terms of design and a scale of approximately 15% of ITER can be used to investigate the chamfering structure design of tungsten modules to reduce heat flow and different tungsten-based materials [20].

2.2.2 Application of tungsten in China's EAST

Tungsten-based materials are gradually used in the EAST device, and the development of PFMs and related components could be divided into several stages: (1) in the first stage (2006–2007), EAST had just been completed and begun being discharged, and its interior was directly connected to the support structure using stainless steel, without a cooling water pipeline; (2) in the second stage (2008–2010), the internal PFM of EAST was graphite with SiC coating on the surface, and the graphite PFM was bolted with

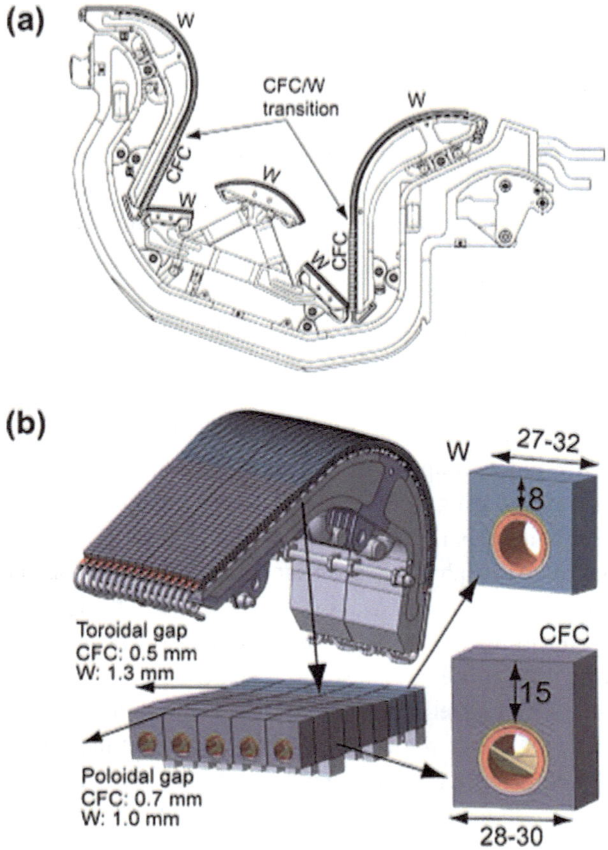

Figure 2.1: (a) Schematic diagram of the structure of the ITER divertor and (b) inner target plate of the divertor [10].

Figure 2.2: Application of tungsten materials in different international tokamak devices: (a) ASDEX All-W metal wall [15]; (b) JET class ITER wall [17]; and (c) WEST W/Cu divertor [18].

the heat sink. Under this condition, EAST obtained a 100-s long pulse and H-mode plasma discharge in the 2010 test; (3) in the third stage (2011–2012), the PFM of the divertor continued to use graphite, and molybdenum was used as the PFM first wall bolted to the heat sink materials. In the experiment in the spring of 2012, EAST achieved a record of 411 s of long pulse and 32 s of H-mode operation; (4) in the fourth phase (2013 onward), attempts have been made to implement a W/Cu divertor and gradually adopt full tungsten as the first wall material. The vertical target parts of the W/Cu divertor will be subjected to a heat load of 7–10 MW/m^2, and the dome and first wall components will endure a heat load of 3–5 MW/m^2 [21].

To meet these demands, the heating power must be increased, with the steady-state heat load at the target plate of the divertor expected to reach 10 MW/m^2. The W/Cu divertor project was initiated in 2010. At present, the upper divertor of EAST has achieved all-tungsten (Figure 2.3) [22], which adopts a modular design, consisting of 80 modules. Each module occupies a horizontal position of 4.5° and is composed of inner and outer target plates and vaults, with mono-block parts for the impact point area of the inner and outer target plates, and flat-type parts for the buffer plate and vault area. For the EAST device, the thermal fatigue standard for mono-block components is 1,000 cycles at 10 MW/m^2 and for flat-type components is 1,000 cycles at 5 MW/m^2. In 2016, EAST obtained a "fully non-inductive current-driven (steady-state) high-confinement mode plasma for over 60s," providing a strong guarantee for tungsten/copper divertor components [20]. This achievement demonstrates that China has the capability to develop tungsten/copper divertors with similarity to those in the ITER, laying the groundwork for the manufacturing of components and materials essential for long-pulse and high-confinement plasma operations.

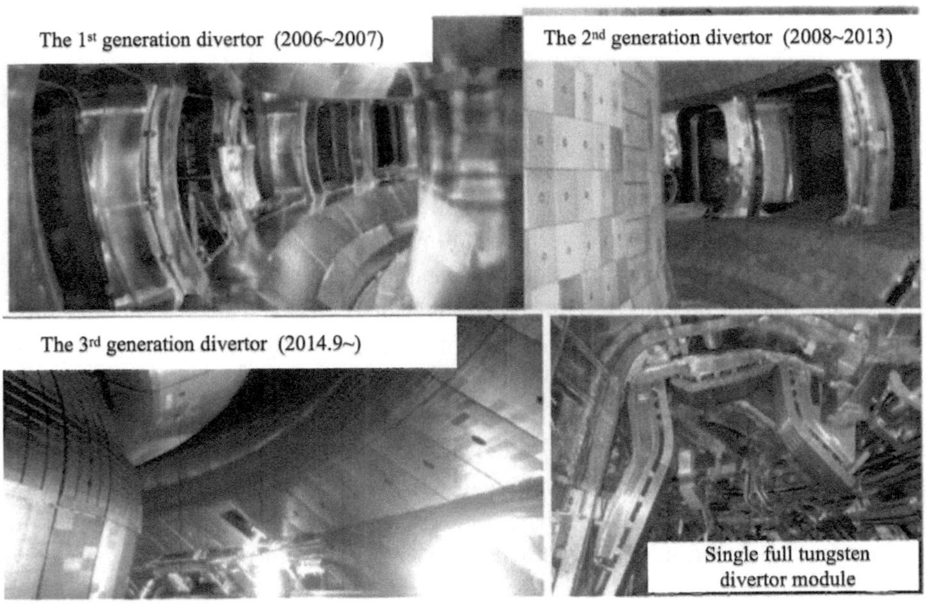

Figure 2.3: Position of EAST third-generation divertor [22].

2.2.3 Applications and challenges of tungsten materials for future fusion reactors

In future fusion experiments, the operating temperature and neutron irradiation intensity will continue to increase, presenting ongoing challenges for material selection and application. Neutron irradiation leads to element activation issues, making low-activation materials the primary choice for DEMO reactors. Tungsten is the preferred material for the first wall of the European DEMO and is exposed to high heat flow and multiple strong particle flow irradiation as shown in Figure 2.4. Plasma etching resistance is the decisive factor for the lifespan of the tungsten first wall [23]. Increasing the tungsten wall thickness will enhance the plasma etching resistance, but excessive thickness will absorb too many neutrons, reducing the tritium breeding ratio in the cladding [24]. Therefore, the tungsten wall thickness needs to balance these two factors. The thickness of the first tungsten wall in the DEMO reactor is ~2 mm [25].

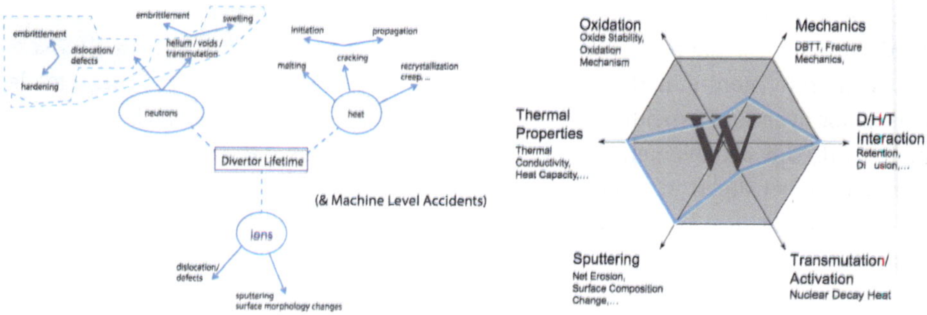

(a) (b)

Figure 2.4: Material problems for plasma components: (a) material-related problems and (b) tungsten as fusion material [23].

At present, DEMO has two divertor design options: water-cooled (~150 °C) and helium-cooled (~600 °C). Water-cooled divertors continue to utilize tungsten as the PFM, with reduced activation ferritic/martensitic steels poised to gradually replace copper alloys as heat sink materials [26]. Helium-cooled divertors will feature tungsten and tungsten alloys, chosen for their high melting points, excellent thermal conductivity, and low neutron activation, as heat sink materials [27].

For future DEMO reactors (Figure 2.5) [28] and commercial reactors, plasma discharge time is expected to exceed 50% of the total operating time of the device in order to ensure sufficient energy output. Therefore, the incident flux of fusion neutrons on wall materials will be substantial. Over 5 years of service, the irradiation damage in iron-based materials can reach 80–100 dpa, and in tungsten materials, about 20–30 dpa [29]. Neutron irradiation causes thermal conductivity loss of the tungsten wall and copper heat sink. If ITER's water-cooled W/Cu mono-block design is adopted, the cooling capacity will be drastically reduced to ~50%. Using low-activation steel instead of copper alloys also leads to a serious reduction in cooling efficiency due to the steel's low thermal conductivity [27]. The melting threshold and melt formation of tungsten are closely related to its thermal conductivity. When subjected to a medium heat flux (~MJ m^{-2}), the tungsten surface is prone to cracking and even local melting. As the heat flux density increases, the tungsten melt begins to evaporate. If the heat flux density continues to rise, the tungsten melt will start to boil violently, resulting in sputtering of molten droplets [30]. In DEMO reactors, plasma vertical displacements events or sudden disruptions events will be entirely mitigated through the optimization of divertor structures (such as Snowflake or Super-X divertors) and techniques to control plasma detachment. These measures aim to maintain the steady-state heat load on the tungsten surface within the range of ≤10 MW/m^2 [29]. However, studying the behavior of tungsten materials under abnormal transient heat flux loading and expanding their tolerance remains crucial.

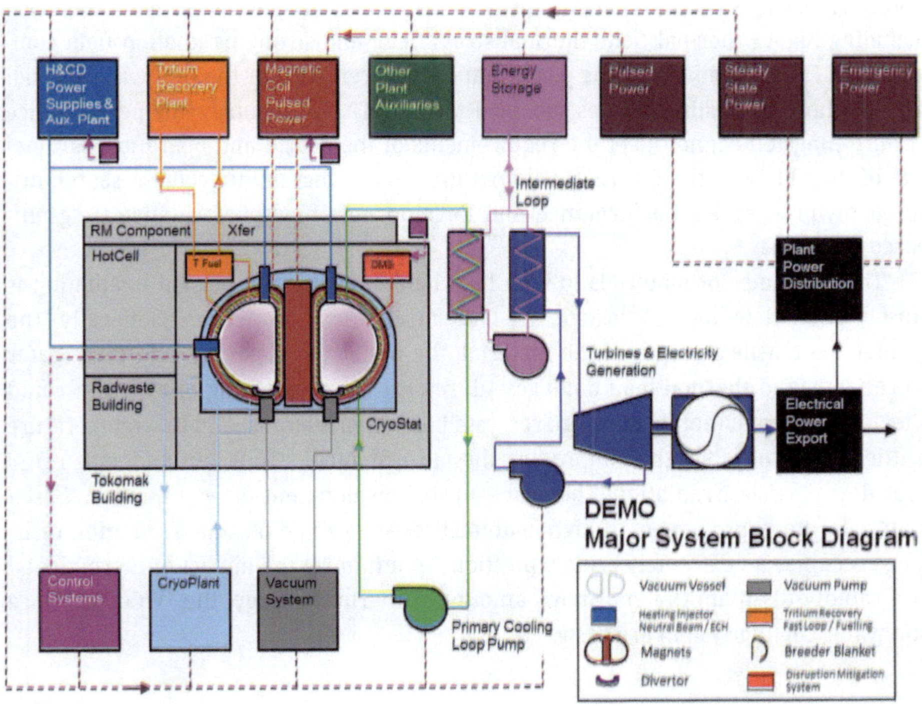

Figure 2.5: Schematic diagram of the DEMO power plant [28].

2.3 Novel tungsten-matrix materials

Despite the theoretical and experimental advancements with tungsten-based materials, fusion materials face complex and evolving service conditions, not a single but a group challenge of many interrelated factors, as shown in Figure 2.6. The synergistic coupling effect of heat, particles, and forces creates complex requirements for tungsten-based materials of the first wall and divertor components: (1) high heat flow requires materials to possess high thermal conductivity, high-temperature strength and thermal shock resistance, and good connection performance with heat sink materials and (2) multiple strong particle flow irradiation requires resistance to D/T/He plasma etching, low fuel particle retention, low neutron irradiation embrittlement, and activity. Therefore, in the operation and service condition of the device, tungsten materials must have excellent comprehensive properties to alleviate or eliminate issues such as fusion neutron irradiation embrittlement, high heat flow irradiation cracking, and plasma-induced blistering and fuzz formation. The first wall material of the China fusion engineering experimental reactor (CFETR) faces more severe service conditions including higher thermal field, thermal stress field, and strong irradiation field compared to ITER. Consequently, the performance requirements for the first wall material surpass those of existing devices such as EAST and ITER. Obviously, the performance of pure tungsten cannot meet the requirements of the design and operation parameters of the CFETR in the future. It is imperative to conduct multifaceted research oriented toward service performance and develop new high-performance tungsten-based materials.

The resistance of materials to high heat load cracking and plasma irradiation is closely related to the microstructure and mechanical properties. Generally, the greater the tensile strength of the material, the better the plasticity, and the greater its resistance to thermal load damage. Addressing the toughening of tungsten could alleviate the irradiation brittleness, recrystallization brittleness, and low-temperature brittleness of tungsten, thus improving the performance of tungsten under neutron/heat flow/particle irradiation. Therefore, in the development of new tungsten-based materials, the improvement of high-heat load resistance and plasma irradiation resistance occupies an extremely critical position. To minimize the impact of thermal load and rapidly dissipate the enormous amount of thermal energy, the W/Cu structure and water cooling system are designed.

2.3.1 Alloyed tungsten

Adding alloying elements to tungsten is a common method to enhance its strength, plasticity, and toughness properties, resulting in significant improvements compared with unalloyed materials.

Figure 2.6: Fusion materials face not a single but many interrelated challenges [23].

2.3.1.1 W-Re alloy

Rhenium (Re) is added to tungsten to improve its strength, plasticity, and toughness. The maximum solubility of Re in tungsten is 37% (at%) at 3,000 °C, decreasing with lower temperature to a minimum of 28% (at%, atomic percent) at 1,600 °C. The rhenium content (wt%) commonly added to the alloy is 3, 5, 10, 25, and 26, which is divided into two categories: low-content W-Re alloy ($Re \leq 5\%$) and high-content W-Re alloy ($Re \geq 15\%$). Higher rhenium content enhances the alloy's high-temperature mechanical properties, with better plasticity observed at the solubility limit of rhenium content. However, when Re content exceeds 26% in the alloy, brittle phases precipitate. The addition of Re raises the recrystallization temperature of tungsten, improves its creep strength, and limits the recrystallization process. This addition significantly enhances not only the high-temperature strength and plasticity of tungsten materials, but also their low-temperature toughness, reducing the ductile-DBTT of tungsten. Therefore, Re expands the stable operating temperature range of tungsten as a plasma-facing material in fusion reactors and improves the service safety of materials and components. Gludovatz et al. [31] studied the fracture toughness K_{Ic} of different tungsten materials (pure W, potassium doped tungsten, W-1wt% La_2O_3, and W26Re) and found that the fracture toughness of all materials increased with rising temperature, but the W-Re alloy exhibited a larger K_{Ic} value than other materials, resulting in significant plastic deformation, as shown in Figure 2.7.

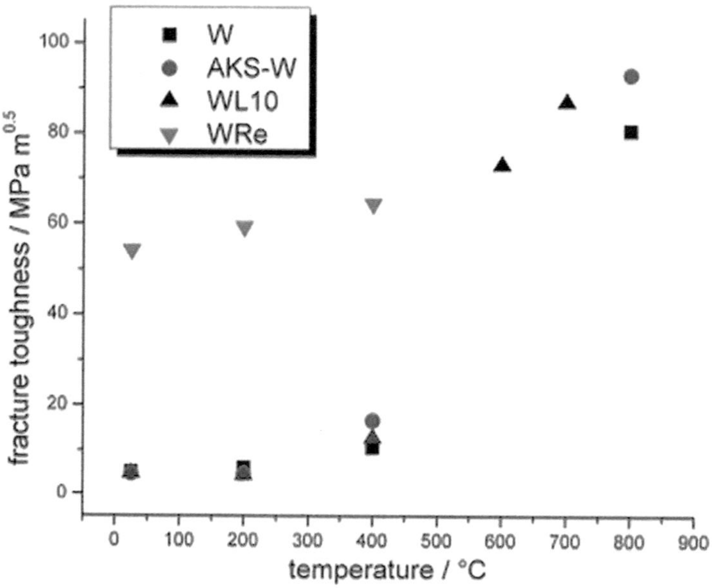

Figure 2.7: Variation of fracture toughness K_Q of pure tungsten, AKS-W (potassium-doped tungsten), WL10 (W-1 wt% La_2O_3), and WRe26 with temperature [31].

Tungsten-rhenium alloy is prone to form twins during plastic deformation, which reduces the dislocation energy of the atomic stacking layer, promotes the improvement of dislocation mobility and the softening of the tungsten solid solution, and consequently increases its plasticity. This phenomenon is called the "rhenium plasticization effect" [32]. Luo et al. [33] found that the addition of iridium (Ir) to tungsten had a more significant solution softening effect than that of Re. Romaner et al. [34] explained the toughening principle by using the dislocation theory, highlighting that the addition of rhenium to tungsten enhances the dislocation properties of 1/2 <111> screw. According to the density function theory, alloying transforms the symmetrical nucleus of the crystal structure into an asymmetric nucleus and simultaneously reduces the resistance of the crystal lattice to the dislocation motion (Peierls-Nabarro force). This increases the number of slip surfaces, reduces stress from plastic deformation, and enhances dislocation mobility, facilitating plastic deformation.

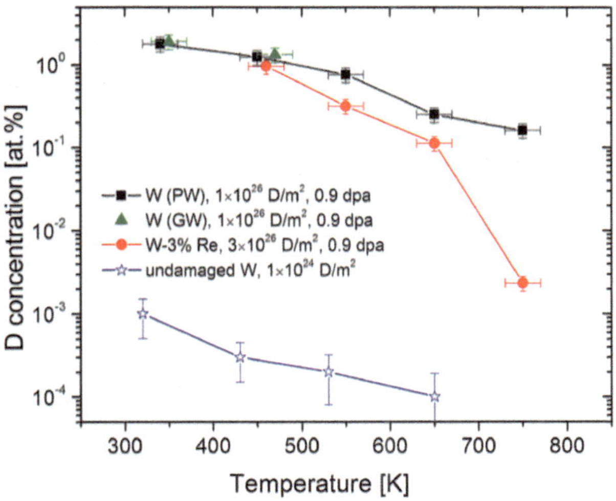

Figure 2.8: Variation curves of deuterium concentration with temperature in the surface depth of 0.2–2 μm after exposure of pure tungsten and tungsten-rhenium alloys measured by NRA after irradiation with 20 MeV tungsten ions [35].

Additionally, the addition of rhenium to tungsten alloys also affects the retention of hydrogen isotopes. Tyburska-Püschel and Alimov [35] studied the deuterium retention of pure tungsten and W-3% Re alloys after ion irradiation of tungsten at different temperatures and found that a small amount of Re can reduce the deuterium retention level in tungsten as shown in Figure 2.8. The deuterium retention in Re-doped tungsten below 650 K was at least 30% lower than that of pure tungsten, and the effect was particularly pronounced at 750 K, where the deuterium concentration was reduced by two orders of magnitude. The underlying reason is that rhenium, acting as a solute atom, recombines with vacancies, reducing the generation rate of high-energy defects. Even at three times the deuterium injection dose compared with pure tungsten, the deuterium retention in W-3% Re alloy was still lower than that of pure tungsten. Therefore, a small amount of transmuted rhenium in tungsten due to neutron irradiation does not increase deuterium retention; instead, it helps to keep the deuterium retention level low, which is highly favorable. Golubeva et al. [36] studied the deuterium retention of tungsten materials doped with different amounts of rhenium and found that the rhenium content did not significantly affect the deuterium retention of tungsten in the range of 1–10% rhenium doping, and there was no significant difference in the deuterium retention of W-(1%, 5%, and 10%) Re alloys, as shown in Figure 2.9. The deuterium retention in tungsten-rhenium alloys was closely related to the material's microstructure, and a great number of cavities appeared on the surface of the material after irradiation with low-energy deuterium ions, particularly noticeable at low rhenium content. The deuterium retention in the W-Re alloys prepared by Plansee exceeded that of polycrystalline tungsten, which was approximately an order of magnitude higher at low doses, with the difference decreasing as the deuterium injection dose increased. Fukuda et al. [37] studied the microstructural evolution of pure tungsten and W-Re alloys under neutron irradiation, and the products of the transmutation of pure tungsten under neutron irradiation were Re and Os, both of which produced brittle precipitates of Re and Os. However, compared with pure tungsten, W-Re alloys exhibited smaller sizes and densities of precipitated phases, with inhibited formation of voids and dislocations, as shown in Figure 2.10. The defective clusters produced by pure tungsten irradiation acted as defect traps and nucleation points, leading to the growth of the precipitated phase.

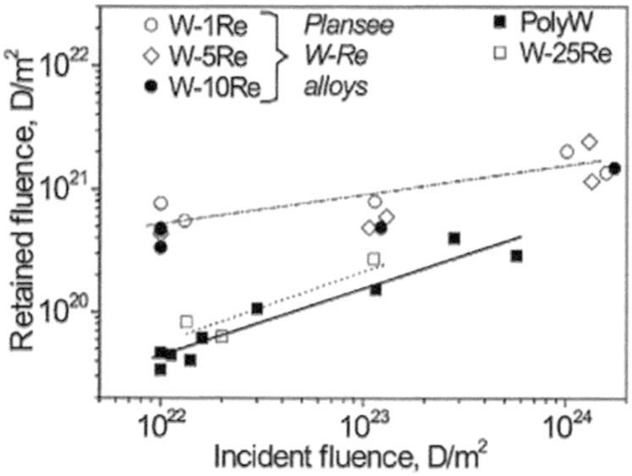

Figure 2.9: Variation curves of deuterium retention with dose of pure tungsten and tungsten-rhenium alloys under deuterium plasma injection with an energy of 200 eV [36].

The addition of alloying elements, including rhenium, to metals generally reduces thermal conductivity. Fujitsuka et al. [38] studied the changes in the thermal conductivity of W-(0%, 5%, 10%, and 25%) Re alloys before and after neutron irradiation at various temperatures. They observed that the thermal diffusion coefficient of the unirradiated W-Re alloy decreased with the increase of rhenium content, a typical outcome of alloying that reduces thermal conductivity. Post-neutron irradiation, the thermal diffusion coefficient of W and W-5%Re alloys decreased compared to preirradiation levels, while that of W-10%Re and W-25%Re alloys increased, especially those of W-25%Re alloys, as shown in Figure 2.11. The thermal diffusion coefficients of W and W-5%Re alloys decreased with rising temperature, whereas those of W-10%Re and W-25%Re alloys increased, as shown in Figure 2.12. The thermal diffusion coefficients of the alloys after neutron irradiation exhibited the same trend as the temperature change. It can be found that adding rhenium reduces tungsten's thermal conductivity, indicating that rhenium addition is generally unfavorable for thermal conductivity. However, alloy compositions beyond 10% can enhance thermal conductivity, possibly due to precipitated phases formed at high temperatures or under irradiation, resulting in a relatively lower rhenium content in the tungsten matrix.

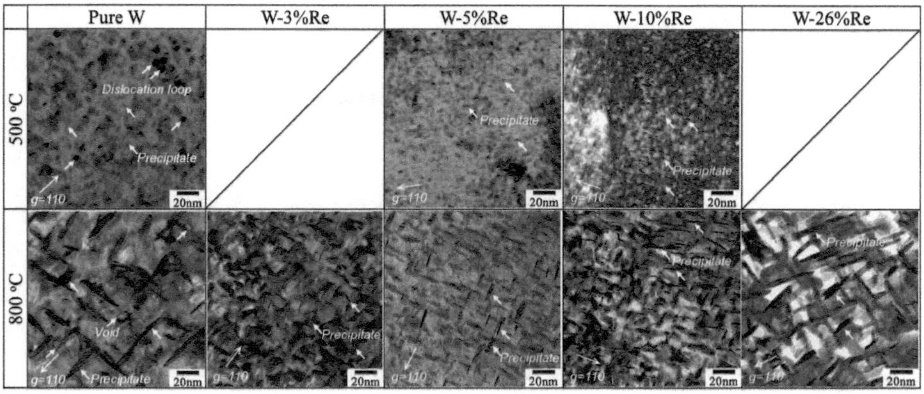

Figure 2.10: TEM plots of pure tungsten and tungsten-rhenium alloys after neutron irradiation at 500 °C, 0.9 dpa and 800 °C, 0.98 dpa [37].

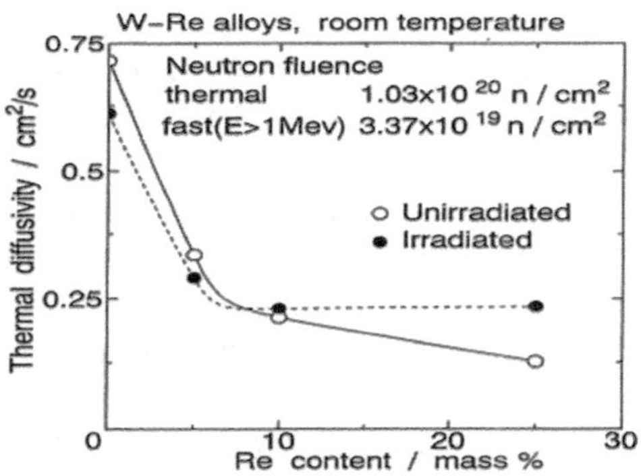

Figure 2.11: Variation of thermal diffusion coefficient of tungsten-rhenium alloy with rhenium content at room temperature [38].

However, as long-term plasma-facing materials in fusion reactors, W-Re alloys tend to form hard and brittle σ phases under neutron irradiation, increasing the brittleness of tungsten-based materials. In addition, owing to the high cost of rhenium and its transmutation after neutron irradiation, the application of W-Re alloys is limited to a certain extent. Therefore, it is advisable to minimize rhenium content in such applications [39].

2.3.1.2 W-Y alloy

In general, the addition of an appropriate amount of rare earth elements can purify tungsten grain boundaries and change the distribution of impurities, reduce gas precipitation, and void formation, consequently increasing the density of materials [40]. Owing to the accumulation of impurities or defects, microcracks are commonly present in tungsten grain boundaries, which propagate under the action of heat and/or

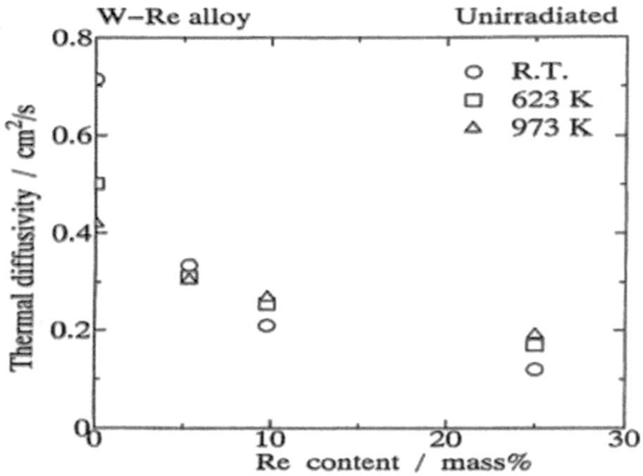

Figure 2.12: Thermal diffusion coefficients of unirradiated tungsten-rhenium alloy at room temperature, 673 K, and 973 K [38].

force, leading to potential fractures in tungsten materials. Rare earth element particles are evenly distributed on tungsten grain boundaries in a near-spherical shape, which can reduce the stress concentration caused by microcracks and alleviate or hinder their propagation [41]. During preparation or processing, the rare earth element particles existing on the grain boundaries limit their movement and hinder grain growth. Consequently, the grain size of tungsten materials containing rare earth elements is much smaller than that of pure tungsten materials. The fracture morphology of tungsten rare earth element materials, as shown in Figure 2.13, indicates that finer grains block the distribution of these microcracks, requiring higher energy for their initiation and propagation, thereby improving the strength of the material [42].

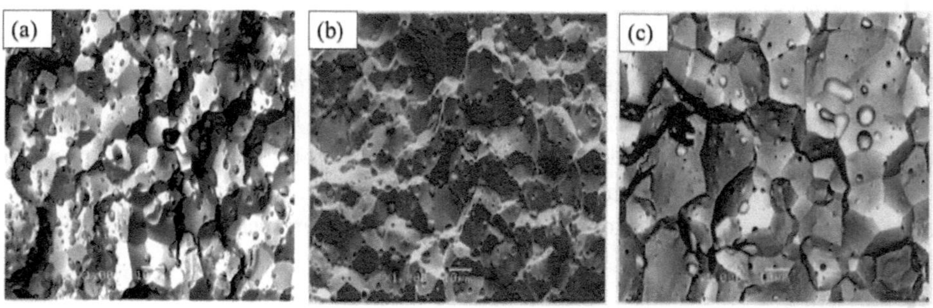

Figure 2.13: SEM diagram of fracture surface of tungsten-rare earth element material: (a) W-Ce; (b) W-La; and (c) W-Y [42].

The strong interaction between rare earth elements and oxygen results in yttrium forming fine oxide particles with oxygen. These particles are diffusely distributed in tungsten grain boundaries and/or intracrystalline, hindering the migration of dislocations and grain boundaries and maintaining a finer grain size at higher temperatures, thereby improving the mechanical properties of tungsten. Although directly adding Y_2O_3 makes the strengthening effect remarkable, it cannot change the distribution of impurity elements in tungsten, thus having little effect on the improvement of material density. However, the brittleness of the material increases significantly when the impurity content is higher. Veleva et al. [43] prepared W-(0.3–2%) Y_2O_3 and W-(0.3–2%)Y alloys by traditional powder metallurgy methods. As shown in Figure 2.14, the density and microhardness of the sintered body and sintered materials of different compositions increased with the Y and Y_2O_3 contents. Among them, W-2%Y alloy had the highest density, while W-2%Y_2O_3 alloy had the highest hardness of 1790 $HV_{0.2}$. This indicates that, compared with Y_2O_3, Y contributes more to the density and has a relatively small effect on the hardness, and the highest hardness is only 1435 $HV_{0.2}$.

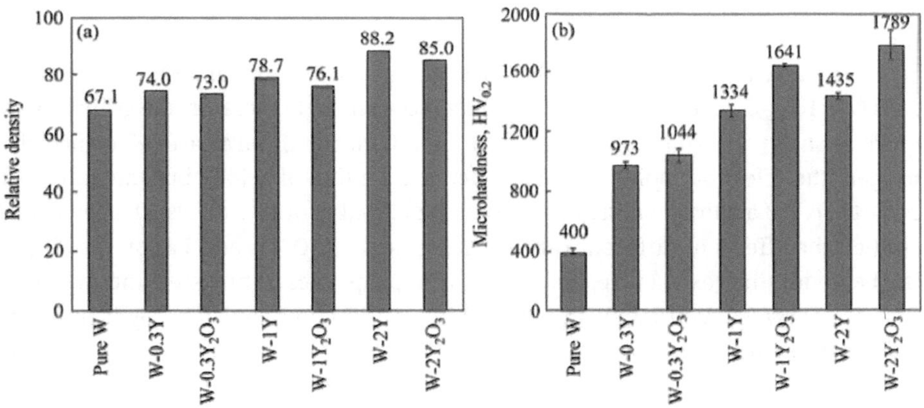

Figure 2.14: Density and microhardness of sintered alloys with different compositions [43].

Veleva et al. [44] prepared W-2%Y alloys via HIP sintering and studied their microstructures and properties. In the ball milling process, Y reacted with the oxygen in tungsten to form fine Y_2O_3 due to its high activity. The diffusion distribution of fine Y_2O_3 second-phase particles played a role in grain refinement and diffusion strengthening, effectively pinning dislocations in the tungsten stretching process, and improving the work hardening ability of the material. The density of W-2%Y alloy reached 97%, and the Y_2O_3 particles were non-uniformly distributed in the tungsten grains and grain boundaries, with sizes ranging from 2 to 20 nm. The tungsten grain sizes were 50 and 150 nm, showing a bimodal distribution, as depicted in Figure 2.15.

Zhao et al. [45] prepared W-(0.25–3%)Y alloys by mechanical alloying combined with SPS technique and found that the lowest density of W-0.25Y was only 95.91%, while W-(0.5–3%)Y was almost fully dense. The EDS results showed the formation of round W-Y-O particles in the W-Y alloy, with the content of the W-Y-O phase increas-

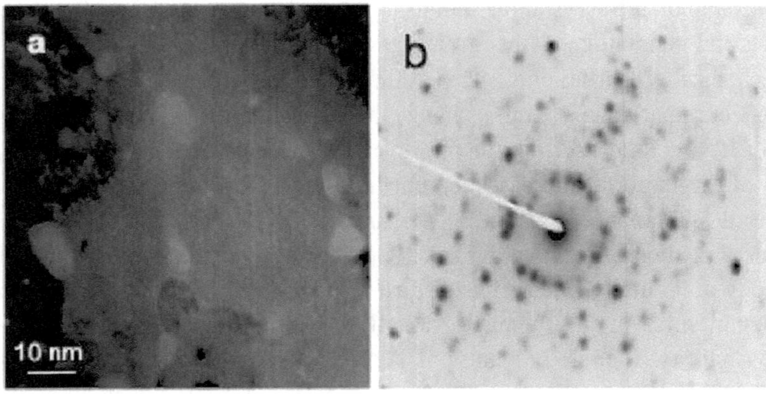

Figure 2.15: (a) TEM dark field image of W-2%Y alloy and (b) the electron diffraction pattern of corresponding selected area [44].

ing with Y content. The average grain size of tungsten decreased with the increase of Y content, and W-1.5%Y alloy had the finest grain size (0.32 μm) and the highest hardness (770 $HV_{0.2}$). However, when the Y content continued to increase, the grains coarsened, as shown in Figure 2.16. In terms of the interaction of rare earth elements with oxygen, the effect of doping with La is better than that of Y [46], but there are few reports on W-La alloys and their properties compared with W-Y alloys. Zhao et al. [47] studied the effects of different rare earth elements (Y_2O_3, Y, and La) on the sintering behavior, microstructure, and mechanical properties of tungsten and prepared W-0.5%Y_2O_3(WYO), W-0.5%Y(WY), and W-0.5% La (WL) tungsten alloys by ball milling and SPS technique. Compared with the other two rare earth elements doped tungsten, Y-doped tungsten had higher density, finer grains, and better mechanical properties. The densities of WY, WYO, and WL were 99.4%, 92.1%, and 88.3%, respectively, and the average grain sizes were 1.10, 2.46, and 4.62 μm, respectively. The hardness and bending strength of W-0.5%Y reached 614.4 $HV_{0.2}$ and 701.0 $HV_{0.2}$, respectively. As shown in Figure 2.17, numerous pores were observed in the individual grains and trident grain boundaries of WYO and WL, which reduced the contact area between tungsten grains, thereby decreasing the bending strength of the material. Moreover, the coarser grains and the non-uniform distribution of oxide particles of WYO and WL were also the reasons for their lower bending strength. Ander et al. [48] analyzed with TDS and found that deuterium retention in W-1%La was significantly lower than that in pure tungsten at injection temperatures below 200 °C.

Lemahieu et al. [49] prepared W-(0.25–1%)Y alloys with different Y contents by SPS and studied their microstructure evolution after transient thermal loading at room temperature and a base low temperature of 400 °C, with an energy density of 0.37–1.14 GW/m^2. With increasing Y content, the average grain size decreased, the hardness increased, and the thermal shock resistance of tungsten was enhanced. Figure 2.18 shows the crack scale after electron beam thermal shock obtained from surface and cross-section SEM, and the cracking thresholds of W-1.00Y and W-0.50Y were found to be between 0.76 and 1.14 GW/m^2, while those of W-0.25Y and W-0.75Y were below 0.38 GW/m^2.

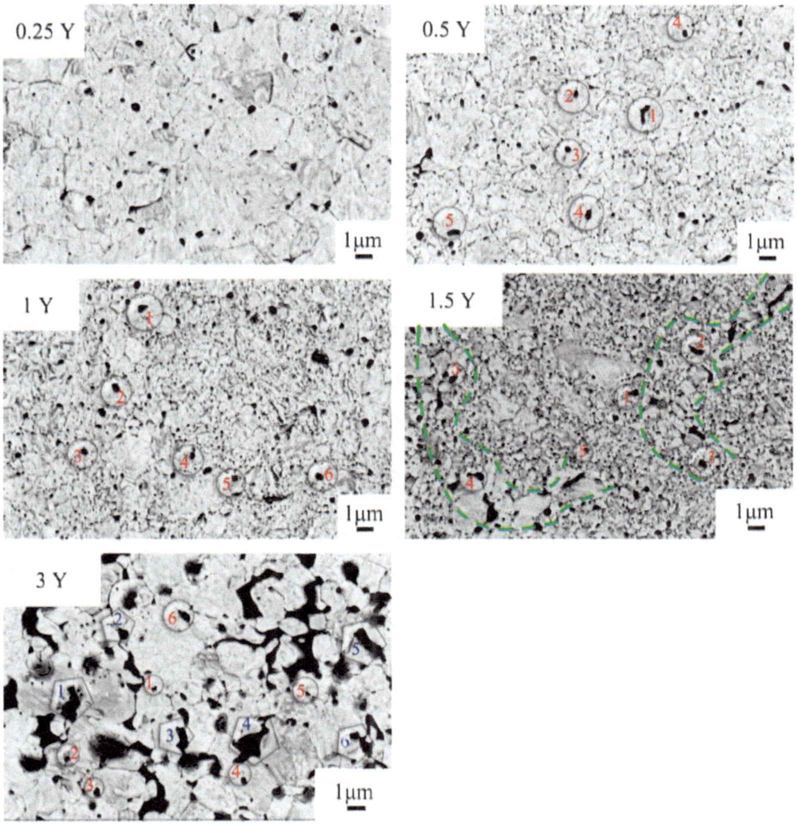

Figure 2.16: Microstructure image of W-(0.25–3%) Y alloy (the dark part is the Y-rich second phase, the red part is the submicroscopic second phase, and the blue part is the microscopic second phase) [45].

2.3.1.3 W-Ta alloy

Tantalum, with its good plasticity, high melting point, fatigue resistance, corrosion resistance, low activation, and resistance to form brittle phases during irradiation, makes W-Ta alloys promising candidates for various helium-cooled divertors in future fusion devices. Tungsten and tantalum exhibit complete mutual solubility across the entire concentration range, both in the liquid and solid states. However, whether tantalum can effectively toughen tungsten is still worth further study. A great number of theoretical calculations show that only rhenium can improve the toughness of tungsten, as the addition of rhenium leads to the transformation of the core structure of the screw dislocation. Alloying tungsten with tantalum depletes the atomic d-orbitals, whereas alloying with rhenium fills the atomic d-orbitals. Li et al. [50] used density

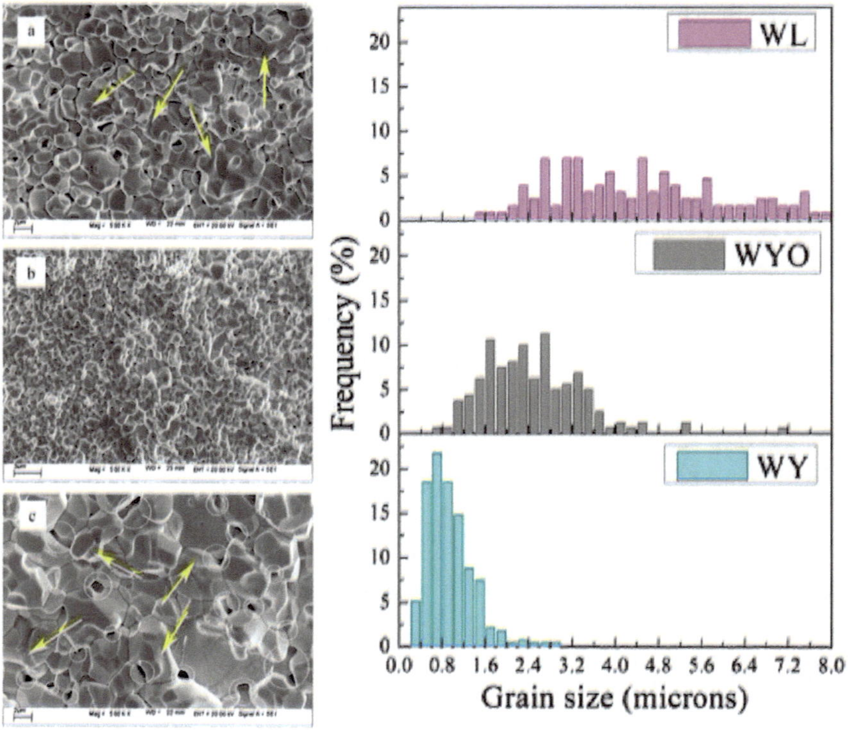

Figure 2.17: SEM image of the fracture surface of rare earth-doped tungsten alloy: (a) WYO; (b) WY; (c) WL; and (d) grain size distribution. Yellow arrows indicate pores at the surface of individual grains, and white-dotted circles indicate pores at trident grain boundaries [47].

functional theory (DFT) to study the effects of tantalum and rhenium on the dislocation of 1/2 <111> screw in tungsten. They found that tantalum-alloyed tungsten maintains a symmetrical dislocation core structure, while rhenium-alloyed tungsten exhibits an asymmetrical structure. Rhenium alloyed tungsten changed the slip surface of tungsten, making it uncertain for W-26%Re, which could be {112} or {123}, but definitely not {110}. In contrast, the slip surface of tungsten tantalum alloys remained {110} (Table 2.1).

Wurster et al. [51] tested a W-Ta alloy and found that its fracture toughness decreased with increasing tantalum content, yielding less than ideal results, as shown in Figure 2.19. The Charpy impact energy of W-1Ta and W-5Ta differed significantly in the high-temperature region. W-5Ta basically remained the same, while W-1Ta increased rapidly at temperatures above 1,200 K, and the Charpy impact energy of W-1Ta was greater than that of pure tungsten in this temperature range. Wurster also studied the fracture toughness of W-Ta alloys prepared by two different methods (industrial hot forging solid solution alloy and powder metallurgy with high-pressure torsion (HPT)) [52] and found that the fracture toughness of industrially

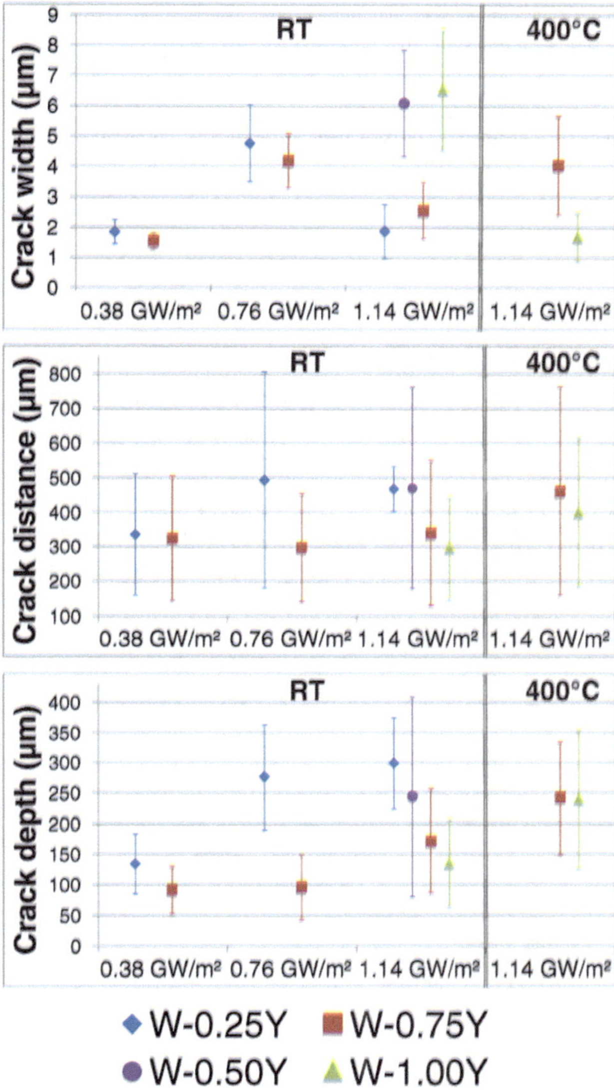

Figure 2.18: The average crack depth, crack width, and crack spacing of the loading surface after electron beam thermal shock [49].

produced W-Ta alloys increased with temperature but decreased with higher tantalum content. The fracture mode of W-10Ta alloy was closely related to tantalum content. For W-10Ta alloy, the maximum test temperature was insufficient to change the fracture mode of the material from transgranular to intergranular, indicating that the increase of tantalum content heightened the internal brittleness of the grain. Compared with pure tungsten, the transgranular fracture toughness of W-Ta alloy was significantly reduced at high temperature. For W-Ta alloys prepared by powder metallurgy, the tantalum volume

Table 2.1: Tungsten alloy cantilever beam bending test specimen dimensions and observed slip surface (*W*, *B*, and *L* represent the width, thickness, and bending length of the cantilever) [50].

Material	*W* (μm)	*B* (μm)	*L* (μm)	Slip surface (μm)
W	5.7	6.0	26.3	{110}
W-24%Ta	4.6	8.6	17.5	{110}
W-24%Ta	4.9	4.5	14.0	{110}
W-26%Re	4.9	6.1	23.1	{112}/{123}

concentration is usually maintained below 10%. Significant interdiffusion occurs between tungsten and tantalum at temperatures greater than 1,500 °C, and if the tantalum content is too high, the Kirkendall effect can lead to defects in the alloy during homogenization. The fracture toughness of W-Ta alloys is anisotropic, with the anisotropy being more pronounced in powder metallurgy-prepared alloys than in those prepared industrially (Figure 2.20). The toughening of W-Ta alloys can be achieved in only one or two principal directions with a suitable microstructure design, although high brittleness at room temperature remains in one direction.

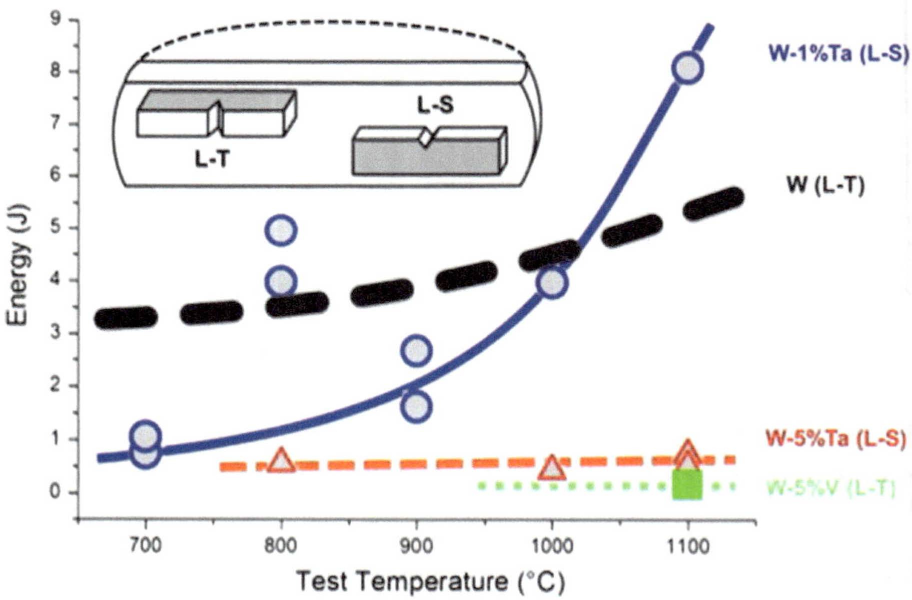

Figure 2.19: Charpy impact test of pure tungsten, W-Ta, and W-V alloys [51].

Mateus et al. [53, 54] used high-energy ball milling combined with the SPS technique to obtain tungsten-tantalum alloys. They studied the radiation damage of tungsten plates, tantalum plates, and tungsten-tantalum alloys by bombardment with single He^+ and D^+

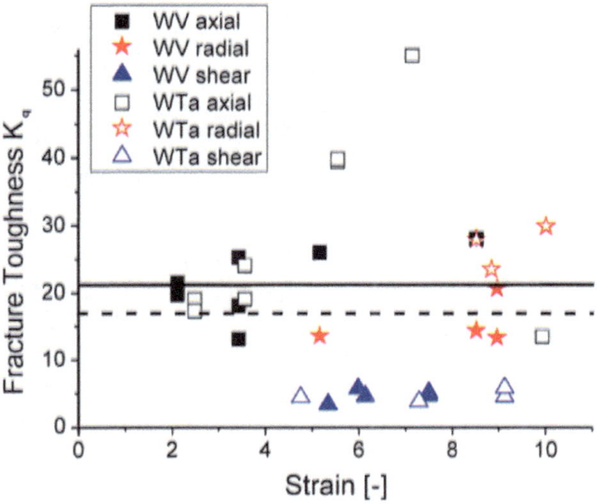

Figure 2.20: Fracture toughness of W-25V and W-30Ta at room temperature under different strains and crack propagation directions [52].

high-energy particles, respectively. The introduction of oxygen as an impurity during the mechanical ball milling process led to more severe surface blistering of the tungsten-tantalum alloy. Dias et al. [55] observed the changes in the microstructure and properties of tungsten-tantalum alloys under the combined effect of helium and deuterium ions by simultaneously injecting He^+ (30 keV, 5×10^{21} at/m^2) and D^+ (15 keV, 5×10^{21} at/m^2) ion beams. The damage varied with the type and combination of the injected ions. Single D^+ injection caused no obvious change on the surface of tungsten-tantalum alloy. However, single He^+ injection led to bubble formation in the tantalum-rich region. When He^+ and D^+ were injected simultaneously, surface blistering was more severe, as shown in Figure 2.21. Zayachuk et al. [56] used TDS to study the deuterium retention behavior of W-Ta alloys and found a correlation between the tantalum content and the amount and location of deuterium retention. W-1%Ta exhibited a single peak at 572 °C, whereas W-5%Ta showed a single peak at 623 °C. The distribution of trapped deuterium between capture points varied with different tantalum alloy compositions. Zayachuk et al. [57] also compared the deuterium retention of pure tungsten, W-1%Ta and W-5% Ta after irradiation with an energy of 50 eV and a flux of 1×10^{24} m$^{-2} \cdot$ s^{-1} deuterium plasma, and found no systematic difference in the total retention between pure tungsten and W-Ta alloys. Compared with pure tungsten, the TDS curves of the W-Ta sample showed no additional desorption peaks, indicating that tantalum alloyed tungsten did not introduce additional defects. As shown in Figure 2.22, all samples exhibited the same low-temperature desorption peak (~500 K). The tungsten sample had a high-temperature peak (600–650 K), while W-Ta sample exhibited a long desorption "tail" extending to the high-temperature region without significant high-temperature peak.

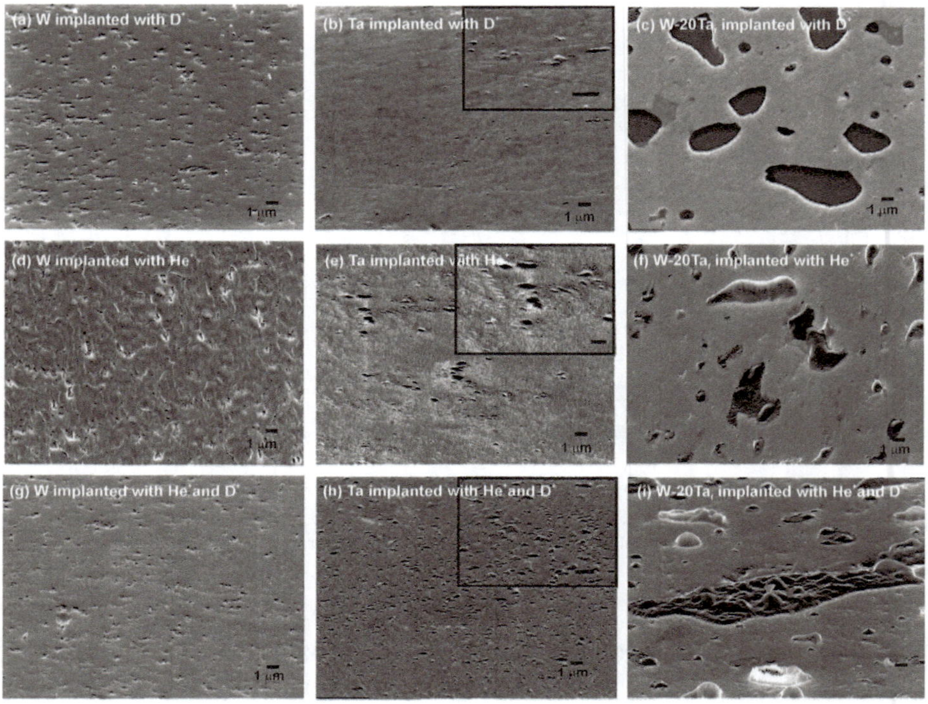

Figure 2.21: Surface microstructure irradiated with tungsten, tantalum, and W-20Ta: (a–c) D^+ ion implantation; (d–f) He^+ ion implantation; (g–i) He^+ and D^+ ions co-implantation [55].

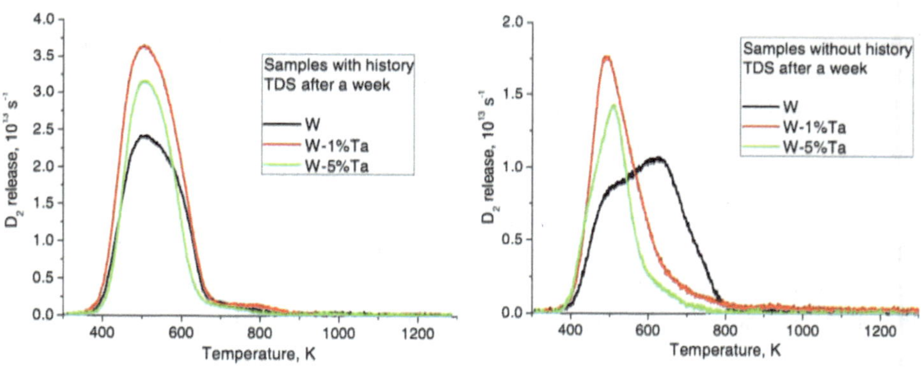

Figure 2.22: TDS profiles of pure tungsten, W-1%Ta, and W-5%Ta irradiated with and without deuterium plasma [57].

2.3.1.4 W-V alloy

Vanadium not only possesses a high melting point and good ductility but also exhibits good solubility with tungsten, forming a continuous solid solution. This capability significantly improves the strength, processability, and corrosion resistance of tungsten-based materials. As shown by the Charpy impact energy values at different temperatures in Figure 2.19, vanadium alloying is not an ideal toughening method for tungsten and cannot reduce the DBTT of the material. DFT calculations also confirm that vanadium alloying cannot change the core structure of screw dislocations, indicating that vanadium is not an ideal toughening element, which is consistent with the experimental results. Arshad et al. [58] prepared W-(1%, 5%, 7%, and 10%) V alloy by ball milling and SPS technique, studied the effect of vanadium content on the microstructure and properties of tungsten, and found that the density, microhardness, and flexural strength of the alloy increased with higher vanadium content, reaching a density of 98.5% for W-10%V, as shown in Figure 2.23. The grain size decreased with the increase of vanadium content, with W-7%V having the smallest grain size at 1 μm. However, abnormal grain growth was noted in W-10%V, possibly due to the reduction in the required sintering temperature caused by vanadium addition. A sintering temperature of 1,600 °C was too high for W-10%V alloy, resulting in grain growth. SEM images revealed vanadium-enriched phases around the tungsten matrix, indicating that a part of the high-plasticity vanadium was not dissolved in tungsten after ball grinding, and filling pores was conducive to increasing density; When distributed uniformly in the matrix, the growth of tungsten grains was hindered. The tungsten grains around the vanadium grains were finer than those in other areas owing to the hindrance and limited growth of the tungsten grains.

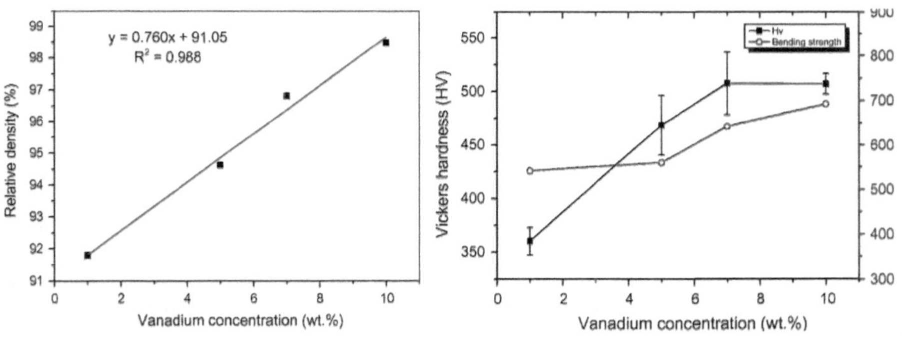

Figure 2.23: Effect of vanadium content on density, hardness, and bending strength of W-V alloy [58].

Palacios et al. [59] prepared pure tungsten, W-4%V, W-1%La_2O_3, and W-4%V-1%La_2O_3 alloys by high-energy ball milling and hot isostatic pressing to explore the effects of alloying and adding oxides on the properties of W. The introduction of vanadium significantly reduced the porosity of W, increased its strength, and improved its fracture toughness. The bending strength reached a maximum value of 850 MPa at 873 K, and the fracture toughness reached a maximum value of 15 MPa $m^{1/2}$ at 673 K. However, the vanadium-doped tungsten increased the DBTT of the alloy (1,073–1,273 K). Vanadium can delay the onset of oxidation; however, once oxidation begins, its severity tends to increase. However, after the addition of La_2O_3, the density and hardness decreased, thereby failing to significantly improve the mechanical properties. The DBTT of W-1%La_2O_3 also increased, although it remained lower than that of W-4%V. W-4%V-1% La_2O_3 alloy had the highest hardness. The strengths of W-4%V-1%La_2O_3 and W-4%V were always higher than those of pure tungsten and W-1%La_2O_3, as shown in Figure 2.24. However, these alloys became more brittle, resulting in lower fracture toughness values, with the DBTT exceeding 1,073 K. In a subsequent study, Palacios et al. [60] also prepared pure tungsten and W-2, 4%V-0.5%Y_2O_3 alloys by high-energy ball milling and hot isostatic pressing and found that the addition of vanadium increased the density of the alloy and refined the grains, with the refinement effect becoming more pronounced as the vanadium content increased. Although ball milling facilitated the alloying of tungsten and vanadium, vanadium segregation still occurred in grain boundary pores. These pores were filled with vanadium, thereby reducing the porosity. The alloy consisted of two microstructures, namely polyhedral grains composed of coarse tungsten grains and "vanadium pools," and nanocomposite structures composed of fine tungsten grains with Y dispersion.

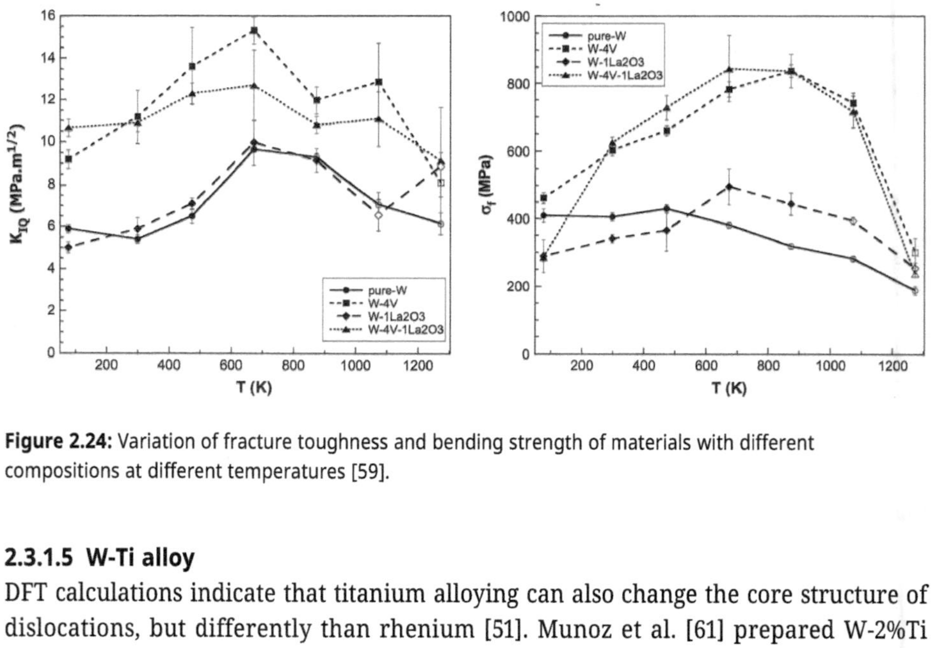

Figure 2.24: Variation of fracture toughness and bending strength of materials with different compositions at different temperatures [59].

2.3.1.5 W-Ti alloy

DFT calculations indicate that titanium alloying can also change the core structure of dislocations, but differently than rhenium [51]. Munoz et al. [61] prepared W-2%Ti alloy and W-1%TiC alloy by mechanical alloying and hot isostatic pressing technology and compared their properties such as strength and plasticity. The results showed that the relative density of W-2%Ti alloy reached 98%, which was higher than that of W-1%TiC, and its microhardness was twice that of the latter, attributed to the solution strengthening of Ti. The stress-strain curves of the two materials were studied through three-point bending tests, revealing that both materials exhibited plasticity only above 1,000 °C. This indicated a significant increase in the DBTT of W after the addition of Ti and TiC. At 1,200 °C, the bending strength of W-2%Ti was 381 MPa compared to 340 MPa for W-1%TiC. This difference was mainly due to the excess residual porosity in W-1%TiC, which resulted in a density of only 95.3% (Figure 2.25).

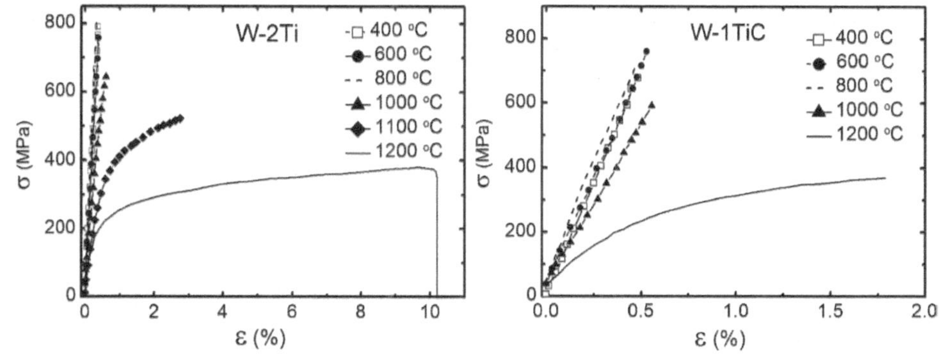

Figure 2.25: Three-point bending stress-strain curves of W-2Ti and W-1TiC at different temperatures [61].

Monge et al. [62] used Ti as the sintering activator and Y_2O_3 particles as the dispersion-strengthened phase to prepare W-x wt% Ti + 0.5 wt%Y_2O_3 (x = 2 and 4) materials by HIP and found that the addition of titanium promoted the densification of the materials, which facilitated the production of fully dense materials. Regardless of whether Y_2O_3 was added, the incorporation of 2% or 4% Ti into the W matrix resulted in a completely dense material. With the increase of titanium content, the hardness improved significantly. Aguirre et al. [63] prepared a small amount of Y_2O_3-doped W-Ti alloy by ball milling and the HIP method and tested the strength and toughness of the material in the range of 25–1,000 °C, finding that the addition of titanium greatly improved the bending strength and fracture toughness as shown in Figure 2.26. However, it also led to a substantial increase in the DBTT, which was raised by at least 600 °C compared to pure tungsten, resulting in increased brittleness. The addition of 0.5 wt% Y_2O_3 improved the oxidation resistance of tungsten and therefore enhanced high-temperature mechanical properties. The addition of Ti not only increased the density of tungsten and inhibited the growth of tungsten grains but also reduced the diffusion of tungsten into Y_2O_3 particles. Y_2O_3 facilitated titanium diffusion into tungsten, forming a Ti-W solid solution. The Y_2O_3 particles in the W matrix began to destabilize above approximately 1,600 °C and became coarse particles containing complex components (W-Y and W-Y-Ti oxides), which deteriorated the mechanical properties of tungsten. Savoini et al. [64] prepared ODS-enhanced W-Ti and W-V alloys with La_2O_3 by mechanical alloying and hot isostatic pressing, respectively, in which Ti and V were enriched between tungsten particles to form Ti and V pools with martensite characteristics, while nano-La_2O_3 particles were evenly distributed in the matrix, providing dispersion strengthening. The hardness of the two components decreased with increasing temperature, and when heated to

473–773 K, the hardness of W-4V-1La_2O_3 remained unchanged, whereas the hardness of W-4Ti-La_2O_3 increased abnormally above 473 K. Since the coefficient of thermal expansion (CTE) of Ti and V is about twice that of W, the martensitic transition occurred under the induction of internal stress with increasing temperature, increasing material hardness. Considering the adverse effect of Ti and V segregation in the voids of tungsten particles on the material properties, the content of Ti or V can be reduced to less than 1 wt% in the design of alloy composition.

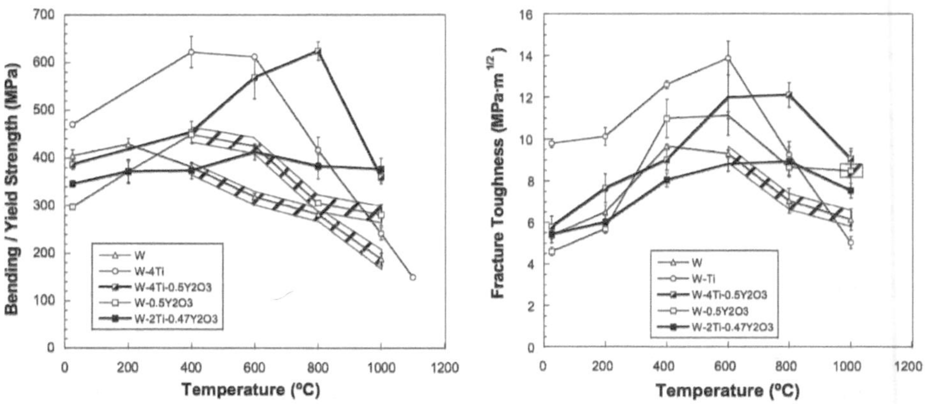

Figure 2.26: Variation of fracture toughness and bending strength of materials with different compositions at different temperatures [63].

2.3.1.6 W-Zr alloy

The strength and toughness of tungsten can be improved by changing the distribution of impurity elements in tungsten or by strengthening grain boundaries. Like rare earth elements La and Y, elements such as Zr and Hf have a strong affinity with C, N, and O, and these elements interact at high temperatures to form high melting point compounds, which effectively reduces the degree and proportion of segregation of impurity atoms in grain boundaries, thus purifying grain boundaries and improving the strength of materials. Xie et al. [65] prepared W-(0.1%, 0.2%, 0.5%, and 1.0%) Zr alloys by SPS, with densities in the range of ~97%. Zr, obtained through the decomposition of ZrH_2, can capture oxygen impurities in tungsten to form nano-sized monoclinic ZrO_2 particles. As the Zr content increased from 0 to 0.2%, the fracture strength at room temperature increased from 154 to 265 MPa, and the fracture energy density increased from 3.73×10^4 J/m^3 to 9.22×10^4 J/m^3. However, further increasing the Zr content led to larger Zr-O particles, which could cause stress concentration, crack initiation, and a decrease in fracture strength and toughness, as shown in Figure 2.27. Xie et al. [66] also prepared W-1%Zr-1%ZrC materials and compared their properties with those of pure W, W-1%Zr, and W-1%ZrC. As shown in the stress-strain curves of the four materials at 400– 700 °C (Figure 2.28), pure tungsten exhibited the lowest strength and the poorest plasticity. Even at 600 °C, pure tungsten still underwent brittle fracture, with a fracture strength of only 309 MPa. In contrast, the W-1%Zr alloy displayed plasticity at 400 °C, with a DBTT that was 200 °C lower than that of pure tungsten, which is advantageous. The elongation of W-1%Zr alloy was higher than that of other materials at various temperatures, indicating that the addition of Zr could effectively improve the plasticity of tungsten. Zr existed in tungsten crystals and grain boundaries in the form of nano-ZrO_2 and W-Zr-O particles, which significantly improved the mechanical properties of the material. W-1%Zr-1%ZrC material had the highest strength, but the plasticity was lower than that of W-1%Zr alloy, and DBTT is 400–500 °C. This decrease in plasticity was due to the excessive number of second-phase particles in the material, which caused a large stress concentration making the plasticity decrease. In terms of improving the strength of the material, the direct addition of ZrC may have a more significant effect.

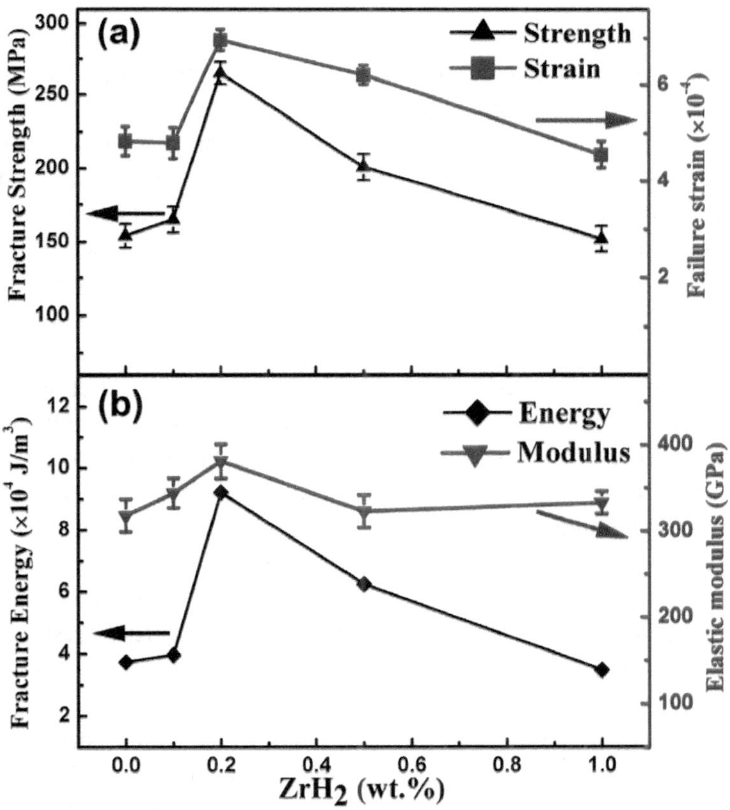

Figure 2.27: Tensile properties of pure tungsten and W-Zr alloys at room temperature: (a) fracture strength and strain with ZrH_2 content and (b) the energy density and elastic modulus at break vary with ZrH_2 content [65].

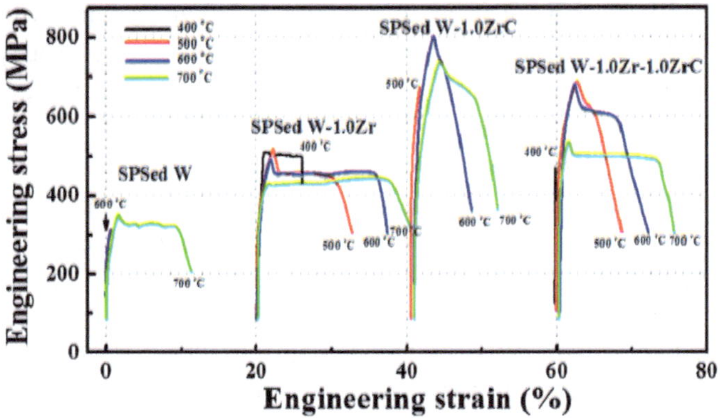

Figure 2.28: Tensile behavior of pure tungsten, W-1%Zr, W-1%ZrC, and W-1%Zr-1%ZrC alloys at 400–700 °C [66].

2.3.1.7 W-K alloy

During the sintering of potassium-doped tungsten, the volatilization of potassium with low-melting point forms micropores in the structure, creating a "pinning" effect on the grain boundary, hindering the growth of grains, and increasing the recrystallization temperature. Consequently, potassium-doped tungsten materials are regarded as potential plasma-facing materials. Potassium, as a non-alloying component, is insoluble in tungsten. It eventually becomes encapsulated in bubbles, with most potassium bubbles and tubes inside tungsten being spherical, rod-shaped with rounded ends, or rugby-shaped. The strengthening mechanism of doped tungsten is "potassium bubble strengthening," and the necessary conditions for potassium bubble strengthening are as follows: (1) Potassium bubbles must exist within the tungsten grains, not at the grain boundaries; otherwise, they will not hinder the dislocation movement and will lose their strengthening effect. (2) During sintering, potassium bubbles form when the internal potassium vapor pressure balances with the grain boundary driving force for tungsten grain growth. Maintaining this equilibrium is crucial; excessive internal potassium vapor pressure can cause bubbles to burst, becoming sources of material cracks. (3) The shape of the potassium bubble must maintain a state of low stress concentration to prevent crack formation. Under conditions of high internal pressure, any irregular shape of the pores inside the material can lead to strong stress concentrations [67].

Preparing high-density potassium-doped tungsten materials with a nanostructure poses challenges due to potassium's unique physical properties (soft, viscous, and low melting point) and high reactivity (susceptibility to oxidation and corrosion), thereby making achieving a uniform distribution of potassium in doped tungsten materials exceedingly difficult. Ensuring potassium remains within tungsten requires the absence of any openings, as potassium has a low boiling point of only 1,043 K, making it prone to escaping through openings. Figure 2.29 illustrates the evolution of potassium bubbles in the preparation of commercial potassium-doped tungsten [68]. The performance of W-K bulk and W-K filaments differs significantly due to variations in their microstructures. In the vertically sintered W-K bulk, the diameter of the potassium bubbles is approximately 200 nm, whereas in W-K filaments obtained by rotary forging and annealing, the potassium bubbles become very small (5–50 nm). This is why W-K is often used as wire, such as in incandescent filaments, rather than in bulk materials. Hence, the current challenge is to achieve nanosized potassium bubbles in W-K bulk materials akin to those found in W-K filaments.

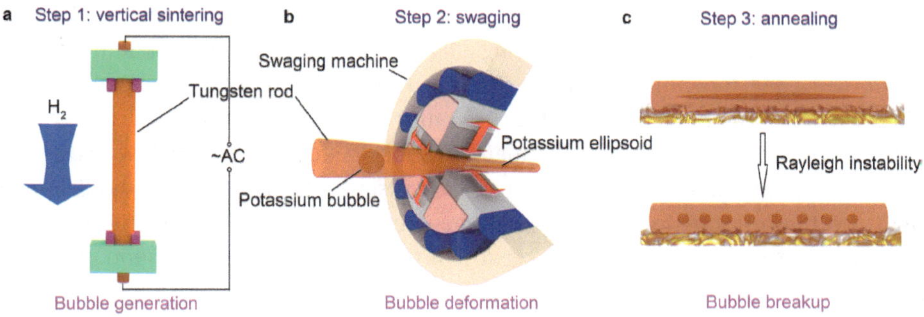

Figure 2.29: Evolution of potassium bubbles in typical commercial potassium-doped tungsten preparation [68].

Japan's ALMT prepared potassium and rhenium-doped tungsten through cold isostatic pressing, sintering, hot rolling, and heat treatment [69]. They also produced tungsten rods with a potassium content of 3.2×10^{-3} wt% by powder metallurgy and rotary swaging [70]. PLANSEE AG in Europe developed WK65 (60–65 ppm K), W vacuum metallizing (30–70 ppm K), and WVM tungsten (WVMW/S-WVMW, 15–40 ppm K) [71]. Schwabmuenchen [72] produced potassium-doped tungsten (6×10^{-3}–7.5×10^{-3} wt%) through drawing processes, and Allied Material Corp. [73] prepared WK bars by powder metallurgy and rotary swaging. Rieth et al. [74] compared WVM rods with pure tungsten and found that the DBTT of WVM was approximately 1,000 °C, which was higher than that of pure tungsten. As indicated in Figure 2.30, only pure tungsten and WVM had an impact absorbed energy plateau (plastic fracture region), starting at 900 and 1,000 °C, respectively. The rod-shaped materials were also found to perform better than the rolled plate, indicating that the microstructure has a significant influence on the plastic transition. Gludovatz et al. [75] investigated the fracture toughness of different tungsten materials, including AKS-W, using different testing methods (three-point bending, double cantilever beam, and compact tension). The fracture toughness of all materials increased with temperature, and AKS-W had a higher fracture toughness value than pure tungsten in the test temperature range (400–800 °C). Sasak et al. [76] prepared potassium-doped tungsten and pure tungsten by powder metallurgy and hot rolling. Potassium-doped tungsten exhibited a higher yield strength and lower DBTT than pure tungsten, with potassium addition refining the grains and enhancing the mechanical properties of tungsten. Figure 2.31 illustrates the stress-strain curves at different temperatures and strain rates, indicating that potassium-doped tungsten exhibited a certain elongation at room temperature, while pure tungsten showed plasticity at 200 °C. Nikolić et al. [77] found that potassium doping could inhibit the recrystallization process of tungsten, as shown in Figure 2.32. The pure tungsten grain grew significantly at 1,600 °C and transformed into equiaxed crystals, while the structure of the W-K filaments remained stable throughout the entire temperature range.

In China, Yan et al. [78] used AKS powder sintering and rotary forging to prepare W-K bulk materials with densities up to 98.2%, exhibiting excellent mechanical properties and thermal shock resistance. Tang et al. [68] prepared nano-potassium bubble-reinforced tungsten bulk materials (KB-W) by combining AKS doping with SPS (Figure 2.33). Nano-potassium bubbles (~50 nm) distributed within tungsten grains and grain boundaries enhanced the material's mechanical properties, lowered tungsten's DBTT, and improved its thermal shock resistance through nanocrystallization. Owing to the introduction of the porous structure, the thermal conductivity of the W-K alloy is relatively reduced, especially in the high-temperature region, necessitating optimization of the porosity and pore size of the W-K alloy.

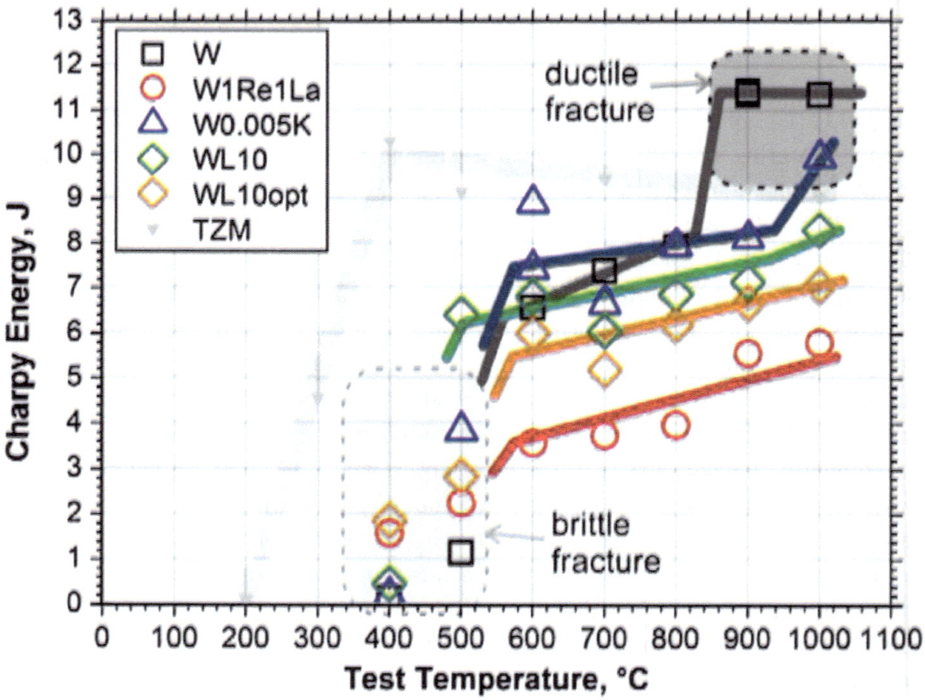

Figure 2.30: Charpy impact test results of tungsten-based alloy rods [74].

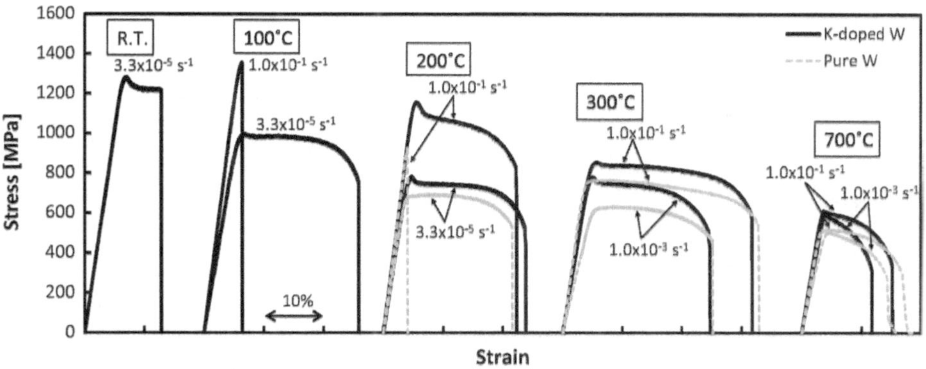

Figure 2.31: Stress-strain curves of pure tungsten and potassium-doped tungsten at different temperatures and strain rates [76].

Figure 2.34 illustrates the thermal conductivity of PW and KB-W against temperature, and the thermal conductivity of KB-W was found to be almost the same as that of PW, indicating that potassium bubbles have better thermal conductivity than ceramic phases, as ceramic phases can damage the thermal conductivity of tungsten alloys.

Both potassium-doped tungsten filaments and W-K bulk exhibited superior high-temperature mechanical properties compared to other tungsten-based materials. The high-temperature mechanical performance of WK-Re, WK-TiC, WK-Y, and WK-Ti prepared through different methods was enhanced by utilizing the strong inhibitory effect of potassium bubbles on dislocation movement [79]. The dispersed potassium bubbles significantly reduced the strain sensitivity of W-3%Re and improved its plasticity and strength at low temperatures; however, the fracture strain of potassium-doped W-3%Re at 474 and 573 K was smaller than that of W-3%Re [69]. Y-doped WK alloy inhibited grain growth and significantly increased the hardness of the material [80]. The addition of TiC impeded the recrystallization process of WK alloy, elevating the recrystallization temperature (1,673–1,873 K), and enhancing the tensile strength at room temperature (from 141 to 353 MPa) [81].

2.3.2 Oxide dispersion-strengthened tungsten (ODS-W)

By means of dispersion intensification, the incorporation of ultrafine oxide into tungsten effectively impedes dislocation movement, significantly enhancing the mechanical properties of tungsten materials at high temperatures and the strength and toughness of tungsten at room temperature. The most widely used dispersed oxides include ThO_2, La_2O_3, Y_2O_3, and Ce_2O_3, known for their high melting points and stable physical and chemical properties, which can effectively hinder the grain growth of materials during sintering and processing. This significantly improves the recrystallization temperature

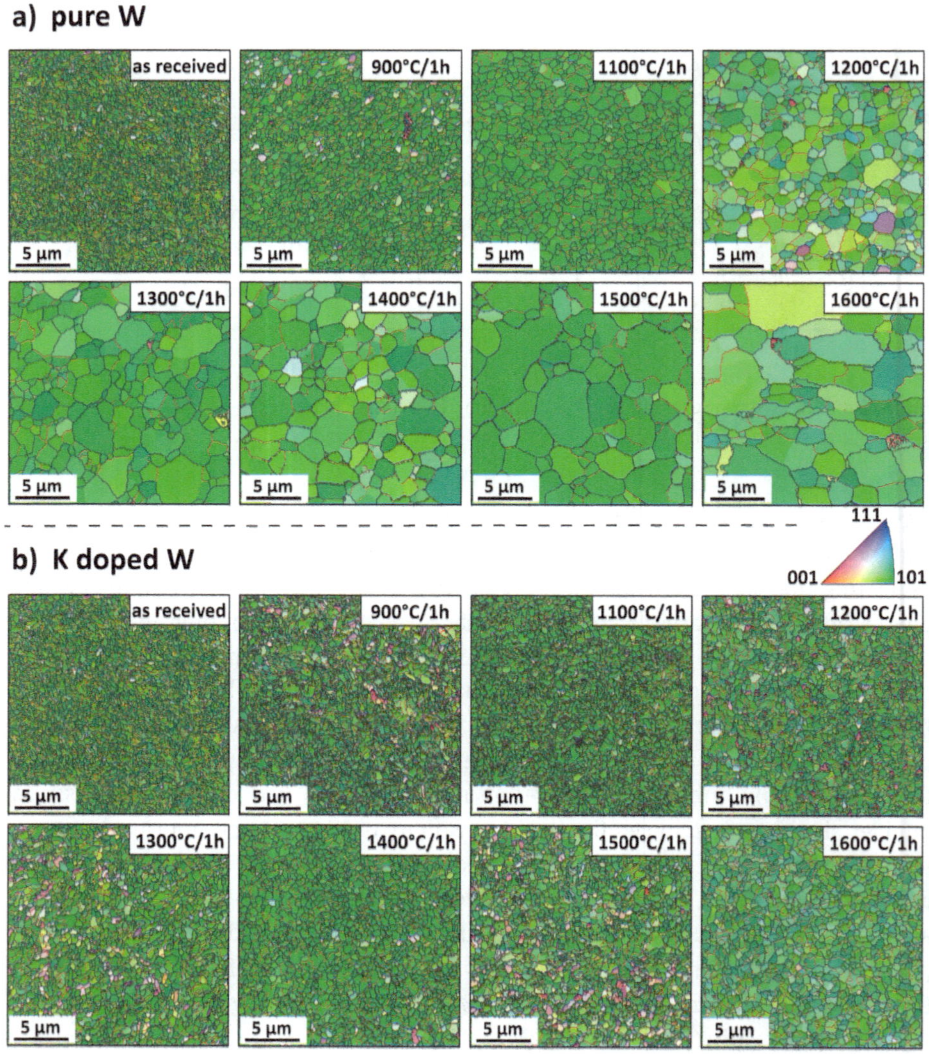

Figure 2.32: EBSD orientation of pure tungsten and potassium-doped tungsten wire cross-sections [77].

and microstructure stability of tungsten-based materials, which is particularly important for applications in high-temperature structural components requiring stable performance. In addition, rare earth elements interact strongly with oxygen at the grain boundary, adsorbing impurity elements, reducing impurity segregation and purifying grain boundaries. Particularly, tungsten-based materials can achieve both low-temperature toughness and high strength after deformation treatments such as rolling and swaging. Simultaneously, the dispersed second phase in tungsten-based materials creates numerous phase interfaces, which can annihilate the point defects caused by irradiation, thereby improving the irradiation resistance of tungsten-based materials.

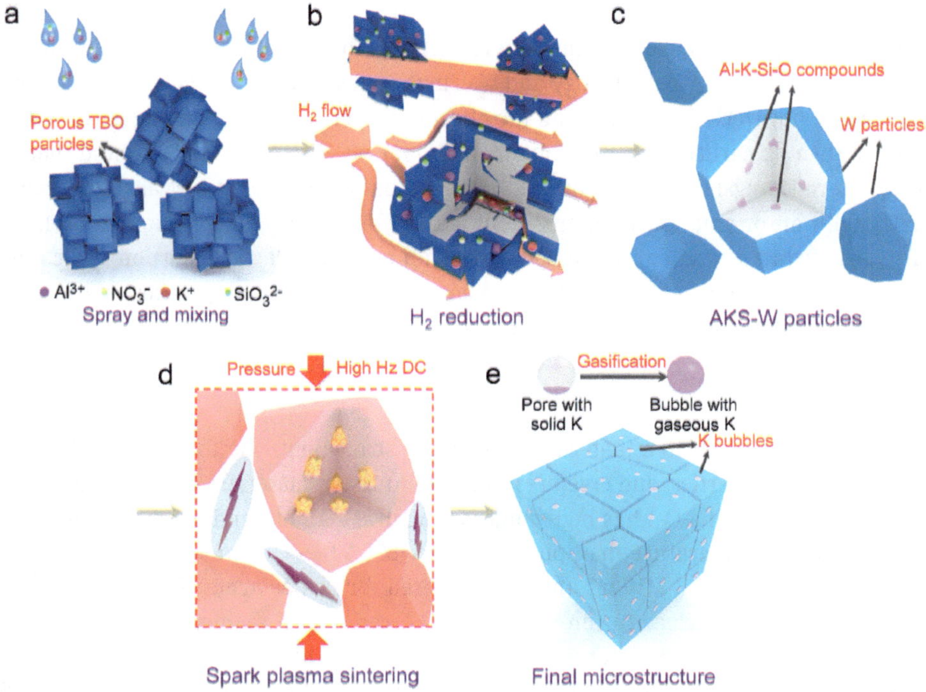

Figure 2.33: AKS-SPS preparation process of potassium bubble-reinforced tungsten alloy [68].

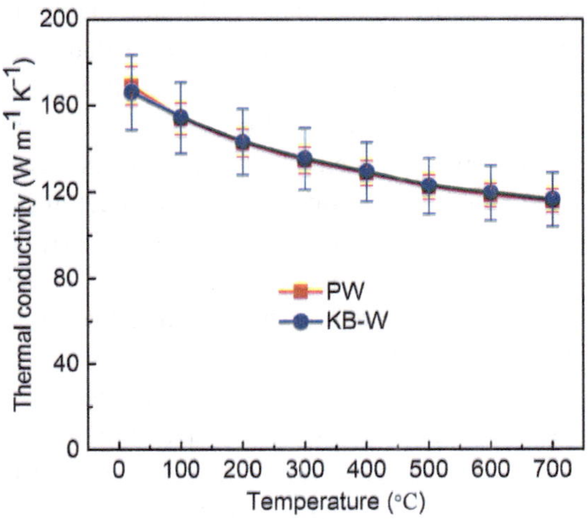

Figure 2.34: Comparison of thermal conductivity between powder metallurgy tungsten (PW) and KB-W [68].

These oxides are typically incorporated into the powder through high-energy ball milling or by reacting rare earth elements with oxygen. They then precipitate in the sintering process, with their morphology and distribution determined by the subsequent sintering and hot working process, and the effect of dispersion strengthening is related to factors such as particle size, distribution, and quantity. Huang et al. [82] found that in the sintered tungsten matrix with SPS, the combination of O and W also generated WO_2, with an orientation relationship of (101)W//(020)WO_2 and a mismatch of 4.9%. This resulted in a coherent phase boundary with a certain elastic distortion, indicating that residual oxygen elements in the tungsten matrices, along with preexisting tungsten oxides, contributed to effective dispersion toughening.

2.3.2.1 W-La_2O_3

The La_2O_3-doped tungsten material significantly strengthens the thermal shock resistance, tensile properties, and creep resistance of tungsten materials. Commercial PLANSEE WL10 (W-1wt%La_2O_3) has better microstructural stability and a 500 °C higher recrystallization temperature than pure tungsten [83]. W-La_2O_3 can withstand a steady-state thermal load of 6 MW/m^2 and is considered for use as armor for divertors and baffles for thermonuclear fusion devices [84]. A rod with a diameter of 10 mm, subjected to final hot-rolled (deformed >80%), contains lanthanum oxide granular fibers approximately 400 nm in diameter and 20 μm in length [85]. Rieth et al. [87] performed Charpy impact tests on commercial pure tungsten and WL10 and found that the DBTT of pure tungsten was 800 ± 50 °C, and the DBTT of WL10 was 950 ± 50 °C, as shown in Figure 2.35. WL10 rods behaved more like unidirectional fiber-reinforcing materials, with cracks propagating along the fibrous La_2O_3, resulting in a higher DBTT than pure tungsten. As a structural material under fusion neutron irradiation, the DBTT of the alloy must be at least 200–400 °C. As a result, neither material is suitable as divertor construction material. Rieth et al. [74] further studied the effects of microstructure characteristics (grain size, texture, anisotropy, and notch) on pure tungsten and WL10 rod materials using Charpy experiments and found that although the addition of La_2O_3 improved tensile strength, creep resistance and inhibition of recrystallization, it did not improve the fracture properties of tungsten. However, the plasticity of the alloy can be enhanced to a certain extent by rationally optimizing the preparation method to achieve a uniform grain size distribution and reduce the size of dispersed particles. Yan et al. [86] studied the effects of deformation process on the fracture behavior and DBTT of pure tungsten and WL10 and found that rotary forging combined with rolling treatment could reduce the DBTT of WL10 and enhance its mechanical properties.

The mechanical properties of molybdenum alloys were significantly improved by effectively introducing the second phase particles of La_2O_3 that are evenly distributed in the crystal and at the grain boundaries [87]. Zhong et al. [88] prepared W-1wt% La_2O_3 powder by a similar method of mechanical ball milling and sintered La_2O_3 reinforced tungsten after cold isostatic pressing. The addition of La_2O_3 inhibited the

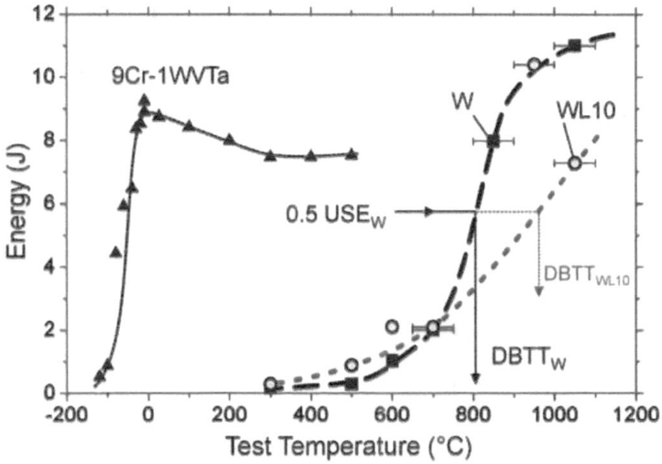

Figure 2.35: Transition from brittle to ductile of pure tungsten and WL10 [85].

growth of tungsten grains during the sintering process, increasing the relative density from 93.2% to 97.8%, and enhancing the bending strength by 35.7%. W-1wt%La_2O_3 also exhibited favorable heat load performance, capable of withstanding a power density of 6 MW/m^2. However, under higher heat loads, the higher surface temperature leads to the melting of La_2O_3 and the appearance of microcracks on the material surface. Cui et al. [89] prepared tungsten alloys in four different states by powder metallurgy including pure tungsten, rolled tungsten, rolled W-1%La_2O_3, and wrought W-1%La_2O_3. As shown in Figure 2.36, pure tungsten exhibited poor density, whereas the density of rolled tungsten remained low owing to its fibrous grains. However, the microstructure and morphology of W-1% La_2O_3 in the rolled and forged states were significantly improved, with increased density and refined grains. Numerous oxide particles were evenly distributed within tungsten grains and at grain boundaries.

Xia et al. [90] prepared high-purity W-1wt% La_2O_3 composites by coprecipitation and SPS, and the cubic crystal La_2O_3 particles were present in tungsten grains and at grain boundaries, with both tungsten and La_2O_3 grains exhibiting narrow size distributions, as shown in Figure 2.37. Yan et al. [91] used hydrothermal-hydrogen reduction method to prepare La_2O_3-doped ultrafine tungsten powders, and the prepared powders were spherical, with a particle diameter of about 700 nm and a narrow size distribution. The precursor had a hexagonal structure of WO_3 nanorods, and the grain size of the reduced powder was determined by the diameter of the WO_3 nanorods. The La_2O_3 phase was confirmed in both the precursor and the reducing powder, indicating the successful doping of La_2O_3 into tungsten. Yar et al. [92] prepared high-purity tungsten powder and W-0.9 wt%La_2O_3 nanocomposite powder by wet-chemical methods, and the tungsten particles were coated with an oxide precipitate phase. After the reduction of the precursor powder, grain growth was inhibited by SPS at 1,300 and 1,400 °C, and the grain size of the sample was related to the sintering temperature and heating rate of SPS, as shown in Table 2.2. The tungsten grains were mainly composed of micron-sized grains and finer subgrains, and the oxide particles were mainly concentrated at the grain boundaries. Tungsten existed in all oxides, indicating the presence of a complex oxide phase (W-O-La) in the samples after sintering.

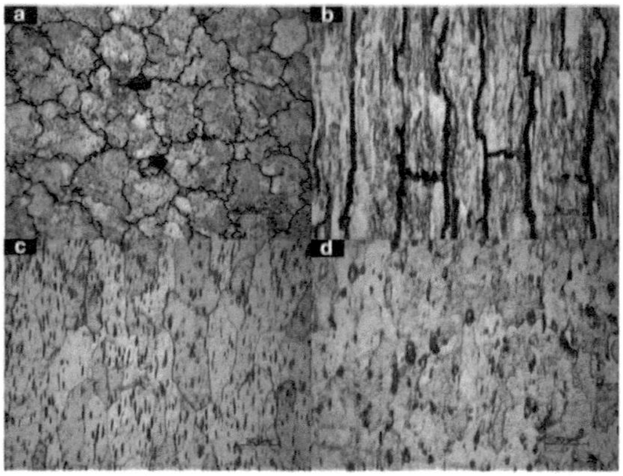

Figure 2.36: Optical microscope image of tungsten alloy: (a) pure tungsten; (b) rolled tungsten; (c) rolled W-1% La_2O_3; and (d) forged W-1% La_2O_3 [91].

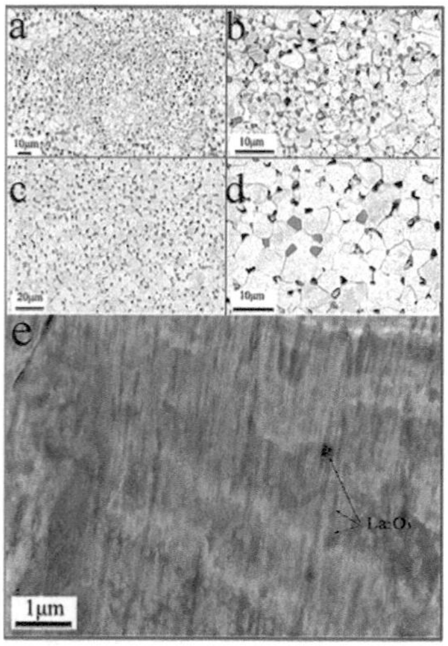

Figure 2.37: SEM-EBSD images of SPS-sintered W material (a, b) commercial La_2O_3/W powder with large grain size distribution and La_2O_3 particles clustered at some grain boundaries; (c, d) La_2O_3/W powder prepared by coprecipitation method, La_2O_3 particles are uniformly distributed at grain boundaries, and the grain size distribution is narrow; and (e) coprecipitation of tungsten nanoparticles with nanoscale tungsten sub-grain boundaries and distribution of ultrafine nano-La_2O_3 particles on a single tungsten grain [92].

Although the original purpose of adding La_2O_3 to tungsten was not to reduce the DBTT of the material, W-La_2O_3 was found to be easier to process at room temperature and therefore reduced processing costs. This is mainly because no micro-cracks formed around La_2O_3 particles during the deformation and processing [86]. However, with the addition of La_2O_3, the hardness increased, making hot processing difficult. The thermal conductivity of W-1wt%La_2O_3 was slightly reduced compared to pure tungsten, and Ghezzi et al. [93] put W-1wt%La_2O_3 armor in a quasi-steady-state plasma accelerator device and exposed to 100 marginal local mode (ELM) transient incidents, observing that the W-La_2O_3 armor performed worse than pure tungsten in the simulated transient event.

Table 2.2: Properties of spark plasma sintering specimens at different temperatures and temperature rates [92].

Components	Heating rate (°C/min)	Holding temperature (°C)	Density/relative density (g/cm^3)/(%RT)	Grain size (μm)	Microhardness (Hv)
Pure tungsten	100	1,300	17.3/91	3.0	317
W-0.9La_2O_3	100	1,300	16.9/89	4.2	327
W-0.9La_2O_3	50	1,400	17.8/94	9.7	406

2.3.2.2 W-Y_2O_3

Y_2O_3 has demonstrated excellent dispersion strengthening effects in tungsten-based materials and has received extensive attention. The fine particles of Y_2O_3 are dispersed in the tungsten matrix, which can significantly refine the tungsten grains, thereby improving the strength and creep performance at room temperature, and substantially increasing the recrystallization temperature of tungsten alloys. In addition, adding a certain amount of Y_2O_3 can improve the densification of sintered tungsten while reducing the sintering temperature. At the same time, compared with pure tungsten, yttrium oxide dispersion-strengthened tungsten-based materials have significantly improved resistance to thermal shock, radiation, and oxidation, making them valuable for plasma-facing material applications. The dispersion of Y_2O_3 can significantly improve the armor-piercing and self-sharpening capabilities of tungsten alloys, which have been used as armor-piercing warheads [94].

Liu et al. [95] prepared pure tungsten, W-1% La_2O_3 and W-1% Y_2O_3 materials by ball milling of nanocomposite powders and microwave sintering at 1,500 °C for 30 min, and the sintering density varied with different components, among which the relative density of pure tungsten and W-1% Y_2O_3 was higher, close to 97%, while the W-1% of La_2O_3 was only 95%, as shown in Table 2.3. Both Y_2O_3 and La_2O_3 nanoparticles could play a role in grain refinement and significantly hinder the growth of tungsten grains during the sintering process, but Y_2O_3 was more pronounced, as shown in Figure 2.38. The hardness and thermal conductivity of W-1% Y_2O_3 were also higher than those of W-1%La_2O_3, but their thermal conductivity was worse than that of pure tungsten. In contrast, the addition of Y_2O_3 has a better effect on the densification, strengthening and grain refinement of tungsten alloy. Liu et al. [96] prepared nanoscale W-1% La_2O_3 and W-1%Y_2O_3 composites by sol-gel and microwave sintering, in which the average size of powder particles was less than 50 nm. Microwave sintering adopts a low sintering temperature (1,500 °C) and a short holding time (30 min) to prevent excessive grain growth and make the nanoscale oxide particles of 10–50 nm evenly distributed in the tungsten matrix. From the comparison of grain size and performance data, the density, average crystal particle size, and Vickers microhardness of W-1%La_2O_3 and W-1% Y_2O_3 after sintering were found to be of 92.4% and 93.6%, 1.1 and 0.7 μm, 4.12 and 6.03 GPa, respectively. This indicates that W-1% Y_2O_3 exhibits better sintering properties, finer grains, and higher hardness than W-1%La_2O_3, as shown in Table 2.4. With ODS-W prepared by the sol-gel method, the solution was mixed uniformly at the molecular level, and the distribution of oxide particles obtained was more uniform than that obtained by the mechanical ball milling method.

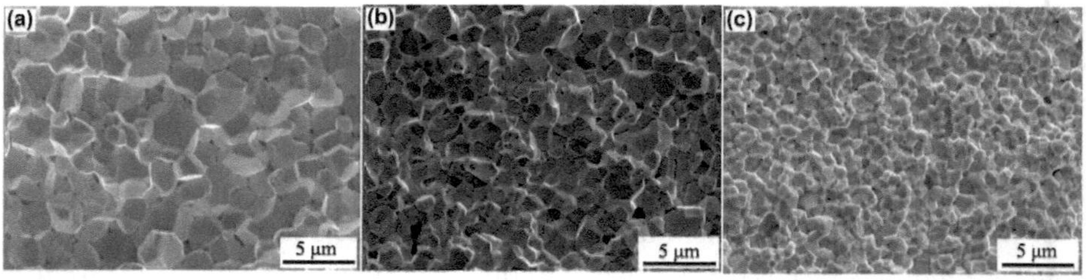

Figure 2.38: SEM diagram of tungsten alloy section: (a) pure W; (b) W-1%La_2O_3; and (c) W-1%Y_2O_3 [95].

Table 2.3: Properties of pure tungsten and ODS-W prepared by ball milling and microwave sintering at 1,500 °C for 30 min [95].

Material	Density (g/cm^3)	Relative density (%)	Grain size (μm)	Microhardness (GPa)	Thermal conductivity (W/m• K)
W	18.7	96.9	3.2	5.04	156
W-1wt%La_2O_3	18.0	95.0	1.4	4.21	121
W-1wt%Y_2O_3	18.2	96.8	0.7	6.91	131

Yar et al. [97] improved the wet-chemical method to prepare W-1wt%Y_2O_3 nanocrystalline powder and then used discharge plasma sintering to obtain W-1wt%Y_2O_3 material. The two-step sintering method was used to explore the effect on the sintering properties of the reduced powder so that the sintered body obtained high density and finer grains. Compared with W-La_2O_3 sintered with SPS, it can be found that W-Y_2O_3 has smaller grains, higher density, and greater microhardness, as well as better sintering performance. Wahlberg et al. [98] used a novel wet-chemical method to pre-

Table 2.4: Properties of ODS-W prepared by sol-gel and ball milling and microwave sintering at 1,500 °C for 30 min [96].

Material	Preparation method	Density (g/cm^3)	Relative density (%)	Grain size (μm)	Microhardness (GPa)
W-1wt%La_2O_3	Sol-gel	17.49	92.4	1.1	4.12
W-1wt%Y_2O_3	Sol-gel	17.60	93.8	0.7	6.03
W-1wt%La_2O_3	Ball milling	18.0	95.0	1.4	4.21
W-1wt%Y_2O_3	Ball milling	18.2	96.8	0.7	6.91

pare W-Y_2O_3 composite powders with highly homogeneous distribution, which were sintered into bulk materials with a density of 88% at SPS at 1,100 °C. Examination of the fracture morphology of the sintered specimen (Figure 2.39) revealed nano-oxide particles distributed on the surface of the tungsten grains, along with pits resembling oxide particles in size and shape. This observation indicated that some oxide particles detached from the fracture surface during the fracture process, highlighting their presence in at least two regions: the surface and the interior of the tungsten grains, with a particular concentration at the grain boundaries and within the grains themselves.

Kim et al. [99] prepared Y_2O_3, HfO_2, and La_2O_3 diffusion-reinforced tungsten matrix composites by mechanical alloying and SPS. Among these oxides, Y_2O_3 was found to be the most effective element in promoting the densification of tungsten during sintering, primarily due to the formation of a new liquid phase consisting of W, Y, and O. During sintering, the W atom can diffuse and move through the grain boundary or the Y_2O_3 phase via the continuous dissolution-reprecipitation process, so W and Y_2O_3 were soluble, thus improving sintering activity of tungsten. This is different from the study reported by Itoh et al. [100] that W-Y_2O_3 composites contain only W and Y_2O_3. With the increase of Y_2O_3 content, the melting point of tungsten decreased, enhancing atomic mobility, and when the content of Y_2O_3 diffuse phase reached 5 wt%, the density of the composite material approached 100%. When the oxide doping amount was 0.5 wt%, the density of the composite was less than that of pure tungsten, regardless of the oxide type. When the doping amount was higher than 1wt%, the density of W-Y_2O_3 composites improved with higher Y_2O_3 content. However, for HfO_2 and La_2O_3 doping, the density of the composite material was always less than that of pure tungsten, as shown in Figure 2.40. The grain sizes of the three ODS-W materials decreased with the increase of oxide content: W-Y_2O_3, W-HfO_2, and W-La_2O_3 decreased from 18.8 μm to 3.7, 11.6, and 9.2 μm, respectively. All oxide particles are capable of preventing tungsten grain growth, but Y_2O_3 is the most effective element to prevent grain coarsening. Additionally, Y_2O_3 proved to be the most effective strengthening element, with the strength of W-Y_2O_3 increasing to 1.5 times the hardness of tungsten as Y_2O_3 content rose. In contrast, adding HfO_2 only slightly increased tungsten strength, while La_2O_3 addition above 2 wt% significantly reduced tungsten hardness.

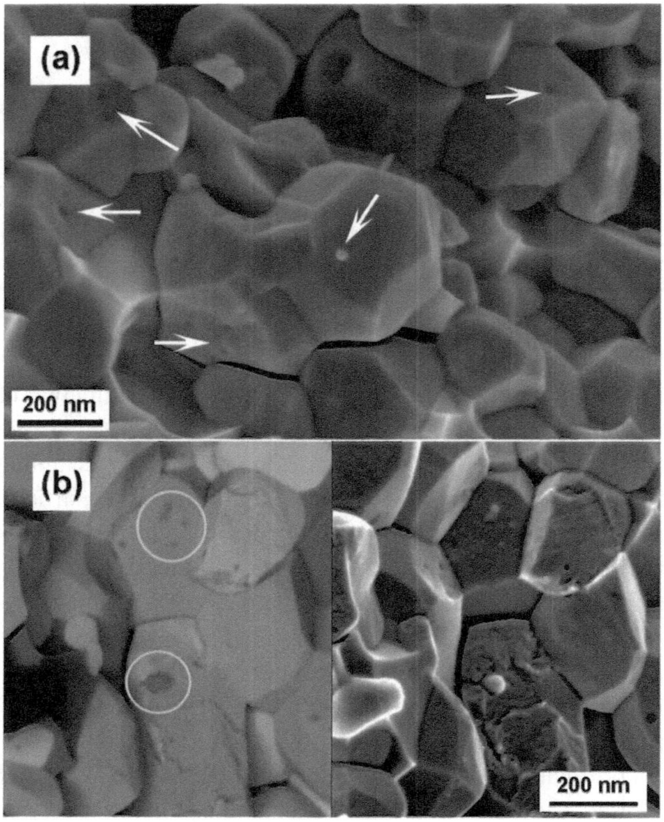

Figure 2.39: (a) The fracture surface of the sintered sample shows pits on the surface of oxide nanoparticles and tungsten grains and (b) backscattered electron image showing black-gray oxide nanoparticles [98].

Dong et al. [101] enhanced the traditional wet-chemical method by adding the surfactant sodium dodecyl sulfate (SDS) and the macromolecular compound polyvinylpyrrolidone (PVP), combined with SPS, to prepare W-Y_2O_3 alloys (Figure 2.41), with fine tungsten grains and uniformly distributed Y_2O_3. As can be seen from Figure 2.42, there were high-density Y_6WO_{12} nanoparticles (10 nm) inside the tungsten grains, which were generated by the diffusion reaction between tungsten atoms and nano-Y_2O_3 particles, consistent with the findings of Kim [101]. The Y_6WO_{12} phase inside the tungsten grain had a coherent relationship with the W matrix, contributing to the material's strengthening. There was a 10-nm thick W atom diffusion layer at the interface between Y_2O_3 and W grains. Due to the limitation of the number of W atom diffusion, no new phase was formed at this interface, and the oxide agglomerated at the tungsten grain boundary remained in the Y_2O_3 phase. Tan et al. [102] prepared W-2vol% Y_2O_3 by wet-chemical method and SPS technique and found that the chemical prepa-

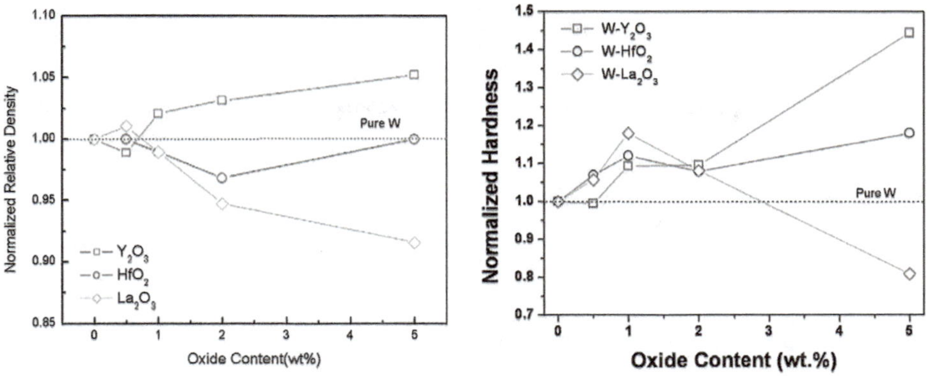

Figure 2.40: Variation of relative density and hardness of ODS-W with oxide content [99].

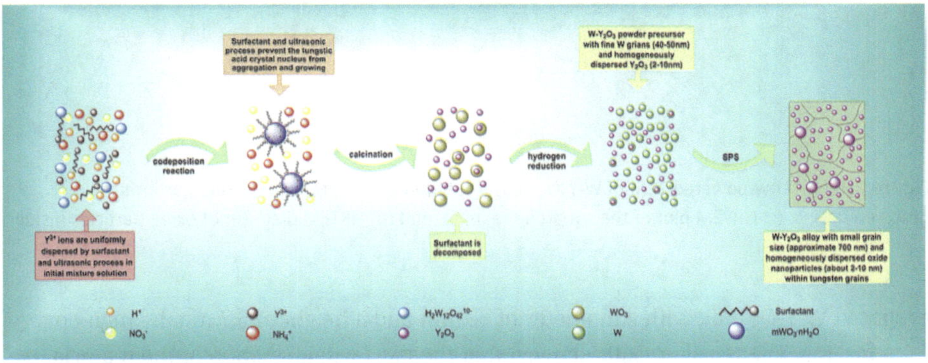

Figure 2.41: Schematic diagram of the process of preparation of W-Y_2O_3 alloy by wet-chemical method and SPS [101].

ration of the powder could refine the grain and particles, improving the mechanical properties of the material.

Battabyal et al. [103, 104] collaborated with Plansee to prepare W-2%Y_2O_3 composites with a relative density of 99.3% by sintering and hot forging. Among them, the average grain size of W matrix was 1–2 μm, and the particle size of Y_2O_3 was 300–1,000 nm. W-2Y_2O_3 began to exhibit plasticity at 400 °C, and its flexural strength gradually decreased with increasing temperature, from ~1.3 GPa at 298 K to ~580 MPa at 1,273 K [103]. As shown in Figure 2.43, the tensile results demonstrated that the material exhibited significant plasticity within the temperature range of 673–1,273 K, with an ultimate tensile strength of 500 MPa at 400 °C and an elongation of about 10%. However, its Charpy impact performance was poor, showing a very low impact absorption work value at 773–1,273 K lower than that of WVM and WL10. The impact energy increased significantly at 773–873 K, indicating that the DBTT of this material was at 773–873 K [106]. Lian et al. [105] prepared W-0.7vol%Y_2O_3 powder by liquid-liquid doping and then

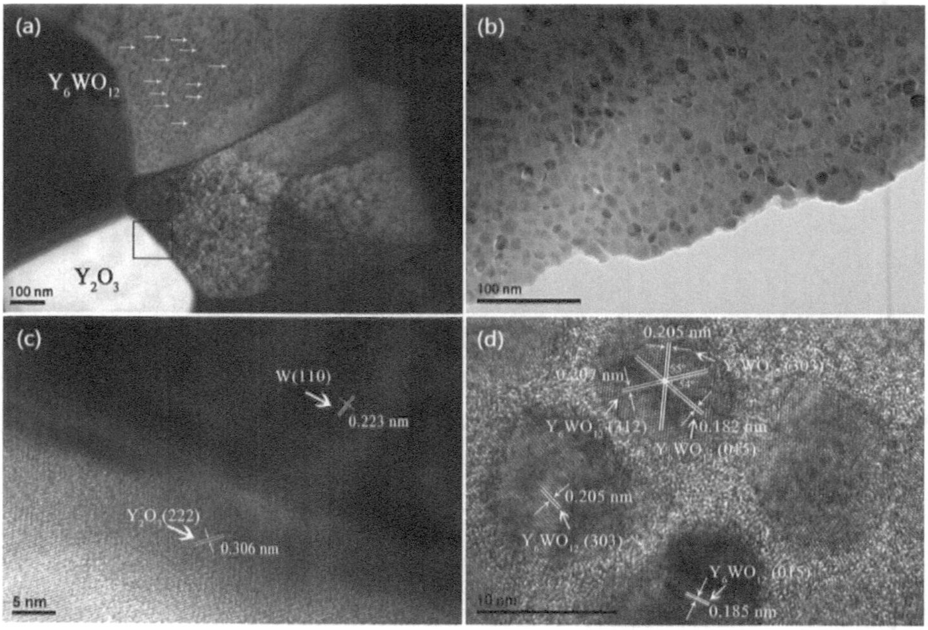

Figure 2.42: (a) Low-powered TEM of W-Y_2O_3 alloy; (b) high-power TEM shows oxide particles inside tungsten grains; (c) HRTEM plot of the square area in (a); and (d) HRTEM diagram of oxide particles inside tungsten grains [101].

prepared W-Y_2O_3 bulk with a diameter of 62 mm and a height of 6.7 mm by high-energy -rate forging. The tensile results showed that the plastic deformation of the material occurred at 100 °C, achieving a of 1,040 MPa with a strain of 2.9%. Xie et al. [106] prepared W-1.0%Y_2O_3 (WY10) materials by SPS and high-temperature rotary swaging and found that the bars after rotary forging underwent tensile plastic deformation at 200 °C, with a strength of about 660 MPa, and tensile plasticity at 150 °C after annealing at 1,300 °C for 1 min. For materials that had not undergone deformation, tensile plastic deformation did not occur until 500 °C. By rotary forging and annealing, the thermal conductivity of W-Y_2O_3 alloy had been greatly improved, and the thermal conductivity of the alloy had reached 198 W/m•K and pure tungsten only 175 W/m•K, as shown in Figure 2.44. Zhao et al. [107] prepared W-Y_2O_3 alloys by SPS and high-temperature sintering plus hot rolling, and the results of three-point bending and high-temperature tensile at room temperature showed that the hot-rolled W-Y_2O_3 samples had higher mechanical strength and toughness. Compared with the samples prepared by SPS, the thermal conductivity of deformed W-Y_2O_3 was increased by nearly 35% to 143.3 W/m•K at room temperature, indicating that the deformation favored the heat dissipation from the loading surface to the matrix. However, the thermal conductivity of both samples was smaller than that of pure tungsten. Compared with SPS-sintered small specimens, the properties of fast forging, rotary forging, or rolling materials had been greatly improved.

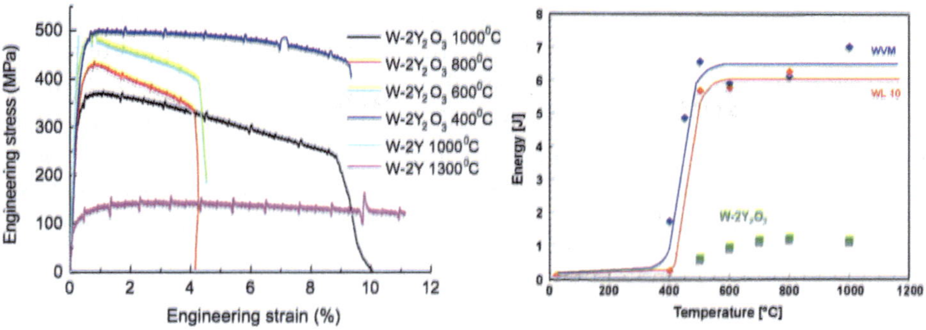

Figure 2.43: Engineering stress-strain curves and charpy impact test results of W-$2Y_2O_3$ alloy [106].

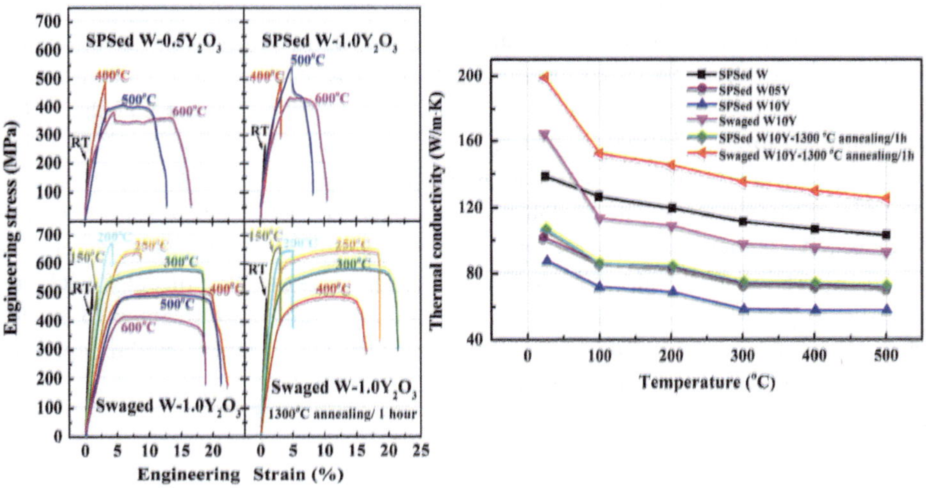

Figure 2.44: Comparison of tensile behavior and thermal conductivity of W-$2Y_2O_3$ alloy [106].

2.3.2.3 W-ThO_2

Since the last century, Sadek et al. [108] have been studying the rare earth oxide ThO_2 and have identified that the addition of ThO_2 to tungsten can effectively increase the yield strength of tungsten and significantly reduce its recrystallization brittleness. Furthermore, they have shown that it also improves the high-temperature strength of tungsten alloys. However, due to its radioactivity after neutron irradiation, ThO_2 is no longer considered a reinforcement phase for tungsten in the PFMs. In addition, Ding et al. [109, 110] also prepared rare earth oxide Pr_2O_3 and CeO_2 diffusion-reinforced tungsten matrix composites by wet chemical method and SPS sintering, which achieved good results in obtaining high-density fine-grained tungsten materials.

Consequently, the incorporation of rare earth oxides into tungsten alloys represents an effective approach for fabricating fine-grained and high-density tungsten-based alloys. Nevertheless, during the actual sintering process of W-La_2O_3/Y_2O_3 powder, the added nanoscale (<50 nm) La_2O_3 or Y_2O_3 particles are prone to agglomeration and growth, ultimately resulting in the formation of micron-sized particles (1–10 μm) La_2O_3 or Y_2O_3 in the material, which weakens the strengthening effect, causes stress concentrations that lead to cracking, and subsequently reduces both the ductility and toughness of the material.

2.3.3 Tungsten-carbide composites (CDS-W)

The melting point of doped oxides is typically considerably lower than that of tungsten itself. In the context of a fusion reactor, the oxides are susceptible to melting before tungsten in the event of an instantaneous thermal shock or local bombardment by high-energy particles. The melting points of La_2O_3 (~2,300 °C) and Y_2O_3 (~2,400 °C) are relatively low; only ThO_2 (~3,050 °C) having a melting point above 3,000 °C. In comparison to oxides, tungsten-based alloys with carbide dispersion strengthening exhibit distinct advantages: TiC (~3,000 °C), ZrC (~3,500 °C), HfC (~3,900 °C), and other carbides have higher melting points and better compatibility with tungsten materials. The incorporation of carbides, such as TiC, ZrC, or HfC, into tungsten through mechanical alloying and deformation processing yields comparable outcomes to those achieved through oxide diffusion strengthening. Moreover, carbides exhibit superior properties at low doses of neutron irradiation.

2.3.3.1 W-TiC

TiC and tungsten exhibit similar thermal expansion coefficients, high melting points, low densities, and robust high-temperature strengths and can form (Ti, W)C solid solution with W, which serves as an effective reinforcing component for tungsten. In a study by Kurishita et al. [111], an ultrafine-grained (UFG) recrystallized (UFGR) W-(0.25–1.5%) TiC alloy, designated UFGR W-TiC, was prepared by mechanical alloying and HIP process. TiC particles were dispersed throughout the grains and at the grain boundaries, exhibiting an average grain size of 15 and 30 nm, respectively. In 2004, Ishijima et al. [112] successfully prepared a fully dense UFGR W-TiC material for the first time, which was sintered by HIP at a temperature of two-fifths of the melting point and achieved complete densification, with minimal grain growth in comparison to the ball milling powder. The UFGR W-TiC alloy had attracted significant attention both domestically and internationally, being the first instance of tungsten-based materials with nanostructures being prepared. UFGR W-TiC exhibited equiaxed crystals in the recrystallized state with an average grain size of 50–200 nm, varying depending on the ball milling atmosphere and TiC content. TiC additions of as little as 0.25 wt%

had been demonstrated to greatly refine the grains, whereas W-TiC/Ar exhibited grain size approximately half that of W-TiC/H [113], a consequence attributed to the pinning effect of nanoscale Ar bubbles on grain boundaries. The proportion of ∑1 to ∑29 b coincidence site lattice (CSL) boundaries was only 20%, with majority of grain boundaries exhibiting high energy and arbitrary orientations (Figure 2.45). Low-energy grain boundaries, such as small-angle grain boundaries and low-∑ coincidence site grain boundaries, exhibited superior mechanical properties, including higher fracture strength and enhanced plasticity, in comparison to high-energy irregular large-angle grain boundaries. Additionally, these low-energy grain boundaries could reduce the vacancy density at the grain boundaries and prevent the segregation of harmful impurity elements in the grain boundaries. As illustrated in Figure 2.46, the fracture strength of W-TiC at room temperature was contingent upon the TiC added, the ball milling atmosphere, and the HIP pressure. Adding 0.25 wt% TiC greatly improved the fracture strength of W-TiC/Ar and W-TiC/H, reaching 1.3 and 2 GPa, respectively, but the fracture strength could not be changed by increasing the TiC content to 1.5 wt% [114]. The fracture strength of W-1.1TiC/Ar reached 4.4 GPa after HIP sintering at 1 GPa pressure and 1,923 K thermoplastic processing, which was the highest fracture strength among tungsten materials reported so far. Both materials had no plasticity at room temperature with DBTT up to 830 K, mainly due to the high yield strength of UFGR W-TiC, whose maximum fracture strength was also much smaller than its yield strength. Figure 2.47 shows a comparison of the CTEs and thermal conductivity of UFGR W-0.5TiC/H and ITER-grade pure tungsten [115], with the findings that the CTEs of W-0.5TiC/H and pure tungsten are basically the same in the test temperature range (300–1,800 K). However, the thermal conductivity of the two materials differs, with W-0.5TiC/H exhibiting a lower thermal conductivity than pure tungsten. Nevertheless, at high temperatures, the thermal conductivity of the two materials approaches a similar value. The thermal conductivity of pure tungsten decreases with increasing temperature, but the thermal conductivity of W-0.5TiC/H basically does not change with temperature, remaining at 100 W/m• K due to its high density of grain boundaries and dispersed phase particles.

Kurishita et al. [116] prepared an UFG near-fully dense W-(0.3–0.7)%TiC alloy by mechanical alloying and HIP, with a grain size of 0.06–0.2 μm and a relative density of 99%, and the degree of grain refinement depended on the TiC content and ball milling atmosphere. W-0.5%TiC was ball-milled in a hydrogen atmosphere, and its three-point flexural fracture strength reached 1.6–2 GPa at room temperature. Figure 2.48 shows the TEM diagram of a TiC particle at the grain boundary after W-0.3TiC neutron irradiation, revealing a K-S orientation relationship between the TiC particle and the tungsten matrix (111)fcc | | (110)bcc,[10]fcc | | [11]bcc. The existence of such a good orientation relationship between the dispersed phase of TiC and tungsten implies that TiC is precipitated from the supersaturated state. The mechanical alloying process helps decompose TiC into Ti and C, forming a W-Ti-C solid solution; HIP sintering also promotes the reaction of Ti and C solutes with the W matrix to form a fine TiC precipitate

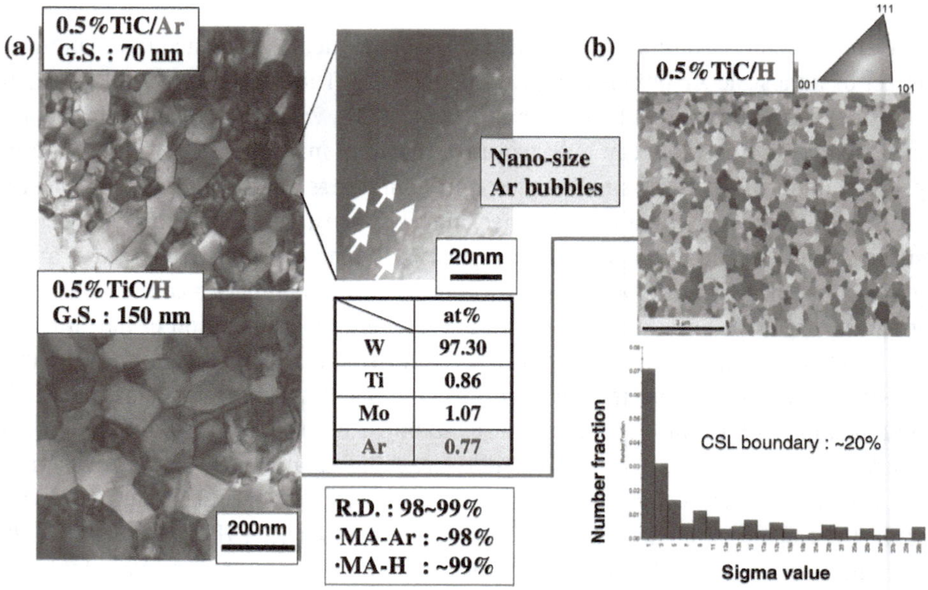

Figure 2.45: (a) TEM bright-field images of W-0.5TiC/H and W-0.5TiC/Ar and (b) EBSP plot showing the grain boundary orientation of W-0.5 TiC/H [113].

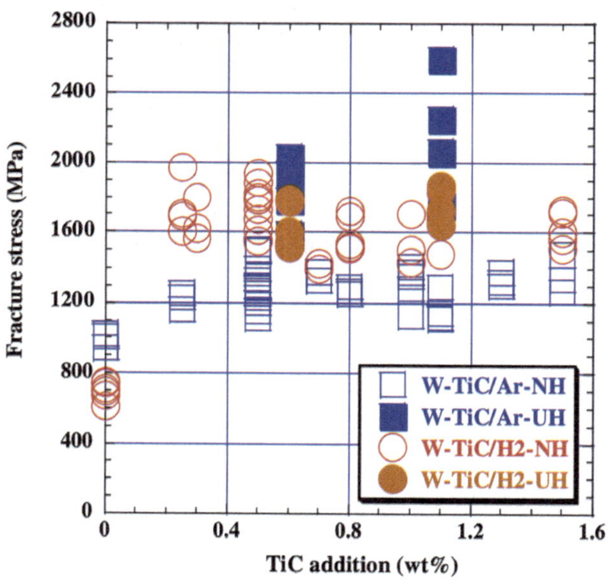

Figure 2.46: The effects of TiC addition, ball mill atmosphere, and HIP pressure on the three-point flexural fracture strength of UFGR W-TiC at room temperature, where NH represents HIP pressure of 0.2 GPa and UH represents ultra-high pressure of 1 GPa [114].

phase. TiC is able to self-regulate lattice constants, forming a solid solution with W, and generating non-stoichiometric TiC_x segregation at grain boundaries, which serves to reinforce these grain boundaries. The K-S relationship is already established and does not depend on neutron irradiation. It can be observed that TiC particles exhibit remarkable stability under neutron irradiation, and the K-S structure has been demonstrated to effectively enhance tungsten's resistance to neutron irradiation.

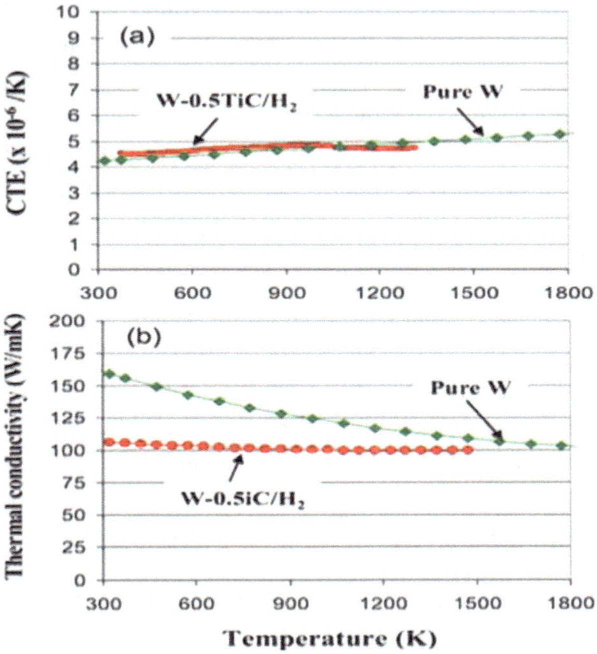

Figure 2.47: (a) Coefficient of thermal expansion and (b) thermal conductivity of W-0.5TiC/H and ITER-grade pure tungsten as a function of temperature [115].

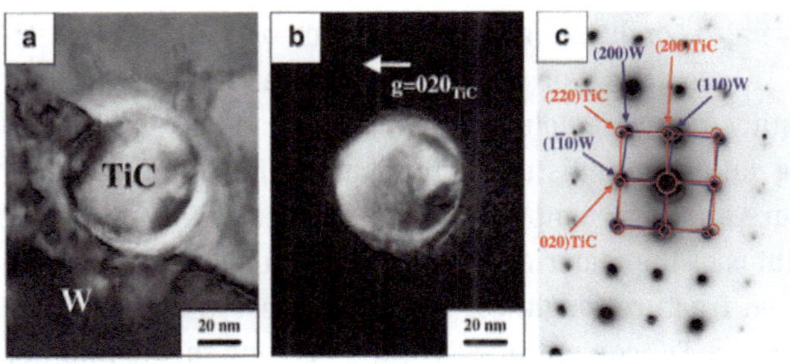

Figure 2.48: TEM diagram of a TiC particle at the grain boundary after W-0.3 TiC neutron irradiation: (a) bright-field image; (b) darkfield image; and (c) diffraction plot of the constituency [116].

Additionally, these UFGR W-TiC composites, despite their enhanced properties, exhibit certain limitations. These include a higher DBTT and a lower thermal conductivity compared to ITER-grade pure tungsten. In their study, Kurishita et al. [111] prepared a W-(0.25–1.5%) TiC composites material, which was toughened, fine-grained, and recrystallized through a process of superplastic deformation using MA-HIP-GSMM (grain boundary sliding-based microstructural modification) method. This process, which activates grain boundary slip and maintains the recrystallized equiaxed crystal state, has the potential to reduce the yield strength by facilitating grain rotation and movement between adjacent grains during superplastic deformation. Consequently, the processed tungsten material exhibited a DBTT lower than room temperature, along with excellent thermal shock and thermal fatigue properties as well as superior resistance to neutron irradiation and plasma etching resistance compared to pure tungsten. The mechanism of strengthening is as follows: Mechanically alloyed tungsten mixed with TiC powder precipitates nano-TiC particles at the grain boundary during HIP sintering, thereby strengthening the grain boundary; following this, superplastic deformation at high temperature densified the sintered billet and continued to precipitate TiC; additionally, the oxygen content was strictly controlled to reduce the formation of the brittle phase W_2C.

For UFGR W-1.1TiC/H, the DBTT was found to exceed 830 K, and the GSMM treatment technology significantly reinforced the grain boundaries, reducing the DBTT to room temperature or even lower and enhancing the fracture strength, as illustrated in Figure 2.49. The strengthening of the grain boundaries was attributed to an increase in the number of TiC dispersed phases at grain boundaries, which changed the structure of the grain boundaries from high-energy irregular boundaries to CSL grain boundaries. The DBTT of W-1.1TiC/H treated by GSMM technology depends on the oxygen and nitrogen content, especially the oxygen content. A large amount of oxygen forms oxides of Ti and W_2C phase, thereby inhibiting the rebonding of TiC decomposed into Ti and C. While W_2C is a brittle phase as a source of cracks. However, implementing the GSMM technology does not improve the thermal conductivity of UFGR W-TiC/H; the thermal conductivity remains largely unchanged.

Domestic researchers have also conducted in-depth research on tungsten materials diffusely reinforced with TiC particles. For instance, Song et al. [118] prepared a series of W-TiC composites with varying TiC volumes (0, 10, 20, 30, and 40 vol%) using ball milling and hot-pressing sintering and studied their mechanical and physical properties at room temperature and high temperature. Their research revealed that the hardness and elastic modulus of the materials increased with the addition of TiC, while the fracture strength and fracture toughness initially increased and then decreased, reaching a maximum at a TiC content of 20 vol%. Following the addition of TiC, a notable decline in thermal conductivity of the material was observed, with the thermal conductivity of the single-crystal tungsten reducing from 153 W/m•K to 27.9 W/m•K (40 vol% TiC), and the expansion coefficient of the material exhibited an increase compared to that of pure tungsten (Figure 2.50). A schematic diagram of the

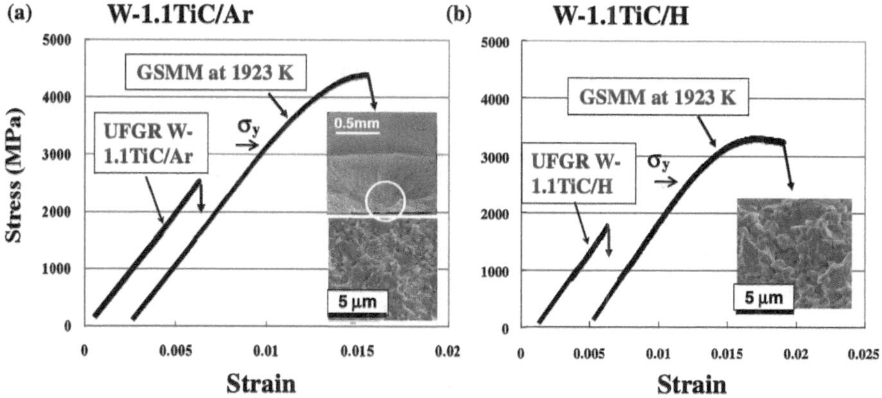

Figure 2.49: Three-point bending stress-strain curves at room temperature before and after the treatment of GSMM technology: (a) W-1.1TiC/Ar-UH and (b) W-1.1TiC/H-NH [114].

(Ti, W)C solid solution formation during the sintering process is presented in Figure 2.51. TiC_{1-x} is a non-stoichiometric compound with a NaCl-type crystal structure, where x ranges from 0 to 0.5. The radius of Ti^{4+} and W^{4+} is 0.068 and 0.070 nm, respectively. At elevated temperatures, solid-state diffusion allows for the replacement of W^{4+} by Ti^{4+} forming a (Ti, W)C solid solution. Additionally, Ti can also diffuse into tungsten. The interdiffusion of Ti and tungsten at the TiC/W interface leads to the formation of a (Ti, W)C solid solution, which enhances the bonding strength of the interface and thus improves the composite's strength and toughness.

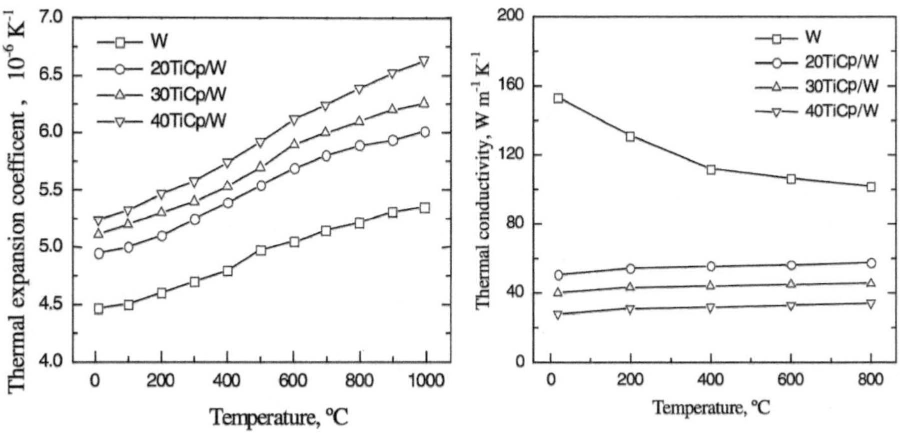

Figure 2.50: Variation of thermal expansion coefficient and thermal conductivity of W-TiC composites with different TiC additions with temperature [118].

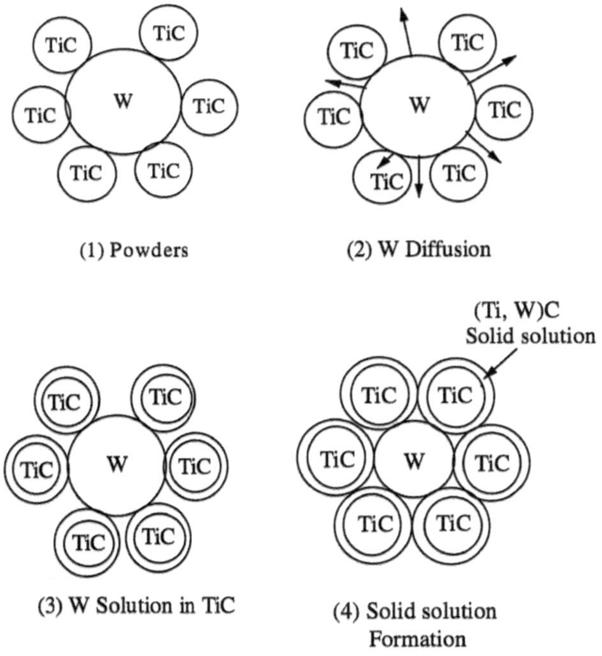

Figure 2.51: Schematic diagram of formation mechanism of (Ti, W)C solid solution in W-TiC composites [118].

Employing a wet chemical method in conjunction with SPS sintering, Yan et al. [119] prepared W-(0–0.9%)TiC alloys and studied the influence of TiC content on the microstructure and mechanical properties of these alloys. The research revealed that the grain size of W-TiC material decreased with increasing TiC content, while the TiC particle size exhibited an opposite trend. The flexural strength of the W-TiC alloy improved with increasing TiC content up to 0.5% at room temperature. At this composition, the density, bending strength, and bending strain reached their peak values at 0.5% TiC, indicating that W-0.5TiC alloy exhibits the optimal mechanical properties, as detailed in Table 2.5. The thermal conductivity of the W-TiC alloy gradually decreased with increasing TiC content. However, for alloys with TiC content between 0.1% and 0.5%, the thermal conductivity was only slightly lower than that of pure tungsten within the tested temperature range (Figure 2.52). The deterioration of the thermal conductivity and mechanical properties of the W-0.7TiC and W-0.9TiC can be attributed primarily to the segregation of TiC particles at grain boundaries. As the TiC content increases from 0 to 0.5%, the number of TiC particles increases, the grain size decreases, that is, the grain boundary density rises, and the electron scattering intensifies. Despite an increase in density and a reduction in porosity and electron scattering, the overall effect was a gradual decrease in thermal conductivity. However, further increases in

Table 2.5: Average grain size, TiC particle size, density, and mechanical properties of pure tungsten and W-(0–0.9) TiC alloys [119].

Sample	Grain size (μm)	TiC size (nm)	Density (%)	Hardness ($Hv_{0.2}$)	Flexural strength (MPa)	Bending strain (%)
Pure tungsten	2.16	–	95.44	469	645.88	0.84
W-0.1TiC	1.76	50	96.58	633.3	812.84	1.16
W-0.3TiC	1.34	60	97.31	670.7	921.12	1.19
W-0.5TiC	0.91	62	97.61	739.1	1065.72	1.23
W-0.7TiC	0.71	71	94.99	843.3	822.47	0.84
W-0.9TiC	0.53	97	93.72	1014.5	843.68	0.9

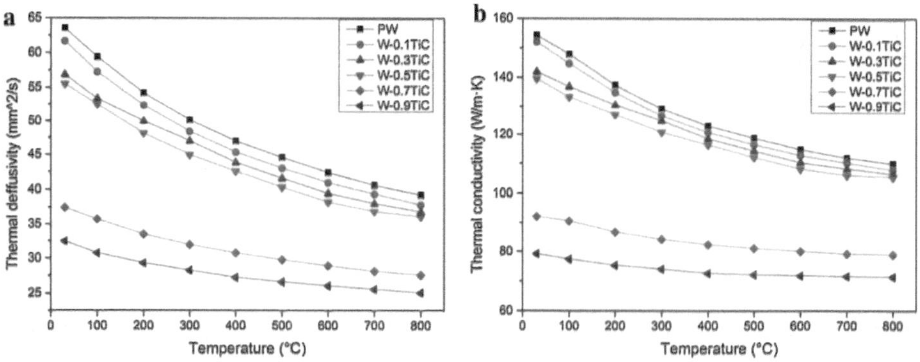

Figure 2.52: Variation of thermal diffusivity and thermal conductivity of pure tungsten and W-(0–0.9) TiC composites from room temperature to 800 °C [119].

TiC content led to a decrease in density, an increase in porosity, and a significant reduction in thermal conductivity.

Miao et al. [120] employed powder metallurgy and a high-temperature rolling process to prepare W-0.5%TiC alloys. Their findings revealed a progressive reduction in the material's hardness and fracture strength due to the recovery process at annealing temperatures up to 1,500 °C which subsequently exhibited a sudden decline due to complete recrystallization at annealing temperatures up to 1,600 °C (Figure 2.53). The rolled W-0.5% TiC exhibits typical brittle fracture at room temperature, but displays ductility at 200 °C, reaching its maximum strength of 789 MPa, with an elongation of 4.8%.

Zhong et al. [121] employed the mechanical alloying method to prepare various W-TiC materials and discovered that the introduction of TiC effectively strengthens the grain boundaries and improves the mechanical properties of alloy materials, as evidenced by significant physical property test outcomes. In particular, the relative density, flexural strength, Vickers microhardness, and Young's modulus of W-1% TiC

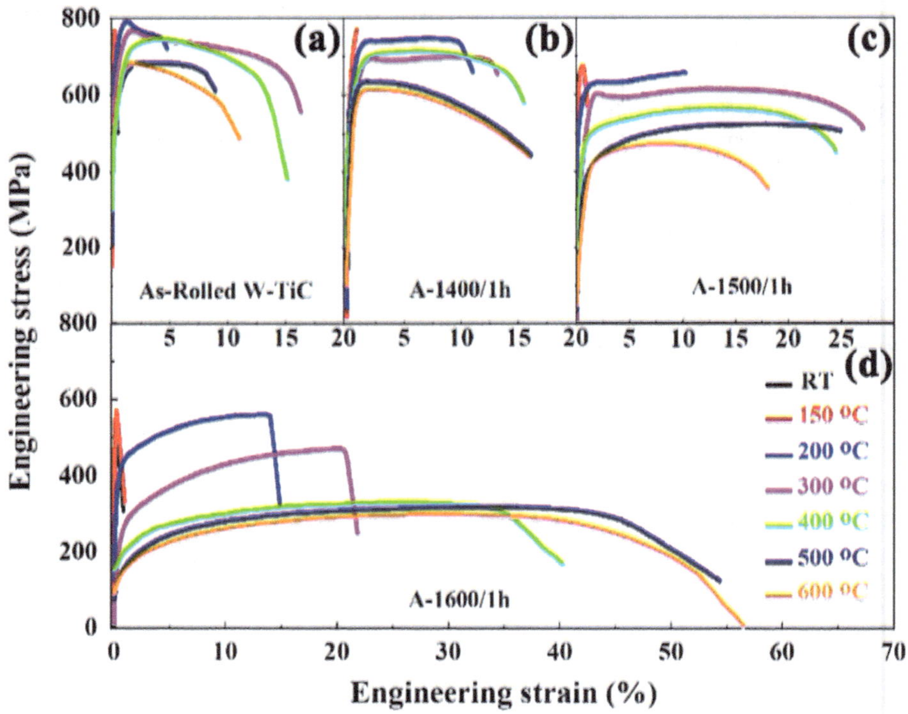

Figure 2.53: Tensile properties at different temperatures of W-TiC plates in the rolled state and annealed at 1,400, 1,500, and 1,600 °C for 1 h [120].

material increased to varying degrees, reaching values of 98.4%, 1,065 MPa, 4.33, and 396 GPa, respectively. Furthermore, TiC was observed to effectively enhance the thermal load-bearing capacity of the material at temperatures below the recrystallization temperature of tungsten alloy.

2.3.3.2 W-ZrC

ZrC has a melting point of 3,510 °C, which is higher than the melting point of tungsten (3,410 °C) and TiC (3,067 °C). Compared to TiC, ZrC exhibits superior high-temperature strength. Additionally, these two materials demonstrate comparable thermal expansion coefficients and favorable compatibility with tungsten. First-principles calculations reveal that the binding energy of Zr-O in the W matrix exceeds that of ZrC. ZrC is added to the W matrix as a diffusion enhancement phase and is decomposed into Zr and C through ball milling. Subsequently, Zr initially reacts with the free oxygen in the matrix to form stable ZrO_y particles, and then the remaining C reacts with the surrounding W matrix to form WC_x. Finally, the ZrO_y, ZrC and WC_x generated in this process form a complex compound, non-stoichiometric, W-Zr-C_x-O_y. The brittle phase W_2C can be avoided by strictly controlling the amount of ZrC. This purification process reduces the embrittlement effect of free oxygen on grain boundaries and an enhancement of the low-temperature plasticity of tungsten. The lattice constant of ZrC (111) is approximately the same as that of W (110) (0.22 nm), allowing for the formation of a coherent or semi-coherent structural relationship at the dispersive phase and the tungsten matrix interface, as shown in Figure 2.54. Therefore, ZrC particles can be stable in the tungsten matrix and are not prone to agglomerate. This coherent phase interface ensures the stability of nano-ZrC particles, preventing them from agglomerating. It also pins dislocations and hinders the grain boundary movement, thereby strengthening the grain and grain boundaries. At the same time, the coherent phase interface may also act as a dislocation slip channel, allowing dislocations to cut through particles, which reduces dislocation plugging and improves material plasticity. The strengthening and toughening mechanism of W-ZrC can be attributed to three aspects [122]. Firstly, the phase boundary strengthened by the coherent structure. Secondly, the grain boundaries strengthened by the semi-coherent structure. Finally, the grain boundaries purified and strengthened by Zr capturing impurities and oxygen elements (Figure 2.55).

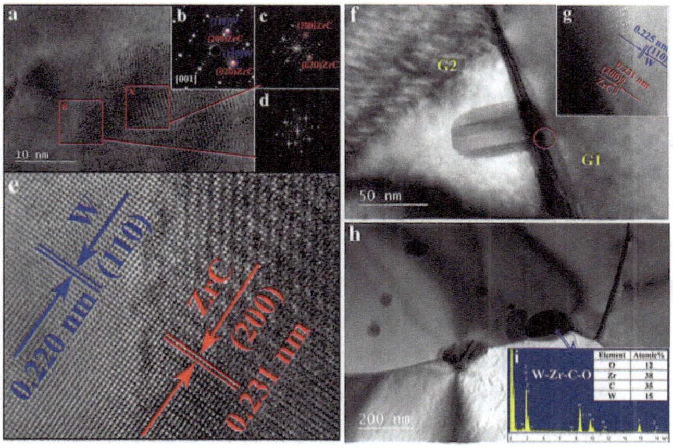

Figure 2.54: Interface relationship between ZrC and W in W-0.5% Zr alloy [122].

Xie et al. [123] prepared a series of W-ZrC with varying ZrC content (0%, 0.2%, 0.5%, 1.0%) by SPS sintering and found that adding a small amount of ZrC could enhance the strength and ductility of tungsten-based materials. The alloy with ZrC contents of 0.2%, 0.5%, and 1.0% exhibited brittle fracture at 500 °C and plasticity at 600 °C, indicating that the DBTT of the alloys is between 500 and 600 °C, which is lower than that of pure tungsten samples sintered under the same conditions. An increase in ZrC content from 0.2% to 1.0% resulted in a reduction in the high-temperature elongation of tungsten-based materials at a consistent temperature. At 700 °C, the tensile strength and total elongation of W-0.5%ZrC alloy were 535 MPa and 24.8%, respectively, which

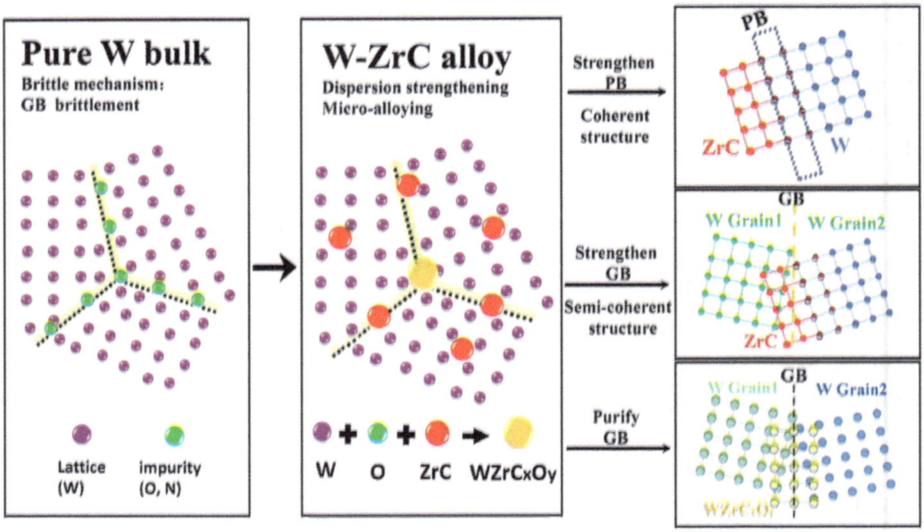

Figure 2.55: Strengthening and toughening mechanism of W-ZrC [122].

were 59% and 114% higher than those of pure tungsten, as shown in Figure 2.56. Based on these results, the W-0.5% ZrC component appears to be the optimal choice in terms of tensile strength and plasticity.

Based on the aforementioned study, W-0.5%ZrC bars with a diameter of 9.1 mm were prepared via pressureless sintering and high-temperature rotary swaging [124]. The mechanical properties of these high-temperature rotary forged W-0.5%ZrC bars exhibited marked improvement compared to samples prepared by SPS sintering. In addition, the thin plate (1 mm thickness) of W-0.5%ZrC, rolled by cold-pressing and high-temperature sintering, demonstrated superior tensile mechanical properties (Figure 2.57). The fracture energy of the rolled W-ZrC alloy at room temperature was 3.23×10^7 J/m^3, which is about 10 times that of the rotary forged W-ZrC alloy. As shown in Table 2.6, the rolled W-ZrC alloy displays enhanced thermal conductivity in comparison to the rotary forged W-ZrC alloy, thereby exhibiting superior thermal shock resistance. As the temperature rises, the thermal conductivity of W-0.5%ZrC declines, being inferior to that of ITER-grade pure tungsten at room temperature. The thermal conductivity of W-0.5%ZrC is enhanced following the annealing process, which can be attributed to the high density of grain boundaries and phase boundaries in the rolled fine grain W-0.5%ZrC, in conjunction with the strong scattering of free electrons at these two boundaries. With the increase in annealing temperature, the grain size gradually increases and the interface scattering decreases, resulting in a gradual increase in thermal conductivity, which remains lower than that of pure tungsten, as shown in Figure 2.58. The tensile and yield strengths of DBTT are 1,413 MPa and 980 MPa, respectively, as measured by the internal friction amplitude effect between

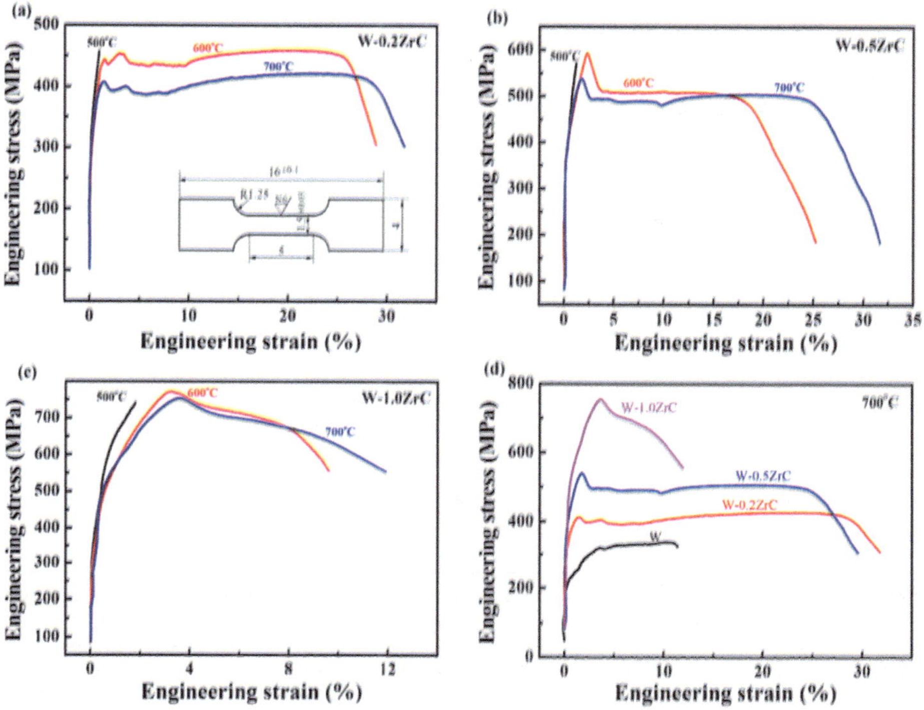

Figure 2.56: Tensile stress-strain curves of W-(0%, 0.2%, 0.5%, 1.0%) ZrC alloy [123].

50 and 80 °C [125]. The 8.5-mm-thick W-0.5%ZrC sheet exhibited better mechanical properties. At room temperature, this material exhibits flexural toughness, with a strain of about 3%, a flexural strength of up to 2.5 GPa, and a three-point bending-calibrated DBTT of 100 °C, which was about 300 °C lower than that of commercial pure tungsten [124]. The picture of the sample obtained after the three-point bending test fully illustrated the excellent plasticity properties of the 8.5-mm-thick W-0.5%ZrC sheet. Seen from the tensile engineering stress-strain curves, the tensile strength is 991 MPa with an elongation of 1.1% at room temperature. At 100 °C, the tensile strength and elongation were 1.1 GPa and 3%, respectively; at 500 °C, the tensile strength was 583 MPa, and the elongation was as high as 45%, as shown in Figure 2.59. The pinning effect of ZrC or W-Zr-C-O particles greatly improves the thermal stability of tungsten-based materials. Figure 2.60 presents a comparative analysis of the hardness evolution of W-0.5%ZrC and pure tungsten after annealing at different temperatures. The data indicates that the recrystallization temperature of W-0.5%ZrC is approximately 1,300 °C, which is 100 °C higher than that observed in hot-rolled pure tungsten [126].

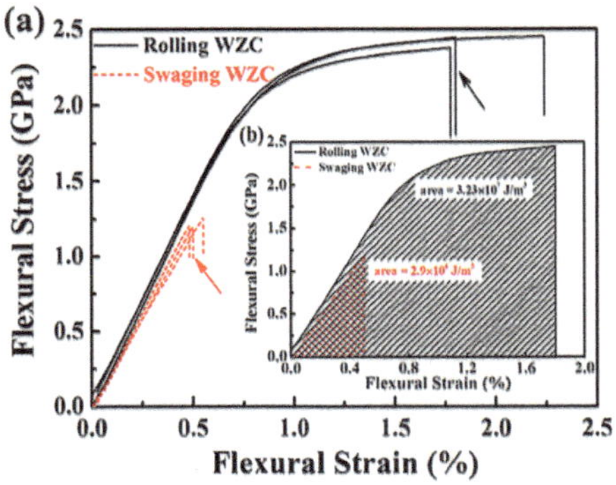

Figure 2.57: (a) Bending stress-strain curves and (b) calculated fracture energies at room temperature for rotary forged ZrC and rolled W-ZrC alloys [126].

Table 2.6: Comparison of properties of rotary forged W-ZrC(S-WZC) and rolled W-ZrC(R-WZC) Alloys at room temperature [124].

Alloy	Density (%)	Grain size (μm)	Hardness (GPa)	Hv 100 g (Hv)	Flexural strength (GPa)	Fracture energy (J/m^3)
S-WZC	99.2 ± 0.2	1.53 ± 0.34	6.1 ± 0.2	478 ± 6	1.21 ± 0.3	2.9×10^6
R-WZC	99.2 ± 0.2	1.46 ± 0.26	6.7 ± 0.2	509 ± 4	2.43 ± 0.2	3.23×10^7

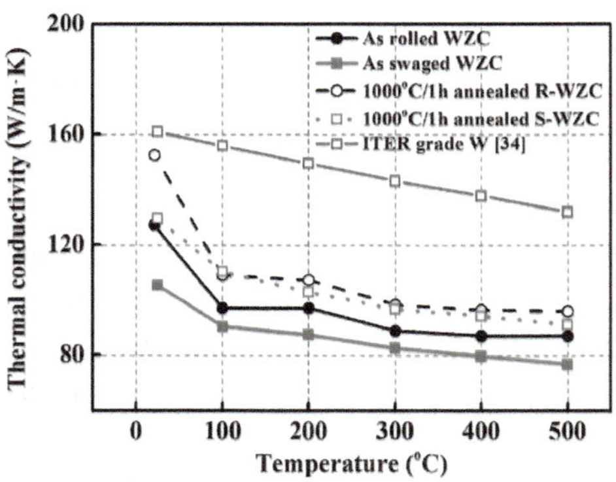

Figure 2.58: Temperature-dependent thermal conductivity of rotary forged/rolled W-ZrC before and after annealing [124].

Song et al. [127] observed that the flexural strength of W-ZrC material increased with increasing temperature, reaching a maximum value of 829 MPa at 1,000 °C, which was 17% higher than that of 705 MPa at room temperature. At high temperatures, the W-ZrC alloy exhibited a fracture process that involved four distinct stages: "micro-crack initiation, crack connection growth, rapid crack propagation and material

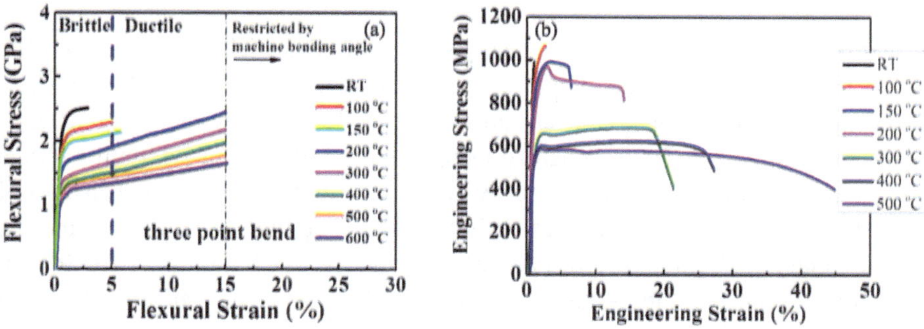

Figure 2.59: Three-point bending stress-strain curves and tensile stress-strain curves of W-0.5%ZrC sheets at different temperatures [122].

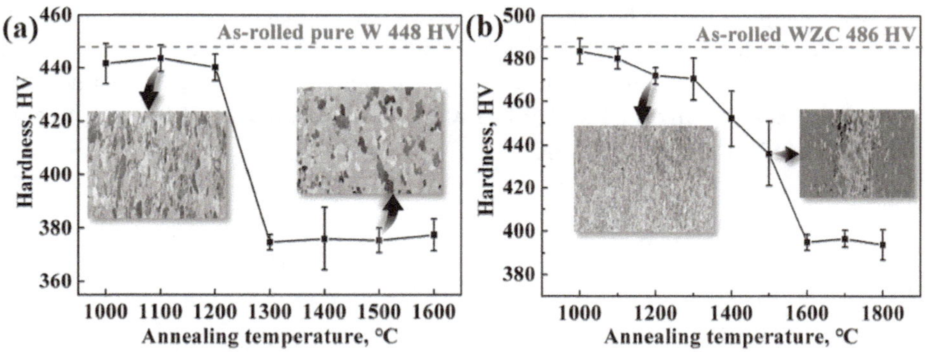

Figure 2.60: Hardness evolution of W-0.5%ZrC and pure tungsten after annealing at different temperatures [126].

fracture." The plastic matrix helped to slow crack propagation and forced the crack to follow a tortuous path, while the tungsten tended ductile fracture. Their study demonstrated that with the increase of temperature, the tungsten matrix undergoes plastic deformation, and the dislocation strengthening and load transfer of particles can be fully exerted, showing the characteristics of improving; In addition, the (Zr, W)C phase is generated at the W-ZrC interface, which improves the strength of the interface. Therefore, the transcrystalline fracture occurs in the W matrix under the action of external force, exerting the high intragranular strength of tungsten grains. Fan et al. [128] employed the method of "sol-heterogeneous precipitation-spray drying-thermal reduction" to prepare W-ZrC (1–4%) composite powder, which was subsequently sintered to produce W-ZrC blocks, and found that the addition of ZrC improved sintering performance. Although increasing ZrC content enhances material strength, a high concentration of second-phase particles reduces the dispersion strengthening effect, increasing material brittleness.

2.3.3.3 W-TaC

TaC exhibits a remarkably high melting point (4,200 K), which surpasses both the melting point of TiC and pure tungsten. The doping of TaC is anticipated to enhance the thermal stability of the material. Kurishita et al. [117] also prepared W-2.2%TaC and W-3.3%TaC with a TFGR structure using the MA-GSMM process, with average grain sizes of 2.0 µm and 1.2 µm, respectively, and found that the volume fractions of 3.3% TaC and 1.1% TiC were approximately the same. The findings indicated that the grain boundary strengthening effect of TaC was less effective than that of TiC. In contrast to the existence of the K-S relationship at the W-TiC interface, the TaC phase did not maintain a specific orientation relationship with the W matrix interface. The DBTT of W-(2.2, 3.3)TaC was higher than that of W-1.1TiC materials, but when the test temperature was slightly increased, the plasticity of W-TaC increased significantly, as shown in Figure 2.61, where the DBTT was defined as the nil ductility temperature (NDT). Precipitated phases such as TaC, TaO_2, and W2C were formed in W-(2.2, 3.3) TaC, with the preferential formation of TaO_2 and W2C precipitated phases due to differences in the ability to form carbides and oxides; this preferential formation results in a reduction of TaC phase content and a strengthening effect on grain boundaries. This partly explains why the DBTT of W-TaC is higher than that of W-TiC. The precipitated phase in W-1.1TiC was only TiC.

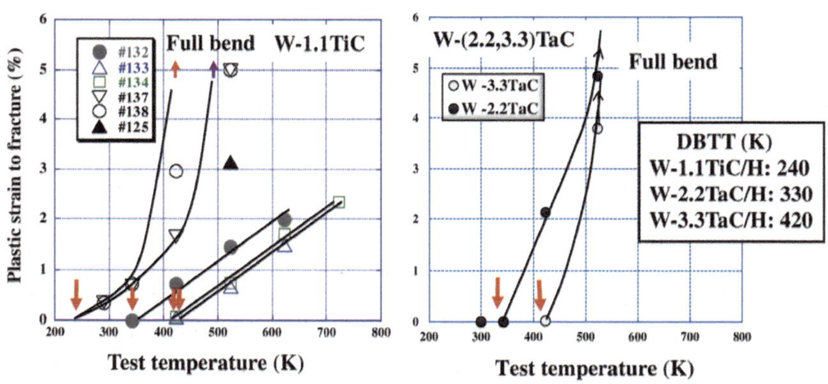

Figure 2.61: The plastic fracture strain versus temperature curves of W-(2.2, 3.3) TaC/H and W-1. 1TiC/H with TFGR structure, and DBTT is defined as nil ductility temperature (NDT); see arrow in the figure [117].

Ueda et al. [129] found that the plasma etching resistance of W-TaC alloys was stronger than that of W-TiC alloys because the melting point of Ta was higher than that of Ti. Oya et al. [130] found that W-TaC alloys exhibit superior hydrogen isotope retention and plasma etching resistance compared with W-TiC alloys, owing to the fact that the trapped hydrogen isotopes react with carbides, with Ta and D forming TaD and Ti and D forming TiD_2. Moreover, the melting point of TaC is higher than that of TiC, making TaC more stable and less prone to form tantalum hydrides, thereby resulting in relatively low hydrogen isotope retention.

Miao et al. [131] added nano-TaC particles (average diameter 80 nm) to the tungsten matrix (average diameter 2 μm), and subsequently, they rolled the W-0.5% TaC sheet at high temperature after the high-energy ball-milling and hot-pressing sintering processes to achieve a DBTT of about 200 °C, a tensile strength of 982 MPa, and an elongation of 12.0% at 200 °C. The tensile strength of the sheet is above 570 MPa between room temperature and 500 °C, and the elongation reaches 40.5% at 400 °C. The hydrogen atmosphere is employed throughout the preparation process, which serves to reduce the impurity oxygen content and, consequently, minimize the formation of submicron oxide particles. This results in a reduction in particle size and an improvement in the low-temperature plasticity of the material, as illustrated in Figure 2.62. In addition, round Ta_2O_5 particles were found at the sub-grain boundaries, formed by the capture of impurity oxygen elements by TaC. Xie et al. [132] also prepared W-1.0% TaC sheets by hot rolling following ball-milling and sintering. The nanoparticles in the tungsten matrix raised the recrystallization temperature of the material to 1,400 °C, and the tensile strength reached 571 MPa at 500 °C. The DBTT of W-1.0%TaC and fully recrystallized W-1.0%TaC was around 200 °C (Figure 2.63).

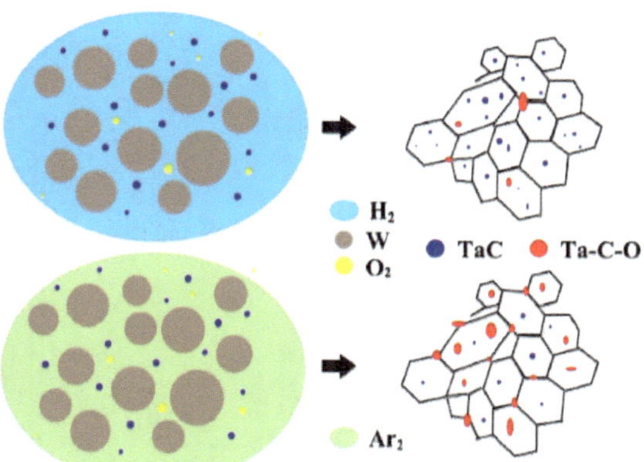

Figure 2.62: Schematic diagram of the reduction of impurity oxygen in the preparation of W-TaC [131].

The directly added TaC was relatively large, and some TaC particles in it underwent a reaction with the O impurity, forming submicron Ta-C-O compounds at the grain boundaries, resulting in a reduction in the mechanical and thermal stability of the material, thereby making W-TaC less robust than W-0.5% ZrC sheets. Given the above reasons, Miao et al. [133] undertook a modification of their process by incorporating nano-Ta and C powders into W powder and utilizing high-energy ball milling to form a solid solution of W-Ta-C. Preferentially forming TaC during sintering resulted in nanoparticle dispersion reinforcement of the obtained W-0.5%Ta-0.01%C (W-05Ta-001C). The tensile strength of W-05Ta-001C, obtained by SPS sintering, reached 565 MPa and the elongation reached 54.0% at 400 °C, as shown in Figure 2.64. The average particle size of TaC was 62 nm, and the particles were uniformly distributed in the grains, while the average size of TaC particles in the W-TaC alloy prepared by direct addition method was 170 nm. As the annealing temperature increased to 1,500 °C, the material's hardness was found to decrease slightly, with its recrystallization temperature being approximately 1,600 °C. In contrast, the W-0.5TaC material underwent annealing at 1,300 °C for 1 h, resulting in a significant reduction in hardness due to the increased grain growth (Figure 2.65). The nano-TaC in W-05Ta-001C effectively impeded the grain boundary migration and grain growth of tungsten, consequently elevating the recrystallization temperature. However, the coarse TaC particles in W-0.5TaC reduced their pinning effect on grain boundaries, thereby lowering the recrystallization temperature.

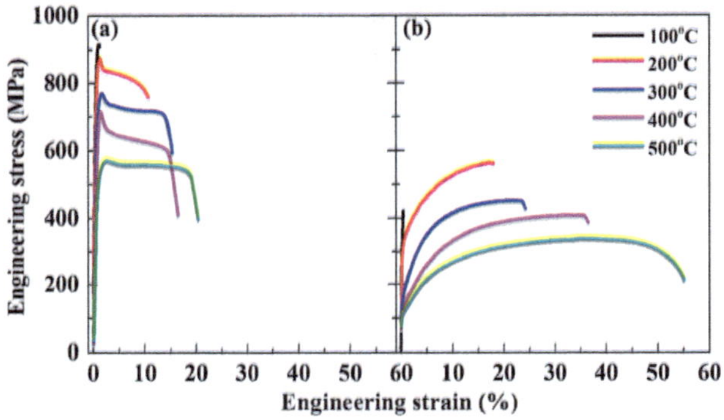

Figure 2.63: Tensile behavior of alloy: (a) W-1.0% TaC and (b) fully recrystallized W-1.0% TaC [132].

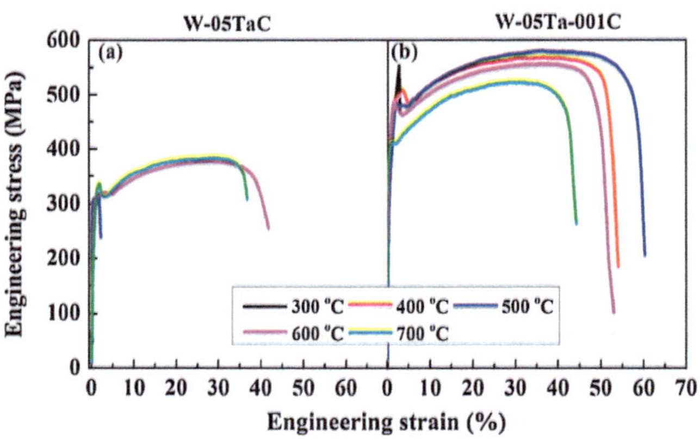

Figure 2.64: High-temperature tensile behavior of an alloy from 300 to 700 °C: (a) W-0.5TaC and (b) W-0.5TaC-0.01C [133].

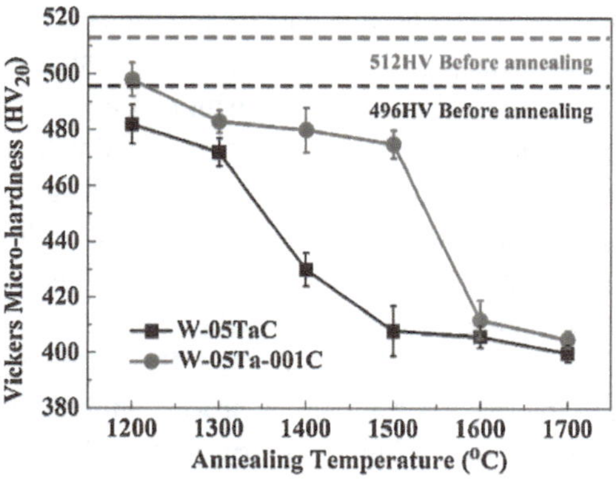

Figure 2.65: Evolution of hardness after isothermal annealing of W-0.5TaC and W-0.5TaC-0.01C [132].

2.3.3.4 W-HfC

HfC exhibits a remarkably high melting point (3,900 °C), surpassing that of both TiC and ZrC. This characteristic suggests that HfC doping could enhance the thermal stability of the material, and W-HfC materials may exhibit resistance to hydrogen isotope retention and plasma etching. Wang et al. [134] prepared small samples of W-(0.1%, 0.2%, 0.5%, 0.8%, 1.0%)HfC with different compositions by SPS sintering to obtain the optimal amount of HfC. As shown in Figure 2.66, W-0.1%HfC and W-0.2%HfC fracture in the elastic deformation region below 600 °C, whereas W-0.5% HfC, W-0.8% HfC, and W-1.0% HfC show obvious plastic deformation at 500 °C, especially W-0.5% HfC, with an elongation rate up to 49.6% at 600 °C. However, as the HfC content increased, the elongation rate exhibited a corresponding decrease. At 600 °C, the W-0.8% HfC and W-1.0%HfC samples exhibited elongation rates of 25% and 29%, respectively. The reason is that excess HfC can form particle clusters, which can lead to stress concentrations as a possible source of cracks. Combined with the results of density and tensile tests, the optimal HfC addition was 0.5 wt%.

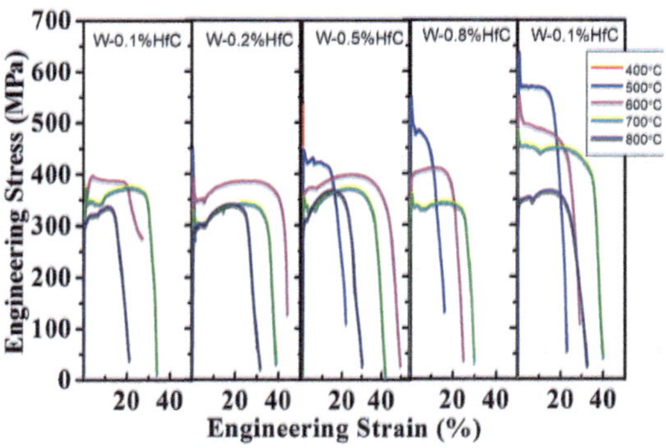

Figure 2.66: Tensile curve of SPS-sintered W-HfC [134].

It is well-established that the low-temperature plasticity of the material and the DBTT of tungsten can be further improved by plastic deformation. Consequently, Wang et al. prepared rod-shaped W-0.5HfC in the rotational calcined state, which exhibited a typical brittle fracture at 200 °C with a tensile strength of 720 MPa and plasticity at 250 °C with a DBTT of 250 °C, which is 200 °C lower than that of W-0.5HfC in the SPS-sintered state. Moreover, the tensile strength of rod-shaped W-0.5HfC is superior to that of the undeformed W-0.5HfC at each test temperature. At 250 °C, the tensile strength can reach 800 MPa and the elongation reaches 4.1%, as shown in Figure 2.67. After annealing at 1,200 °C for 1 h, the DBTT of it was reduced by 50 °C, its elongation reached 15% at 200 °C, and the tensile strength reached 840 MPa; as the result the annealing process eliminated the residual stresses and defects present in the sample following rotary swaging. What's more, after annealing at 1,200 °C, the tensile strength of W-0.5HfC remained above 550 MPa in the range of 300–800 °C, indicating excellent high-temperature thermal stability. The thermal conductivity of W-0.5HfC reached 174 W/m•K after annealing, remaining above 137 W/m•K from room temperature to 500 °C. No matter W-0.5% TaC (124 or 70 nm) or W-0. 5% HfC (105 nm), the average size of the dispersive phase in W substrate was much larger than that of W-ZrC (50 nm), so its strength was lower than that of W-0.5% ZrC plate. Additionally, due to its large diffusive phase and average grain size, the defect trap per unit volume was lower than that of W-0.5%ZrC, leading to reduced self-healing capabilities and inferior irradiation resistance.

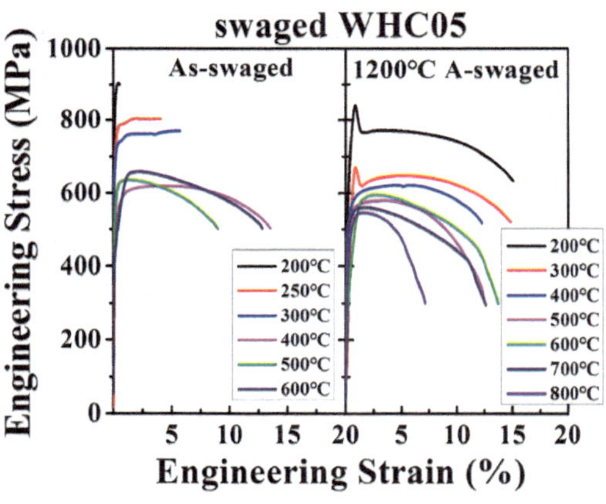

Figure 2.67: Elongation curves of rotary forged W-0.5HfC and W-0.5HfC after annealing at 1,200 °C [134].

2.3.4 Ultrafine-grained/nanocrystalline tungsten

UFG crystals and nanocrystals (NCs) exhibit excellent ductility and strong resistance to irradiation, swelling, and embrittlement. The self-healing mechanism of nanomaterials provides a promising avenue for enhancing their resistance to irradiation. Figure 2.68 presents a schematic diagram of the principle of self-healing of NC materials [135]. These materials demonstrate the capacity to absorb defects (such as interstitial atoms and vacancies) caused by irradiation due to the presence of a large number of grain boundaries. Therefore, selecting UFG and NC tungsten in an appropriate manner can enhance the ductility and irradiation resistance of tungsten and leads to improvements in other properties. These improvements ultimately meet the requirements of the operating conditions of PFMs in fusion reactors.

Currently, many viable methods for preparing bulk ultrafine/nanocrystalline metals are being considered, which can be classified into two categories: top-down and bottom-up approaches [136]. Top-down processing involves the direct refinement of coarse-grained metals through large plastic deformation, also known as the SPD method. Conversely, bottom-up processing begins with the production of nanoscale particles, which are subsequently sintered, as in powder metallurgy. The regulation of grain growth is of paramount importance in this process. When sintering is to be conducted via the SPS method, a low temperature, short duration, and rapid heating must be employed to prevent the excessive growth of grains. The so-called large plastic deformation method is employed to achieve strong grain refinement of the material by generating SPD, and the resulting refined grains are typically at the submicron or even nanometer level. UFG and NC materials can be prepared from bulk coarse-grained materials by different SPD processes. Although the mechanisms by which their grains and microstructures are refined differ, analysis has revealed that these materials possess the characteristics of a clean interface and a dense structure. It can be concluded that the obtained bulk nano-SPD materials are suitable for providing large-scale samples for scientific research purposes and can also be fabricated into commercial structural parts [137].

The two well-established processes for SPD are equal channel angular pressing (ECAP) and HPT [138]. The high pressure of several gigapascal applied during HPT enables grain refinement to reach the nanometer level. Faleschini et al. [139] performed HPT treatment on three commercial tungsten-based materials, namely pure tungsten (99.98% purity), WL10 (W-1wt%La_2O_3), and WVMW (0.005% potassium-doped tungsten alloy). The results indicated that HPT treatment of these three commercial tungsten-based materials significantly improved their fracture toughness at room temperature, as shown in Figure 2.69 (where the gray area indicates that the linear fracture mechanism is no longer effective). Wei et al. [140] conducted HPT treatment on tungsten at 500 °C, producing fully dense NC tungsten with grain boundaries exhibiting large-angle misorientations and a high density of blade-shaped dislocations. Furthermore, the concentration of impurities at the grain boundaries was significantly reduced. This observation indicates that the ductility of tungsten increases at room temperature.

Deep plastic deformation can be achieved using the equal-channel angular pressing process, resulting in bulk UFG tungsten material with high density, large volume, and strong toughness. In the study by Wei et al. [141], an UFG tungsten material (~500 nm) was prepared by ECAP, which led to both enhanced strength and notable

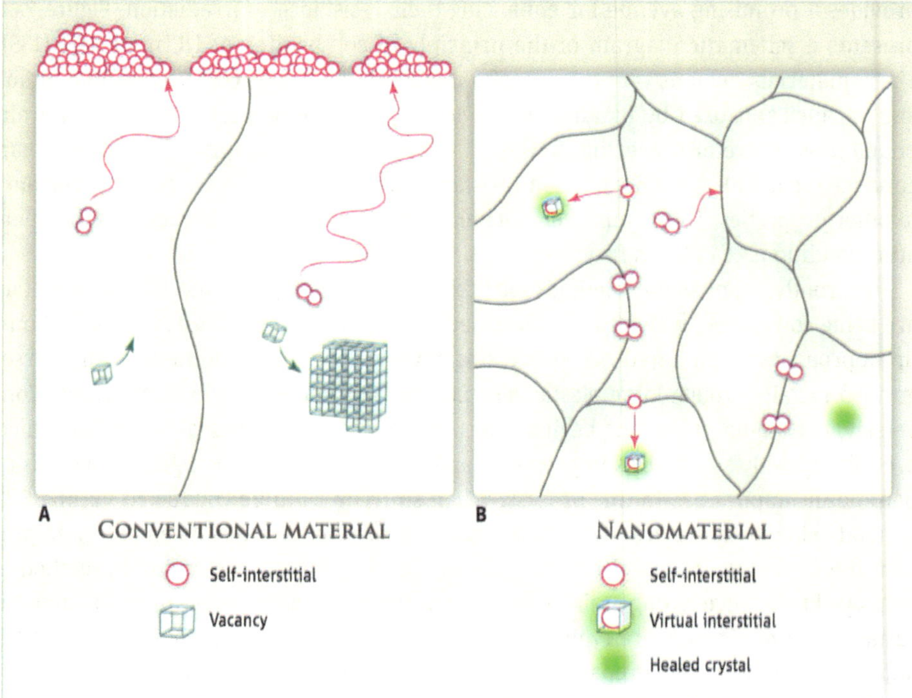

Figure 2.68: Self-healing mechanism of irradiation damage in nanocrystalline materials: (a) traditional materials and (b) nanocrystalline materials [135].

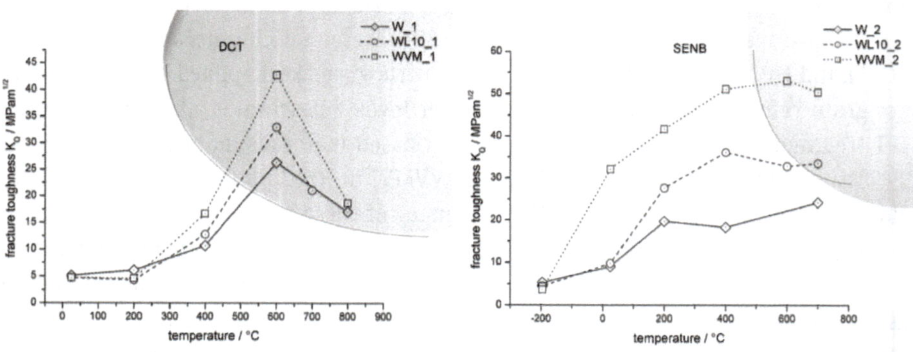

Figure 2.69: The fracture toughness curves of three commercial tungsten alloys in the sintered state and in the rolled state (60% deformation, SENB) after sintering, where SENB represents a unilateral notched bending specimen, prepared by HPT process [139].

increases in ductility. Zhang et al. [142] obtained ultrafine crystalline materials with a grain size of about 0.9 μm and a DBTT below 350 °C by applying ECAP treatment to commercial pure tungsten, while the DBTT of coarse-grained tungsten without ECAP treatment was higher than 483 °C, indicating that ECAP treatment not only obtains UFG bulk tungsten but also significantly lowers its DBTT compared to coarse-grained tungsten. Hao et al. [143] prepared UFG tungsten by ECAP at relatively low temperatures and found that ECAP processes at 800 and 950 °C could refine the grain size of tungsten from microns to UFG scales, with grain sizes of 0.3–1.5 and 0.5–2 μm, respectively (Figure 2.70), while improving the strength and toughness of the material.

Figure 2.70: TEM diagram of ultrafine-grained tungsten prepared by ECAP: (a) two-pass pressing at 800 °C and (b) three-pass pressing at 950 °C [143].

HPT enables the deformation of material by increasing pressure, thus allowing the processing of challenging materials, such as refractory metals, even at lower temperatures. Lower temperatures can lead to a greater degree of material refinement. For instance, pure tungsten treated with HPT typically results in NC tungsten, whereas ECAP produces only ultrafine crystalline tungsten. To achieve UFGs, the material must undergo cold-rolling at a certain temperature after ECAP to further plastically deform. HPT is only used for producing thin (≤1 mm) bulk products, while ECAP can be employed to prepare large-scale bulk UFG/NC tungsten, which has high density, superior toughness, and enhanced brittleness. ECAP the potential to play a pivotal role in the advancement and utilization of fusion materials.

Additionally, UFG tungsten can be prepared by powder metallurgy, which involves initially producing the requisite precursor powder and subsequently sintering the powder using an appropriate method. The high melting point and low diffusion coefficient of pure tungsten make sintering an extremely challenging process. Sintering typically needs to be conducted at the highest possible temperature, which results in the formation of coarse grains. Currently, special sintering methods, such as HIP, SPS, and ultra-high-voltage electrification sintering, are predominantly used to prepare ultrafine crystalline tungsten.

2.4 Summary

Various tungsten-based materials have been adjusted and optimized through different ways such as dispersion strengthening, alloying, and compositing, but there are still no tungsten-based materials and technologies that can fully meet the requirements of

PFMs in fusion reactors. Future efforts will focus on the application of an integrated approach, utilizing a combination of preparation techniques for tungsten-based materials, to enhance the microstructural design and control of these materials, thus optimizing their comprehensive performance characteristics. At the international level, alloying/diffuse particle doping and other methods have been employed with the objective of enhancing the thermal/mechanical and anti-irradiation properties of W-PFM. To enhance the composition and properties of tungsten materials, domestic research mainly focuses on dispersion strengthening. Researchers are encouraged to compare the existing domestic diffusion-reinforced tungsten systems and select one or two large-scale preparation and engineering application solutions for systematic research and development, which should be based on a full demonstration of each solution and should consider the potential of exploring other advanced tungsten-based materials.

References

[1] A.R. Raffray, R. Nygren, D.G. Whyte, S. Abdel-Khalik, R. Doerner, F. Escourbiac, T. Evans, R.J. Goldston, D.T. Hoelzer, S. Konishi, P. Lorenzetto, M. Merola, R. Neu, P. Norajitra, R.A. Pitts, M. Rieth, M. Roedig, T. Rognlien, S. Suzuki, M.S. Tillack, C. Wong. High heat flux components-readiness to proceed from near term fusion systems to power plants [J]. Fusion Engineering and Design, 2010, 85(1): 93–108.

[2] C. Linsmeier, M. Rieth, J. Aktaa, T. Chikada, A. Hoffmann, J. Hoffmann, A. Houben, H. Kurishita, X. Jin, M. Li, A. Litnovsky, S. Matsuo, A. von Müller, V. Nikolic, T. Palacios, R. Pippan, D. Qu, J. Reiser, J. Riesch, T. Shikama, R. Stieglitz, T. Weber, S. Wurster, J.-H. You, Z. Zhou. Development of advanced high heat flux and plasma-facing materials [J]. Nuclear Fusion, 2017, 57: 092007.

[3] R.A. Pitts, S. Carpentier, F. Escourbiac, T. Hirai, V. Komarov, A.S. Kukushkin, S. Lisgo, A. Loarte, M. Merola, R. Mittear, A.R. Raffray, M. Shimada, P.C. Stangeby. Physics basis and design of the ITER plasma-facing components [J]. Journal of Nuclear Materials, 2011, 415(1): 957–964.

[4] N. Noda, V. Philipps, R. Neu. A review of recent experiments on W and high Z materials as plasma-facing components in magnetic fusion devices [J]. Journal of Nuclear Materials, 1997, 241–243: 227–243.

[5] J.W. Davis, V.R. Barabash, A. Makhankov, L. Plöchl, K.T. Slattery. Assessment of tungsten for use in the ITER plasma facing components [J]. Journal of Nuclear Materials, 1998, 258–263: 308–312.

[6] I. Smid, M. Akiba, G. Vieider, L. Plöchl. Development of tungsten armor and bonding to copper for plasma-interactive components [J]. Journal of Nuclear Materials, 1998, 258–263: 160–172.

[7] M. Roedig, W. Kuehnlein, J. Linke, M. Merola, E. Rigal, B. Schedler, E. Visca. Investigation of tungsten alloys as plasma facing materials for the ITER divertor [J]. Fusion Engineering and Design, 2002, 61–62: 135–140.

[8] S.N. Mathaudhu, A.J. Derosset, K.T. Hartwig, L.J. Kecskes. Microstructures and recrystallization behavior of severely hot-deformed tungsten [J]. Materials Science and Engineering: A, 2009, 503(1–2): 28–31.

[9] Q. Wei, L.J. Kecskes. Effect of low-temperature rolling on the tensile behavior of commercially pure tungsten [J]. Materials Science and Engineering: A, 2008, 491(1–2): 62–69.

[10] R.A. Pitts, S. Carpentier, F. Escourbiac, T. Hirai, V. Komarov, A.S. Kukushkin, S. Lisgo, A. Loarte, M. Merola, R. Mitteau, A.R. Raffray, M. Shimada, P.C. Stangeby. Physics basis and design of the ITER plasma-facing components [J]. Journal of Nuclear Materials, 2011, 415(1): S957–S964.

[11] M. Shimada, A.E. Costley, G. Federici, K. Loki, A.S. Kukushkin, V. Mukhovatov, A. Polevoi, M. Sugihara. Overview of goals and performance of ITER and strategy for plasma-wall interaction investigation [J]. Journal of Nuclear Materials, 2005, 337–339: 808–815.

[12] H. Bolt, V. Barabash, G. Federici, J. Linke, J. Loarte, J. Roth, K. Sato. Plasma facing and high heat flux materials-needs for ITER and beyond [J]. Journal of Nuclear Materials, 2002, 307–311: 43–52.

[13] H. Bolt, V. Barabash, W. Krauss, J. Linke, R. Neu, S. Suzuki, N. Yoshida, ASDEX Upgrade Team. Materials for the plasma-facing components of fusion reactors [J]. Journal of Nuclear Materials, 2004, 329–333: 66–73.

[14] L. Guanghong, L. Guangnan, L. Jiangang. Research progress on plasma-wall interaction of magnetic confinement nuclear fusion tokamak [J]. Materials in China, 2010, 29(7): 42–48.

[15] R. Neu, V. Bobkov, R. Dux, J.C. Fuchs, O. Gruber, A. Herrmann, A. Kallenbach, H. Maier, M. Mayer, T. Putterich, V. Rohde, A.C.C. Sips, J. Stober, K. Sugiyama, ASDEX Upgrade Team. Ten years of W programme in ASDEX upgrade-challenges and conclusions [J]. Physica Scripta, 2009, T138: 014038.

[16] C. Ruset, E. Grigore, H. Maier, R. Neu, H. Greuner, M. Mayer, G. Matthews. Development of W coatings for fusion applications [J]. Fusion Engineering and Design, 2011, 86(9–11): 1677–1680.

[17] G.F. Matthews, M. Beurskens, S. Brezinsek, M. Groth, E. Joffrin, A. Loving, M. Kea, M.L. Mayoral, R. Neu, P. Prior, V. Riccardo, F. Rimini, M. Rubel, G. Sips, E. Villedieu, P. de Vries, M.L. Watkins, EFDA-JET Contributors. JET ITER-like wall: Overview and experimental programme [J]. Physica Scripta, 2011, T145: 014001.

[18] M. Missirlian, J. Bucalossi, Y. Corre, F. Ferlay, M. Firdaouss, P. Garin, A. Grosman, D. Guilhem, J. Gunn, P. Languille, M. Lipa, M. Richou, E. Tsitrone. The WEST project: Current status of the ITER-like tungsten divertor [J]. Fusion Engineering & Design, 2014, 89(7–8): 1048–1053.

[19] M. Richou, M. Missirlian, E. Tsitrone, J. Bucalossi, C. Desgranges, M. Firdaouss, D. Guilhem, H. Greuner, C. Hernandez, M. Lipa, G. Pintsuk, F. Samaille. The WEST project: Validation program for WEST tungsten coated plasma facing components [J]. Physica Scripta, 2015, T167: 014029.

[20] L. Feng, L. Guangnan, L. Qiang, W. Wanjing. Research on the application of tungsten in nuclear fusion reactor [J]. China Tungsten, 2017, 32(2): 41–48.

[21] L. Qiang. Research on material design and component preparation of tungsten copper deflector in EAST[D]. Ph.D., Beijing: University of Chinese Academy of Sciences, 2012.

[22] L. Cao, Z. Zhou, D. Yao. EAST full tungsten divertor design [J]. Journal of Fusion Energy, 2015, 34(6): 1451–1456.

[23] J.W. Coenen, S. Antusch, M. Aumann, W. Biel, J. Du, J. Engels, S. Heuer, A. Houben, T. Hoeschen, B. Jasper, F. Koch, J. Linke, A. Litnovsky, Y. Mao, R. Neu, G. Pintsuk, J. Riesch, M. Rasinski, J. Reiser, M. Rieth, A. Terra, B. Unterberg, T. Weber, T. Wegener, J.-H. You, C. Linsmeier. Materials for DEMO and reactor applications-boundary conditions and new concepts [J]. Physica Scripta, 2016, T167: 014002.

[24] M.E. Sawan, M.A. Abdou. Physics and technology conditions for attaining tritium self-sufficiency for the DT fuel cycle [J]. Fusion Engineering and Design, 2006, 81: 1131–1144.

[25] I. Yu, R. Fetzer, B. Bazylev. Effect of design geometry of the demo first wall on the plasma heat load [J]. Nuclear Materials and Energy, 2016, 9: 560–564.

[26] P. Norajitra, S.I. Abdel-Khalik, L.M. Giancarli, T. Ihli, G. Janeschitz, S. Malang, I.V. Mazul, P. Sardain. Divertor conceptual designs for a fusion power plant [J]. Fusion Engineering and Design, 2008, 83 (7–9): 893–902.

[27] M.R. Gilbert, S.L. Dudarev, S. Zheng, L.W. Packer, S. J.-Ch. An integrated model for materials in a fusion power plant: Transmutation, gas production, and helium embrittlement under neutron irradiation [J]. Nuclear Fusion, 2012, 52: 083019.

[28] G. Federici, W. Biel, M.R. Gilbert, R. Kemp, N. Taylor, R. Wenninger. European DEMO design strategy and consequences for materials [J]. Nuclear Fusion, 2017, 57: 092002.

[29] J.H. You. A review on two previous divertor target concepts for DEMO: Mutual impact between structural design requirements and materials performance [J]. Nuclear Fusion, 2015, 55: 113026.

[30] G. Pintsuk. Tungsten as a plasma-facing material [A]. R.J.M. Konings Comprehensive Nuclear Materials [C]. Amsterdam: Elsevier, 2012: 551–581.

[31] B. Gludovatz, S. Wurster, A. Hoffmann, R. Pippan. Fracture toughness of polycrystalline tungsten alloys [J]. International Journal of Refractory Metals and Hard Materials, 2010, 28(6): 674–678.

[32] J.J. Park. Creep strength of a tungsten–rhenium–hafnium carbide alloy from 2200 to 2400 K [J]. Materials Science and Engineering: A, 1999, 265(1–2): 174–178.

[33] A. Luo, D.L. Jacobson, K.S. Shin. Solution softening mechanism of iridium and rhenium in tungsten at room temperature [J]. International Journal of Refractory Metals and Hard Materials, 1991, 10(2): 107–114.

[34] L. Romaner, C. Ambrosch-Draxl, R. Pippan. Effect of rhenium on the dislocation core structure in tungsten [J]. Physical Review Letters, 2010, 104: 195503.

[35] B. Tyburska-Püschel, V.K. Alimov. On the reduction of deuterium retention in damaged Re-doped W [J]. Nuclear Fusion, 2013, 53: 123021.

[36] A.V. Golubeva, M. Mayer, J. Roth, V.A. Kurnaev, O.V. Ogorodnikov. Deuterium retention in rhenium-doped tungsten [J]. Journal of Nuclear Materials, 2007, 363–365: 893–897.

[37] M. Fukuda, K. Yabuuchi, S. Nogami, A. Hasegawa, T. Tanaka. Microstructural development of tungsten and tungsten–rhenium alloys due to neutron irradiation in HFIR [J]. Journal of Nuclear Materials, 2014, 455(1–3): 460–463.

[38] M. Fujitsuka, B. Tsuchiya, I. Mutoh, T. Tanabe, T. Shikama. Effect of neutron irradiation on thermal diffusivity of tungsten-rhenium alloys [J]. Journal of Nuclear Materials, 2000, 283–287: 1148–1151.

[39] Z. Lingxu, G. Shuangquan, Z. Yu, G. Changchun. Research progress of new tungsten-based plasma-oriented materials [J]. Materials Reports, 2011, 25(8): 42–45.

[40] T. Dunqiang, L. Yalei, Y. Xin, L. Lei, L. Deping. Effect of impurity elements on structure and performance of tungsten products [J]. Materials Reports, 2013, 27(9): 98–100.

[41] Y. Shimin, H. Guanghua. Effect of sintering process on the properties of cerium tungsten [J]. Tungsten Carbide, 2014, 31(4): 241–246.

[42] Y. Shimin, F. Jinglian. Discussion on the strengthening mechanism of cerium, lanthanum, yttrium, rhenium and potassium on tungsten materials [J]. Tungsten Carbide, 2015, 32(1): 19–23.

[43] L. Veleva, Z. Oksiuta, U. Vogt, N. Baluc. Sintering and characterization of W-Y and W-Y_2O_3 materials [J]. Fusion Engineering and Design, 2009, 84(7–11): 1920–1924.

[44] L. Veleva, R. Schäublin, T. Plocinski, M. Walter, N. Baluc. Processing and characterization of a W-2Y material for fusion power reactors [J]. Fusion Engineering and Design, 2011, 86(9–11): 2450–2453.

[45] M. Zhao, Z. Zhou, Q. Ding, M. Zhong, J. Tan. The investigation of Y doping content effect on the microstructure and microhardness of tungsten materials [J]. Materials Science and Engineering: A, 2014, 618: 572–577.

[46] P.H. Brown, A.H. Rathjen, R.D. Graham, D.E. Tribe. Chapter 92 rare earth elements in biological systems [J]. Handbook on the Physics and Chemistry of Rare Earths, 1990, 13: 423–452.

[47] M. Zhao, Z. Zhou, Q. Ding, M. Zhong, K. Arshad. Effect of rare earth elements on the consolidation behavior and microstructure of tungsten alloys [J]. International Journal of Refractory Metals and Hard Materials, 2015, 48: 19–23.

[48] R.A. Anderl, R.J. Pawelko, S.T. Schuetz. Deuterium retention in W, W1%La, C-coated W and W_2C [J]. Journal of Nuclear Materials, 2001, 290–293: 38–41.

[49] N. Lemahieu, J. Linke, G. Pintsuk, G. Van Oost, M. Wirtz, Z. Zhou. Performance of yttrium doped tungsten under 'edge localized mode'-like loading conditions [J]. Physica Scripta, 2014, T159: 014035.

[50] H. Li, S. Wurste, C. Motz, L. Romaner, C. Ambrosch-Draxl, R. Pippan. Dislocation-core symmetry and slip planes in tungsten alloys: Ab initio calculations and microcantilever bending experiments [J]. Acta Materialia, 2012, 60(2): 748–758.

[51] S. Wurster, N. Baluc, M. Battabyal, T. Crosby, J. Du, C. García-Rosales, A. Hasegawa, A. Hoffmann, A. Kimura, H. Kurishita, R.J. Kurtz, H. Li, S. Noh, J. Reiser, J. Riesch, M. Rieth, W. Setyawan, M. Walter, J.-H. You, R. Pippan. Recent progress in R&D on tungsten alloys for divertor structural and plasma facing materials [J]. Journal of Nuclear Materials, 2013, 442(1–3): S181–S189.

[52] S. Wurster, B. Gludovatz, A. Hoffmann, R. Pippan. Fracture behaviour of tungsten-vanadium and tungsten-tantalum alloys and composites [J]. Journal of Nuclear Materials, 2011, 413: 166–176.

[53] R. Mateus, M. Dias, J. Lopes, J. Rocha, N. Catarino, P. Duarte, R.B. Gomes, C. Silva, H. Fernandes, V. Livramento, P.A. Carvalho, E. Alves, K. Hanada, J.B. Correia. Blistering of W-Ta composites at different irradiation energies [J]. Journal of Nuclear Materials, 2013, 438: S1032–S1035.

[54] R. Mateus, M. Dias, J. Lopes, J. Rocha, N. Catarino, N. Franco, V. Livramento, P.A. Carvalho, J.B. Correia, K. Hanada, E. Alves. Effects of helium and deuterium irradiation on SPS sintered W-Ta composites at different temperatures [J]. Journal of Nuclear Materials, 2013, 442(1–3): S251–S255.

[55] M. Dias, R. Mateus, N. Catarino, N. Franco, D. Nunes, J.B. Correia, P.A. Carvalho, C. Sârbu, K. Hanada, E. Alves. Synergistic helium and deuterium blistering in tungsten-tantalum composites [J]. Journal of Nuclear Materials, 2013, 442(1–3): 69–74.

[56] Y. Zayachuk, G. Bousselin, J. Schuurmans, Y. Gasparyan, I. Uytdenhouwen, G. Van Oost. Design of a planar probe diagnostic system for plasmatron VISIONI and its application for the study of deuterium retention in W-Ta alloy [J]. Fusion Engineering and Design, 2011, 86: 1153–1156.

[57] Y. Zayachuk, M.H.J. Hoen, P.A. Zeijlmans van Emmichoven, I. Uytdenhouwen, G. van Oost. Deuterium retention in tungsten and tungsten–tantalum alloys exposed to high-flux deuterium plasmas [J]. Nuclear Fusion, 2012, 52: 103021.

[58] K. Arshad, M.Y. Zhao, Y. Yuan, Y. Zhang, Z.H. Zhao, B. Wang, Z.J. Zhou, G.H. Lu. Effects of vanadium concentration on the densification, microstructures and mechanical properties of tungsten vanadium alloys [J]. Journal of Nuclear Materials, 2014, 455(1–3): 96–100.

[59] T. Palacios, J.Y. Pastor, M.V. Aguirre, A. Martín, M.A. Monge, A. Muñóz, R. Pareja. Mechanical behavior of tungsten-vanadium-lanthana alloys as function of temperature [J]. Journal of Nuclear Materials, 2013, 442(1–3): S277–S281.

[60] T. Palacios, M.A. Monge, J.Y. Pastor. Tungsten-vanadium-yttria alloys for fusion power reactors (I): Microstructural characterization [J]. International Journal of Refractory Metals and Hard Materials, 2016, 54: 433–438.

[61] A. Muñoz, B. Savoini, E. Tejado, M.A. Monge, J.Y. Pastor, R. Pareja. Microstructural and mechanical characteristics of W-2Ti and W-1TiC processed by hot isostatic pressing [J]. Journal of Nuclear Materials, 2014, 455(1–3): 306–310.

[62] M.A. Monge, M.A. Auger, T. Leguey, Y. Ortega, L. Bolzoni, E. Gordo, R. Pareja. Characterization of novel W alloys produced by HIP [J]. Journal of Nuclear Materials, 2009, 386–388: 613–617.

[63] M.V. Aguirre, A. Martín, J.Y. Pastor, J.L. Lorca, M.A. Monge, R. Pareja. Mechanical properties of tungsten alloys with Y_2O_3 and titanium additions [J]. Journal of Nuclear Materials, 2011, 417(1–3): 516–519.

[64] B. Savoini, J. Martínez, A. Muñoz, M.A. Monge, R. Pareja. Microstructure and temperature dependence of the microhardness of W-4V-1La_2O_3 and W-4Ti-1La_2O_3 [J]. Journal of Nuclear Materials, 2013, 442(1–3): S229–S232.

[65] Z.M. Xie, R. Liu, Q.F. Fang, Y. Zhou, X.P. Wang, C.S. Liu. Spark plasma sintering and mechanical properties of zirconium micro-alloyed tungsten [J]. Journal of Nuclear Materials, 2014, 444(1–3): 175–180.

[66] Z.M. Xie, T. Zhang, R. Liu, Q.F. Fang, S. Miao, X.P. Wang, C.S. Liu. Grain growth behavior and mechanical properties of zirconium micro-alloyed and nano-size zirconium carbide dispersion strengthened tungsten alloys [J]. International Journal of Refractory Metals and Hard Materials, 2015, 51: 180–187.
[67] T. Renzheng. Fundamentals of physical metallurgy [M]. Beijing: Metallurgical Industry Press, 1997: 242–243.
[68] B. Huang, J. Tang, L.Q. Chen, X.L. Yang, Y.Y. Lian, L. Chen, X. Liu, X.D. Cui, L. Gu, C.T. Liu. Design of highly thermal-shock resistant tungsten alloys with nanoscaled intra- and inter-type K bubbles [J]. Journal of Alloys and Compounds, 2019, 782: 149–159.
[69] M. Fukuda, T. Tabata, A. Hasegawa, S. Nogami, T. Muroga. Strain rate dependence of tensile properties of tungsten alloys for plasma-facing components in fusion reactors [J]. Fusion Engineering and Design, 2016, 109–111: 1674–1677.
[70] S. Nogami, W.H. Guan, M. Fukuda, A. Hasegawa. Effect of microstructural anisotropy on the mechanical properties of K-doped tungsten rods for plasma facing components [J]. Fusion Engineering and Design, 2016, 109–111(B): 1549–1553.
[71] G. Pintsuk, I. Uytdenhouwen. Thermo-mechanical and thermal shock characterization of potassium doped tungsten [J]. International Journal of Refractory Metals and Hard Materials, 2010, 28(6): 661–668.
[72] D. Terentyev, W. Van Renterghem, L. Tanure, A. Dubinko, J. Riesch, S. Lebediev, T. Khvan, K. Verbeken, J.W. Coenen, E.E. Zhurkin. Correlation of microstructural and mechanical properties of K-doped tungsten fibers used as reinforcement of tungsten matrix for high temperature applications [J]. International Journal of Refractory Metals and Hard Materials, 2019, 79: 204–216.
[73] W.H. Guan, S. Nogami, M. Fukuda, A. Hasegawa. Tensile and fatigue properties of potassium doped and rhenium containing tungsten rods for fusion reactor applications [J]. Fusion Engineering and Design, 2016, 109–111(B): 1538–1542.
[74] M. Rieth, A. Hoffmann. Influence of microstructure and notch fabrication on impact bending properties of tungsten materials [J]. International Journal of Refractory Metals and Hard Materials, 2010, 28(6): 679–686.
[75] B. Gludovatz, S. Wurster, A. Hoffmann, R. Pippan. Fracture toughness of polycrystalline tungsten alloys [J]. International Journal of Refractory Metals and Hard Materials, 2010, 28(6): 674–678.
[76] K. Sasaki, K. Yabuuchi, S. Nogami, A. Hasegawa. Effects of temperature and strain rate on the tensile properties of potassium-doped tungsten [J]. Journal of Nuclear Materials, 2015, 461: 357–364.
[77] V. Nikolić, J. Riesch, R. Pippan. The effect of heat treatments on pure and potassium doped drawn tungsten wires: Part I-Microstructural characterization [J]. Materials Science and Engineering: A, 2018, 737: 422–433.
[78] X. Zhang, Q. Yan, S. Lang, M. Xia, C. Ge. Basic thermal-mechanical properties and thermal shock, fatigue resistance of swaged^{+} rolled potassium doped tungsten [J]. Journal of Nuclear Materials, 2014, 452(1–3): 257–264.
[79] X. Yang, W. Qiu, L. Chen, J. Tang. Tungsten–potassium: A promising plasma-facing material [J]. Tungsten, 2019, 1: 141–158.
[80] B. He, B. Huang, Y. Xiao, Y. Lian, X. Liu, J. Tang. Preparation and thermal shock characterization of yttrium doped tungsten-potassium alloy [J]. Journal of Alloys and Compounds, 2016, 686: 298–305.
[81] K. Shi, B. Huang, B. He, Y. Xiao, X. Yang, Y. Lian, X. Liu, J. Tang. Room-temperature tensile strength and thermal shock behavior of spark plasma sintered W-K-TiC alloys [J]. Nuclear Engineering and Technology, 2019, 51(1): 190–197.
[82] D. Nunes, V. Livramento, U.V. Mardolcar, J.B. Correia, P.A. Carvalho. Tungsten–nanodiamond composite powders produced by ball milling [J]. Journal of Nuclear Materials, 2012, 426(1–3): 115–119.

[83] V. Livramento, D. Nunes, J.B. Correia, P.A. Carvalho, U. Mardolcar, R. Mateus, K. Hanada, N. Shohoji, H. Fernandes, C. Silva, E. Alves. Tungsten–microdiamond composites for plasma facing components [J]. Journal of Nuclear Materials, 2011, 416(1–2): 45–48.

[84] L. Huang, L. Jiang, T.D. Topping, C. Dai, X. Wang, R. Carpenter, C. Haines, J.M. Schoenung. In situ oxide dispersion strengthened tungsten alloys with high compressive strength and high strain-to-failure [J]. Acta Materialia, 2017, 1: 19–31.

[85] G. Vieider, M. Merola, J.P. Bonal, P. Chappuis, D. Duglué, R. Duwe, F. Escourbiac, M. Fèbvre, L. Giancarli, M. Grattarola, G. LeMarois, H.D. Pacher, A. Pizzuto, L. Plöchl, B. Riccardi, M. Rödig, J. Schlosser, I. Smid, P. Tramier, C.H. Wu. European development of the ITER divertor target [J]. Fusion Engineering and Design, 1999, 46(2): 221–228.

[86] I. Smid, M. Akiba, G. Vieider, L. Plöchl. Development of tungsten armor and bonding to copper for plasma-interactive components [J]. Joumal of Nuclear Materials, 1998, 258: 160–172.

[87] M. Rieth, B. Dafferner. Limitations of W and W-1%La_2O_3 for use as structural materials [J]. Journal of Nuclear Materials, 2005, 342(1–3): 20–25.

[88] Q. Yan, X. Zhang, T. Wang, C. Yang, C. Ge. Effect of hot working process on the mechanical properties of tungsten materials [J]. Journal of Nuclear Materials, 2013, 442(1–3): S233–S236.

[89] G. Liu, G.J. Zhang, F. Jiang, X.D. Ding, Y.J. Sun, J. Sun, Y. Ma. Nanostructured high-strength molybdenum alloys with unprecedented tensile ductility [J]. Nature Materials, 2013, 12: 344–350.

[90] Y.C. Mana, Y. Wu, J. Chen. La_2O_3dispersion reinforced tungsten alloy for plasma and its high thermal load performance [J]. Rare Metal Materials and Engineering, 2009, 27(3): 415–417.

[91] K. Cui, Y. Shen, J. Yu, B. Ji. Microstructural characteristics of commercial purity W and W-1%La_2O_3 alloy [J]. International Journal of Refractory Metals and Hard Materials, 2013, 41: 143–151.

[92] M. Xia, Q. Yan, L. Xu, H. Guo, L. Zhu, C. Ge. Bulk tungsten with uniformly dispersed La_2O_3 nanoparticles sintered from co-precipitated La_2O_3/W nanoparticles [J]. Journal of Nuclear Materials, 2013, 434(1–3): 85–89.

[93] L. Xu, Q. Yan, M. Xia, L. Zhu. Preparation of La_2O_3 doped ultra-fine W powders by hydrothermal-hydrogen reduction process [J]. International Journal of Refractory Metals and Hard Materials, 2013, 36: 238–242.

[94] M. Yar, S. Wahlberg, H. Bergqvist, H.G. Salem, M. Johnsson, M. Muhammed. Chemically produced nanostructured ODS-lanthanum oxide-tungsten composites sintered by spark plasma [J]. Journal of Nuclear Materials, 2011, 408: 129–135.

[95] F. Ghezzi, M. Zani, S. Magni, G.M. Vanacore, A. Tagliaferri. Surface and bulk modification of W-La_2O_3 armor mock-up [J]. Journal of Nuclear Materials, 393(3): 522–526.

[96] H.J. Ryu, S.H. Hong. Fabrication and properties of mechanically alloyed oxide-dispersed tungsten heavy alloys [J]. Materials Science and Engineering: A, 2003, 363(1–2): 179–184.

[97] R. Liu, Y. Zhou, T. Hao, T. Zhang, X.P. Wang, C.S. Liu, Q.F. Fang. Microwave synthesis and properties of fine-grained oxides dispersion strengthened tungsten [J]. Journal of Nuclear Materials, 2012, 424 (1–3): 171–175.

[98] R. Liu, X.P. Wang, T. Hao, C.S. Liu, Q.F. Fang. Characterization of ODS-tungsten microwave-sintered from sol-gel prepared nano-powders [J]. Journal of Nuclear Materials, 2014, 450(1–3): 69–74.

[99] M.A. Yar, S. Wahlberg, H. Bergqvist, H.G. Salem, M. Johnsson, M. Muhammed. Spark plasma sintering of tungsten-yttrium oxide composites from chemically synthesized nanopowders and microstructural characterization [J]. Journal of Nuclear Materials, 2011, 412(2): 227–232.

[100] S. Wahlberg, M.A. Yar, M.O. Abuelnaga, H.G. Salem, M. Johnsson, M. Muhammed. Fabrication of nanostructured W-Y_2O_3 materials by chemical methods [J]. Journal of Materials Chemistry, 2012, 22: 12622–12628.

[101] Y. Kim, K.H. Lee, E. Kim, D. Cheong, S.H. Hong. Fabrication of high temperature oxides dispersion strengthened tungsten composites by spark plasma sintering process [J]. International Journal of Refractory Metals and Hard Materials, 2009, 27(5): 842–846.

[102] Y. Itoh, Y. Ishiwata. Strength properties of yttrium-oxide-dispersed tungsten alloy [J]. JSME International Journal. Ser. A, Mechanics and Material Engineering, 1996, 39: 429–434.

[103] Z. Dong, Z. Ma, C. Liu, Q. Guo, Y. Yamauchi, H.R. Alamri, Z.A. Alothman, M.S.A. Hossain, Y. Liu. Synthesis of nanosized composite powders via a wet chemical process for sintering high performance W-Y_2O_3 alloy [J]. International Journal of Refractory Metals and Hard Materials, 2017, 69: 266–272.

[104] X. Tan, L. Luo, H. Chen, X. Zhu, X. Zan, G. Luo, J. Chen, P. Li, J. Cheng, D. Liu, Y. Wu. Mechanical properties and microstructural change of W-Y_2O_3 alloy under helium irradiation [J]. Scientific Reports, 2015, 5: 12755.

[105] M. Battabyal, R. Schäublin, P. Spätig, N. Baluc. W-2wt.% Y_2O_3 composite: Microstructure and mechanical properties [J]. Materials Science and Engineering: A, 2012, 538: 53–57.

[106] M. Battabyal, R. Schäublin, P. Spätig, M. Walter, M. Rieth, N. Baluc. Microstructure and mechanical properties of a W-2wt.%Y_2O_3 composite produced by sintering and hot forging [J]. Journal of Nuclear Materials, 2013, 442(1–3): S225–S228.

[107] Y. Lian, X. Liu, F. Feng, J. Song, B. Yan, Y. Wang, J. Wang, J. Chen. Mechanical properties and thermal shock performance of W-Y_2O_3 composite prepared by high-energy-rate forging [J]. Physica Scripta, 2017, T170: 014044.

[108] Z.M. Xie, R. Liu, S. Miao, T. Zhang, X.P. Wang, Q.F. Fang, C.S. Liu, G.N. Luo. Effect of high temperature swaging and annealing on the mechanical properties and thermal conductivity of W-Y_2O_3 [J]. Journal of Nuclear Materials, 2015, 464: 193–199.

[109] M. Zhao, Z. Zhou, M. Zhong, J. Tan, Y. Lian, X. Liu. Thermal shock behavior of fine grained W-Y_2O_3 materials fabricated via two different manufacturing technologies [J]. Journal of Nuclear Materials, 2016, 470: 236–243.

[110] A.A. Sadek, M. Ushio, F. Matsuda. Effect of rare earth metal oxide additions to tungsten electrodes [J]. Metallurgical Transactions A, 1989, 21(12): 3221–3236.

[111] X.Y. Ding, L.M. Luo, Z.L. Lu, G.N. Luo, X.Y. Zhu, J.G. Cheng, Y.C. Wu. Chemically produced tungsten-praseodymium oxide composite sintered by spark plasma sintering [J]. Journal of Nuclear Materials, 2014, 454: 200–206.

[112] X.Y. Ding, L.M. Luo, H.Y. Chen, X.Y. Zhu, X. Zan, J.G. Cheng, Y.C. Wu. Chemical synthesis and oxide dispersion properties of strengthened tungsten via spark plasma sintering [J]. Materials, 2016, 9(11): 879.

[113] H. Kurishita, H. Arakawa, S. Matsuo, T. Sakamoto, S. Kobayashi, K. Nakai, G. Pintsuk, J. Linke, S. Tsurekawa, V. Yardley, K. Tokunaga, T. Takida, M. Katoh, A. Ikegaya, Y. Ueda, M. Kawai, N. Yoshida. Development of nanostructured tungsten based materials resistant to recrystallization and/or radiation induced embrittlement [J]. Materials Transactions, 2013, 54(4): 456–465.

[114] Y. Ishijima, H. Kurishita, H. Arakawa, M. Hasegawa, Y. Hiraoka, T. Takida, K. Takebe. Microstructure and bend ductility of W-0.3 mass %TiC alloys fabricated by advanced powder-metallurgical processing [J]. Materials Transactions, 2005, 46(3): 568–574.

[115] H. Kurishita, S. Matsuo, H. Arakawa, S. Kobayashi, K. Nakai, T. Takida, K. Takebe, M. Kawai. Superplastic deformation in W-0.5 wt.% TiC with approximately 0.1 μm grain size [J]. Materials Science and Engineering: A, 2008, 477(1–2): 162–167.

[116] H. Kurishita, S. Matsuo, H. Arakawa, T. Sakamoto, S. Kobayashi, K. Nakai, T. Takida, M. Kato, N. Yoshida. Development of re-crystallized W-1.1%TiC with enhanced room-temperature ductility and radiation performance [J]. Journal of Nuclear Materials, 2010, 398(1–3): 87–92.

[117] H. Kurishita, S. Matsuo, H. Arakawa, T. Hirai, J. Linke, M. Kawai, N. Yoshida. Development of nanostructured W and Mo materials [J]. Advanced Materials Research, 2009, 59: 18–30.

[118] H. Kurishita, Y. Amano, S. Kobayashi, K. Nakai, H. Arakawa, Y. Hiraoka, T. Takida, K. Takebe, H. Matsui. Development of ultra-fine grained W-TiC and their mechanical properties for fusion applications [J]. Journal of Nuclear Materials, 2007, 367–370: 1453–1457.

[119] H. Kurishita, S. Matsuo, H. Arakawa, T. Sakamoto, S. Kobayashi, K. Nakai, H. Okano, H. Watanabe, N. Yoshida, Y. Torikai, Y. Hatano, T. Takida, M. Kato, A. Ikegaya, Y. Ueda, M. Hatakeyama, T. Shikama. Current status of nanostructured tungsten-based materials development [J]. Physica Scripta, 2014, T159: 014032.

[120] G.M. Song, Y.J. Wang, Y. Zhou. Thermomechanical properties of TiC particle-reinforced tungsten composites for high temperature applications [J]. International Journal of Refractory Metals & Hard Materials, 2003, 21: 1–12.

[121] S. Lang, Q. Yan, N. Sun, X. Zhang, C. Ge. Effects of TiC content on microstructure, mechanical properties, and thermal conductivity of W-TiC alloys fabricated by a wet-chemical method [J]. Fusion Engineering and Design, 2017, 121: 366–372.

[122] S. Miao, Z.M. Xie, T. Zhang, X.P. Wang, Q.F. Fang, C.S. Liu, G.N. Luo, X. Liu, Y.Y. Lian. Mechanical properties and thermal stability of rolled W-0.5 wt% TiC alloys [J]. Materials Science & Engineering A, 2016, 671: 87–95.

[123] Kinds Of Mana, Y. Fuwen, C. Junling. Experimental study on the thermal load of W-TiC alloy on plasma materials and their electron beams [J]. Rare Metal Materials and Engineering, 2010, 39(4): 750–752.

[124] Z.M. Xie, R. Liu, S. Miao, X.D. Yang, T. Zhang, X.P. Wang, Q.F. Fang, C.S. Liu, G.N. Luo, Y.Y. Lian, X. Liu. Extraordinary high ductility/strength of the interface designed bulk W-ZrC alloy plate at relatively low temperature [J]. Scientific Reports, 2015, 5: 16014.

[125] Z.M. Xie, R. Liu, Q.F. Fang, T. Zhang, Y. Jiang, X.P. Wang, C.S. Liu. Microstructure and mechanical properties of nano-size zirconium carbide dispersion strengthened tungsten alloys fabricated by spark plasma sintering method [J]. Plasma Science and Technology, 2015, 17(12): 1066–1071.

[126] Z.M. Xie, R. Liu, S. Miao, X.D. Yang, T. Zhang, Q.F. Fang, X.P. Wang, C.S. Liu, Y.Y. Lian, X. Liu, G.N. Luo. High thermal shock resistance of the hot rolled and swaged bulk W-ZrC alloys [J]. Journal of Nuclear Materials, 2016, 469: 209–216.

[127] H.L. Ding, Z.M. Xie, Q.F. Fang, T. Zhang, Z.J. Cheng, Z. Zhuang, X.P. Wang, C.S. Liu. Determination of the DBTT of nanoscale ZrC doped W alloys through amplitude-dependent internal friction technique [J]. Materials Science & Engineering A, 2018, 716: 268–273.

[128] H.W. Deng, Z.M. Xie, Y.K. Wang, R. Liu, T. Zhang, T. Hao, X.P. Wang, Q.F. Fang, C.S. Liu. Mechanical properties and thermal stability of pure W and W-0.5 wt%ZrC alloy manufactured with the same technology [J]. Materials Science & Engineering A, 2018, 715: 117–125.

[129] S. Guiming, B. Houshan, Z. Yu. High-temperature fracture behavior of ZrC particle-reinforced tungsten matrix composites [J]. Rare Metal Materials and Engineering, 2000, 29: 101–104.

[130] J. Fan, Y. Han, P. Li, Z. Sun, Q. Zhou. Micro/nano composited tungsten material and its high thermal loading behavior [J]. Journal of Nuclear Materials, 2014, 455(1–3): 717–723.

[131] Y. Ueda, M. Oya, Y. Hamaji, H.T. Lee, H. Kurishita, Y. Gasparyan, O.V. Ogorodnikova, A. Manhard, A. Pisarev. Surface erosion and modification of toughened, fine- grained, recrystallized tungsten exposed to TEXTOR edge plasma [J]. Physica Scripta, 2014, T159: 014038.

[132] M. Oya, H.T. Lee, Y. Ohtsuka, Y. Ueda, H. Kurishita, M. Oyaidzu, T. Yamanishi. Deuterium retention in various toughened, fine-grained recrystallized tungsten materials under different irradiation conditions [J]. Physica Scripta, 2014, T159: 014048.

[133] S. Miao, Z.M. Xie, L.F. Zeng, T. Zhang, X.P. Wang, Q.F. Fang, C.S. Liu. Mechanical properties, thermal stability and microstructure of fine-grained W-0.5 wt.% TaC alloys fabricated by an optimized multi-step process [J]. Nuclear Materials and Energy, 2017, 13: 12–20.

[134] Z.M. Xie, S. Miao, T. Zhang, R. Liu, X.P. Wang, Q.F. Fang, T. Hao, Z. Zhuang, C.S. Liu, Y.Y. Lian, X. Liu, L.H. Cai. Recrystallization behavior and thermal shock resistance of the W-1.0 wt% TaC alloy [J]. Journal of Nuclear Materials, 2018, 501: 282–292.

[135] G. Ackland. Controlling Radiation Damage [J]. Science, 2010, 327(5973): 1587–1588.

[136] R.Z. Valiev, I.V. Alexandrov, Y.T. Zhu, T.C. Lowe. Paradox of strength and ductility in metals processed by severe plastic deformation [J]. Journal of Materials Research, 2002, 17(1): 5–8.
[137] Z. Xin, G. Yuwei, N. Yun, J. Tianfu. Large plastic deformation technology for the preparation of bulk nano/ultrafine grain materials [J]. Materials Reports, 2003, 17(12): 5–8.
[138] R. Valiev. Nanostructuring of metals by severe plastic deformation for advanced properties [J]. Nature Materials, 2004, 3: 511–516.
[139] M. Faleschini, H. Kreuzer, D. Kiener, R. Pippan. Fracture toughness investigations of tungsten alloys and SPD tungsten alloys [J]. Journal of Nuclear Materials, 2007, 367–370: 800–805.
[140] Q. Wei, H.T. Zhang, B.E. Schuster, K.T. Ramesh, R.Z. Valiev. Microstructure and mechanical properties of super-strong nanocrystalline tungsten processed by high-pressure torsion [J]. Acta Materialia, 2006, 54: 4079–4089.
[141] Q. Wei, T. Jiao, K.T. Ramesh, E. Ma, L.J. Kecskes, L. Magness, R. Dowding, V.U. Kazykhanov, R.Z. Valiev. Mechanical behavior and dynamic failure of high-strength ultrafine grained tungsten under uniaxial compression [J]. Acta Materialia, 2006, 54: 77–87.
[142] Y. Zhang, A.V. Ganeev, J.T. Wang, J.Q. Liu, I.V. Alexandrov. Observations on the ductile-to-brittle transition in ultrafine-grained tungsten of commercial purity [J]. Materials Science and Engineering: A, 2009, 503(1–2): 37–40.
[143] T. Hao, Z.Q. Fan, T. Zhang, G.N. Luo, X.P. Wang, C.S. Liu, Q.F. Fang. Strength and ductility improvement of ultrafine-grained tungsten produced by equal-channel angular pressing [J]. Journal of Nuclear Materials, 2014, 455: 595–599.

Chapter 3
Preparation and processing of tungsten matrix materials for fusion reactors

3.1 Introduction

Although pure tungsten has a series of advantages, it still exhibits significant performance drawbacks when used as a plasma-facing material (PFM). These include a high ductile-to-brittle transition temperature (DBTT) of around 400 °C, a low recrystallization temperature of approximately 1,200 °C, a sharp increase in self-sputtering rate under high-energy irradiation, and an increase in DBTT following neutron irradiation. Additionally, under high-energy plasma irradiation, the interaction between plasma particles (D, T, He) and tungsten can cause surface blistering and eventual surface layer exfoliation. Under certain conditions, tungsten can form a surface nanoporous structure due to plasma irradiation, which can trap particles and subsequently affect its thermal conductivity. Therefore, the research objective of fusion energy materials is to obtain the expected ideal microstructure and properties by different preparation methods for tungsten and tungsten alloy materials to solve the embrittlement behavior of tungsten-based materials, improve toughness, reduce DBTT, and enhance irradiation resistance, etc. Ultrafine-grained (UFG)/nanocrystalline tungsten-based materials have been found to possess excellent toughness while maintaining adequate plasticity, exhibiting good resistance to irradiation-induced embrittlement and swelling under specific conditions. The development of nanoscale fine-grained tungsten-based materials has become a critical issue in overcoming application bottlenecks for PFM tungsten-based materials in fusion energy.

3.2 Preparation of tungsten composite powders with different nanostructures

The preparation of tungsten-based materials first requires consideration of powder preparation. Tungsten powder serves as the raw material for tungsten bulk materials and components, playing a crucial role in determining the final performance of the material parts. Properties such as purity and particle size play a decisive role in the preparation of high-performance tungsten materials with fine-grained and fully dense sintered structures. Nanomaterials exhibit significant enhancements in performance compared with traditional materials due to their unique advantages. Nanopowders are expected to refine tungsten grains, thereby improving the material's mechanical properties, such as strength, hardness, and ductility. At present, with the increasing adoption of nanotechnology, there are many methods for preparing tung-

 | https://doi.org/10.1515/9783111556987-003

sten nanocomposite powders [1], providing a wider range of applications for the prepared tungsten-based materials:

(1) Mechanical alloying
Mechanical alloying (MA), also known as mechanical milling (MM), involves milling the elemental metal powders of the alloy to be prepared in a high-energy ball mill equipped with stirring, planetary, or rotor mechanisms. During the milling process, gases such as argon or nitrogen are introduced to prevent powder oxidation. In the process of ball milling, the high-speed collisions between the metal balls and the powder generate huge energy, refining the powder and resulting in pre-alloyed mixed powders with fine-grained or nanocrystalline characteristics. Simultaneously, the repeated mixing and collisions during milling cause the temperature to rise, leading to repeated cold welding and fracturing, which results in a highly uniform mixture of elemental powders, with interdiffusion between the particles. Consequently, a supersaturated solid solution or/and an amorphous phase with a certain degree of solubility or even a high degree of solubility is formed. Due to the simplicity and ease of operation of the equipment used in this technique, MA is the most widely used and popular method for preparing ultrafine (nanocrystalline) alloy powders. The main drawbacks are the tendency to introduce impurities, serious powder agglomeration, and wall sticking issues.

(2) Spray drying method
Spray drying, also known as thermal chemical synthesis, is a method that combines chemical and physical processes, which atomizes the solution through physical means to obtain ultrafine particles in three stages: preparation and mixing of the original solution, spray drying, and fluidized bed conversion. Firstly, various metal salts are mixed in proportion to obtain a mixed solution. Then, ammonium paratungstate (APT), ammonium metatungstate (AMT), and other metal brine solutions are mixed and sent to the atomizer, where they are sprayed at high speed into a drying chamber through a nozzle to obtain fine particles of mixed metal salts. After the collection and calcination, ultrafine particles of nanocrystalline oxide composite powder precursors are obtained, resembling shell-like spherical shapes. Controlling the reduction process of the precursors is also crucial. By reducing or carbonizing the precursor powder under specific conditions, the desired single-component or multi-component alloy composite powders or carbides can be obtained. The spray drying process is simple, making it the most suitable for large-scale production without introducing other heterogeneous impurities.

(3) Condensing drying method
The process begins with the preparation of a mixed solution from multiple metal salt precursors, employing the precipitation method to form a hydroxide solution. When the aqueous material freezes, the solid-phase particles can maintain a uniform state in the water. Upon sublimation of the ice, the solid particles do not approach each other

too closely and thus avoid agglomeration. The condensing drying method offers distinct advantages, including simple equipment, low cost, uniform particle composition, and large batch production capacity-characteristics particularly advantageous for the preparation of ultrafine particles in large-scale factories. However, a notable limitation involves potential contamination by impurities such as sulfur (S) and oxygen (O).

(4) Vacuum plasma spray deposition
Vacuum plasma spray deposition can be used to prepare single-component or multi-component tungsten, as well as tungsten alloy powders or coatings. According to the composition ratio for alloy preparation, tungsten and tungsten alloy mixed powders are first obtained. Next, the high-melting-point mixed powders are melted and sprayed as a single entity. Finally, tungsten and its alloys are prepared using vacuum plasma deposition.

The preparation of nanomaterials can be divided into two categories: one is the top-down approach, and the other is the bottom-up approach. The bottom-up approach, compared to the top-down approach, allows for the production of powders with precise and controllable composition, high purity, and excellent uniformity. It results in smaller grain sizes and requires less energy consumption, thus offering better economic efficiency [2].

3.2.1 Top-down technology

Materials can be shaped and sized to meet specifications using external mechanical tools or methods such as cutting or grinding, which are typically categorized as top-down techniques, with MA being a commonly employed approach. Using a grinder or high-energy ball mill, larger-sized powders undergo repeated deformation, cold welding, tearing, and fracturing processes. Ultimately, this leads to a reduction in grain size and the achievement of solid-state alloying. Through thorough grinding for sufficient time and energy, the grain size can be refined from the micrometer scale to the nanometer scale [3–5], as shown in Figure 3.1. MM equipment is relatively well-equipped, and the process is simple. After designing the composition of tungsten alloy, tungsten powder and other alloy powders are weighed according to the proportions, placed in a ball mill jar, and thoroughly mixed through ball milling to obtain the alloy powder.

Mechanical ball milling can be classified into horizontal and vertical ball milling, depending on the orientation of the ball mill jar. Due to the high density of tungsten powder and its tendency to settle, horizontal ball milling is generally used. Based on the energy input during ball milling, the technique can be divided into conventional ball milling, high-energy ball milling, and the newly developed plasma-assisted ball milling method [6, 7]. However, during the ball milling process, collisions and wear between the ball milling jar and the milling balls can lead to deformation and fracture. Additionally, the protective atmosphere required during grinding can introduce significant impurities, resulting in compositional impurities and minor variations. These impurities, which have low solubility in the tungsten lattice, tend to segregate at grain boundaries during ball milling, thereby weakening the integrity at those boundaries. Although these effects can be mitigated by optimizing the ball milling process through factors such as selecting appropriate milling media and employing inert gas protection [8], ball milling also introduces a large amount of energy into the powder, placing it in a high-energy activation state. This can lead to significant grain growth during the sintering process, and at the same time, the sintering billet is more prone to cracking due to the high internal stress [9]. To obtain high-purity and uniformly mixed tungsten-based alloy powders, it is necessary to select suitable ball milling conditions, such as the material of the ball milling jar, ball-to-powder ratio, ball milling speed, ball milling time, and ball milling atmosphere. In addition to high-energy ball milling, ultra-fine-grained or nanocrystalline materials can also be obtained by severe plastic deformation and equal-channel angular pressing (ECAP) methods.

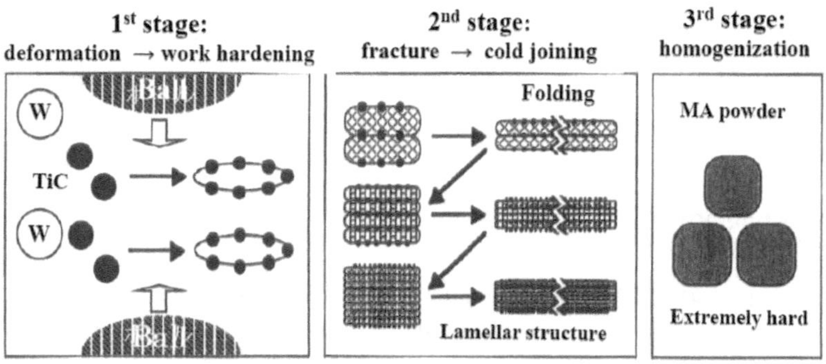

Figure 3.1: Schematic diagram of the mechanical alloying process [5].

3.2.2 Bottom-up technology

The synthesis of nanostructured materials through atomic or molecular-level reactions occurring in the gas phase or solution phase is known as the bottom-up approach [10–12]. As a method for the preparation of ultrafine powders, the wet chemical method overcomes the shortcomings of MA, and the powder composition obtained is accurate and controllable, with high purity and good uniformity [13–15]. In the solution state, the desired dopants are uniformly mixed according to specified measurements, and this method is simple to operate and suitable for large-scale industrial production. Especially in the preparation of second-phase dispersion-strengthened tungsten-based material powder, the above-mentioned method can yield composite powders with uniformly dispersed nanosized second-phase particles. By reacting tungsten and second-phase metal dopants in solution, the two components are mixed at the molecular level, resulting in very uniform second-phase metal-doped tungsten powders. These tungsten powders are subsequently reduced in a high-temperature hydrogen atmosphere to form nano-scale tungsten powders with uniformly distributed second-phase particles.

Mamoun Muhammed et al. [16, 17] in Sweden suspended APT in a rare earth nitrate solution, added nitric acid, precipitated after vigorous stirring, and then used the conventional reduction process to obtain doped tungsten powder, which was sintered into tungsten blocks using SPS. The fracture surface, as shown in Figure 3.2 [16], reveals a transgranular fracture with second-phase oxide nanoparticles uniformly distributed at the tungsten grain boundaries and inside the grains. In China, M. Xia et al. [18, 19] prepared tungsten powder doped with TiC and La_2O_3 by chemical precipitation using AMT as raw material and sintered it. As shown in Figure 3.3, the TEM observation of the TiC/W "core-shell" structured powder prepared by the precipitation-coating method shows that the TiC particles are coated with a W layer [19]. Liu et al. [14] used APT and yttrium nitrate as raw materials to prepare tungsten-containing sols/gels, which were dehydrated and dried, followed by calcination and reduction. Subsequently, microwave sintering was employed to produce La_2O_3 and Y_2O_3-doped W. As shown in Figure 3.4, the doped La_2O_3 and Y_2O_3 oxide particles can be seen evenly distributed within the tungsten matrix.

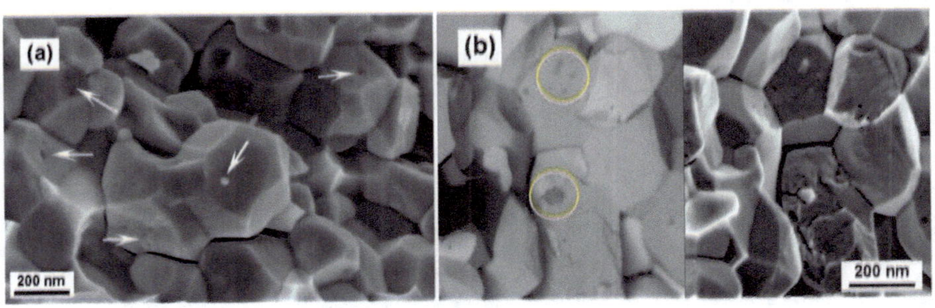

Figure 3.2: Fracture morphology of a sintered sample [16].

Sun Jun et al. [20] used the "liquid-liquid doping" method to design and control the microstructure, and made significant progress in the preparation of high-performance molybdenum alloys, as shown in Figure 3.5. By regulating the proportion and size distribution of the second phase within the grains and at the grain boundaries, the ultrafine second-phase particles occupying the grain boundaries can effectively inhibit the growth trend caused by sintering, which significantly reduces the extent of intergranular fracture. The dispersed second-phase particles within grains promote notable dislocation accumulation during material stretching (Figure 3.6), leading to enhanced work-hardening behavior (as seen in the true stress-true strain curve) and the consequent prevention of premature local softening. The combined effect of the second phase at two locations results in Mo materials exhibiting ultra-high elongation during

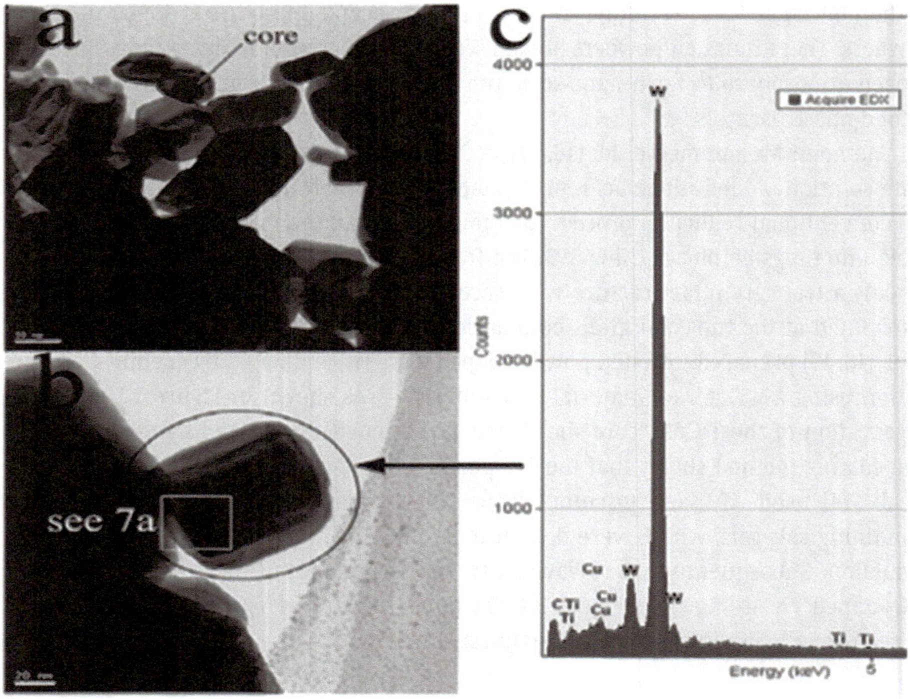

Figure 3.3: TEM diagram of a TiC/W "core-shell" structured powder prepared using the precipitation-coating method [19].

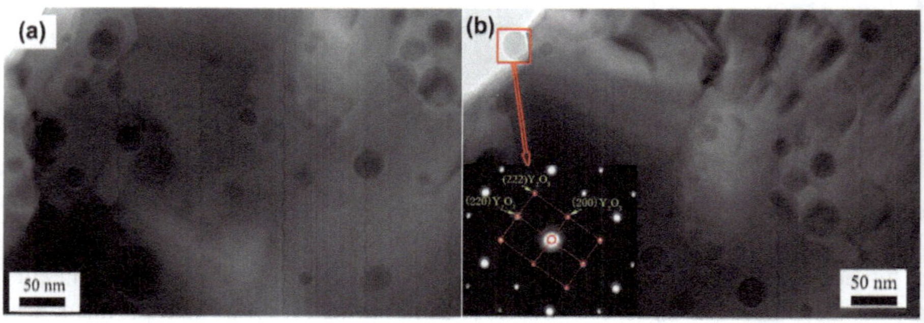

Figure 3.4: TEM diagram of a W-1%Y_2O_3 powder microwave-sintered sample prepared by the sol-gel method [14].

room-temperature tensile tests (Figure 3.7). Given the many similarities between W and Mo, integrating liquid-phase doping to prepare W composite powders with sintering and plastic processing methods holds promise for new breakthroughs in the strengthening and toughening of W materials.

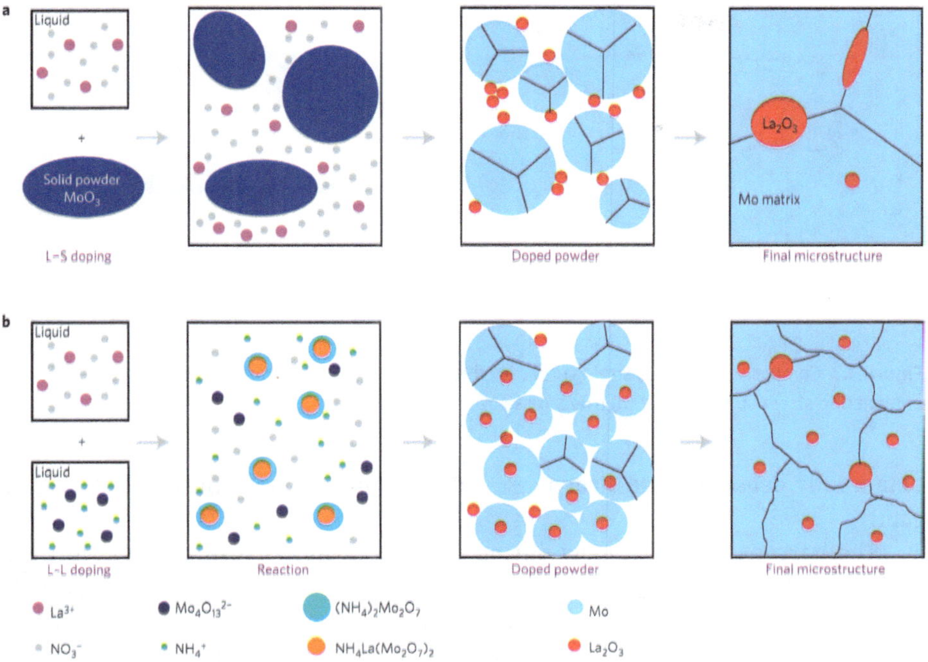

Figure 3.5: Schematic diagram of liquid-solid doping and liquid-liquid doping [20].

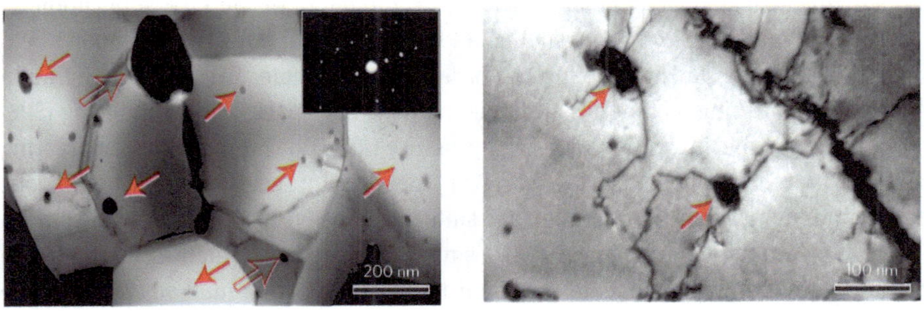

Figure 3.6: TEM plots of NS-Mo before (left) and after (right) tensile testing [20].

3.3 Sintering densification of tungsten alloy

An effective sintering process must be adopted to limit grain growth after the preparation of nano W powder for tungsten materials to maintain the size of nano or ultrafine grains. Compared with coarse-grained powders, nano powders possess higher surface energy, which facilitates grain growth during sintering [21, 22]. Due to the high melting point and low self-diffusion coefficient of W, the sintering temperature must range from 2,000 to 2,500 °C, and the sintering duration is typically prolonged to

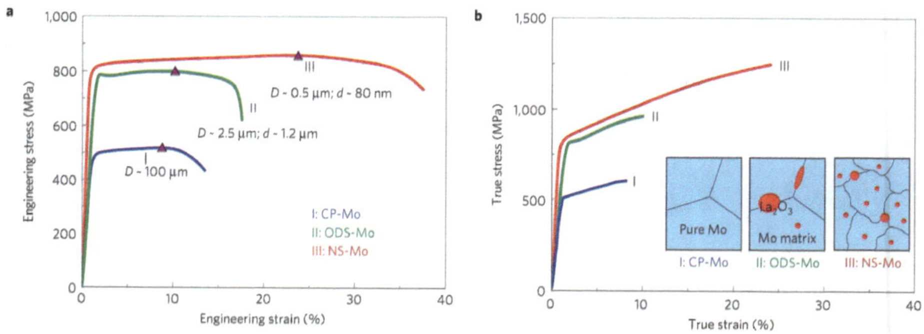

Figure 3.7: Comparison of stress-strain curves for three different Mo alloys at room temperature: (a) engineering stress-strain curves; (b) true stress-strain curves [20].

obtain fully dense sintered structures, and the grain will undoubtedly grow significantly.

The narrow kinetic window during sintering (where densification occurs but grain growth is suppressed) presents a challenge in achieving the desired goals. In this context, materials obtained can be described as "under-sintered" due to insufficient sintering time and lower temperatures, which limit inter-particle diffusion and affect grain growth. Although the addition of components with low melting points, such as nickel (Ni), can promote densification, it also causes grain growth. Moreover, Ni is a highly activated element and is strictly prohibited in fusion PFM applications.

Ensuring high density in tungsten as a PFM is essential. Effective methods to enhance the sintering density of tungsten-based alloys include: first, employing high-energy input sintering techniques, utilizing methods like spark plasma sintering (SPS) [23], resistance sintering under ultra-high pressure (RSUHP) [24], hot isostatic pressing (HIP) [25], and microwave sintering (MS) [26]. These techniques involve instantaneous heating and high-energy input to improve sintering density. Second, enhancing sintering activity can be achieved through pre-sintering treatments, such as adding trace amounts of activating elements or mechanical activation. These methods increase the surface activity of the sintering powder, effectively lowering the required sintering temperature.

3.3.1 Conventional sintering

The traditional sintering method mainly uses thermal conductivity to heat and sinter materials, such as atmosphere sintering and hot-pressing sintering. This heating method generally features a slow heating rate, a long holding time, and a high sintering temperature. The traditional sintering method usually requires holding at high temperatures for several hours, achieving sintered densities of approximately 90%.

However, prolonged high-temperature sintering can lead to grain growth in materials, affecting mechanical and radiation resistance properties. The advantage of this method is its ability to produce large-sized tungsten-based materials suitable for engineering applications, making it a widely used method in tungsten industry production. However, its disadvantage lies in the relatively lower sintered density achieved, which may not be directly suitable for immediate application. Furthermore, due to the high sintering temperature, tungsten grains grow and need to be processed by thermo-mechanical processing to enhance density and refine the grains to compensate for the limitations caused by a sintering process.

3.3.2 Special sintering technologies

Pressureless sintering, which is commonly used in industry, usually at temperatures above 2,000 °C, typically leads to the growth of tungsten grains, which exacerbates its brittleness. To address these issues and improve density while refining the grain structure, alternative sintering methods are being explored. These include methods such as SPS, RSUHP, HIP, and MS. These methods offer advancements toward achieving higher density and finer grain structures compared with traditional sintering approaches:

(1) Spark plasma sintering
SPS, also known as spark plasma sintering or plasma-activated sintering, belongs to an innovative powder metallurgy sintering method that combines electric energy activation, resistance heating, and pressure. This advanced sintering method offers significant advantages in the fabrication of nanomaterials and composite materials. It is characterized by rapid heating, short sintering times, controllable microstructure of sintered bodies, uniform grain size distribution, and high density. Currently, SPS is widely applied in fabricating metals, ceramics, and composite materials, as well as nano, amorphous bulk materials, and gradient materials [27–34].

Unlike conventional heating methods, SPS uses direct current pulses that directly act on the pre-sintered powder, which fully utilizes the resistive heat generated between powder particles, causing the particle surfaces to become plasmified and produce high heat instantly. This process promotes the volatilization of surface impurities, thereby achieving higher sintering activity. Therefore, it is possible to obtain fine-grained sintered bodies with high density at lower temperatures and shorter sintering times. When comparing the heating rates, SPS can achieve the fastest heating rate of up to 1,000 °C/min, whereas hot pressing sintering typically only reaches 50–80 °C/min [35]. Such a difference in heating rates profoundly influences the sintering process. SPS mostly uses conductive materials for the punch and die. Typically, for low-temperature sintering (<800 °C), high-strength tungsten carbide molds are used, which can provide uniaxial sintering pressure of 500 to 1,000 MPa. For high-temperature sintering (1,000–2,300 °C), graphite molds are commonly used, which can provide uniaxial pressure of approximately 60 MPa. The sintering pressure is also closely related to the sintering density and grain growth. Due to the high melting points of tungsten and its alloys, graphite molds are often chosen for SPS because they can withstand the high temperatures required without reacting with the material being sintered.

Kim et al. [36] used SPS to sinter at 1,700 °C to obtain oxide-dispersion-strengthened tungsten-based composites. Due to the fine grain size, the mechanical properties of the tungsten material were significantly improved. Tan Jun et al. [37] used the SPS process to sinter at 1,700 °C for 1 min and obtained UFG W-TiC composites with well-bound sintered particles and a high density of approximately 98.6%. Special sintering techniques also face challenges, such as the difficulty in producing large-sized products and the relatively high cost.

(2) Ultra-high-pressure electric current sintering
The ultra-high pressure electric current sintering, a technique of directly heating the green compact by applying ultra-high pressure while passing a large current, features rapid heating and cooling, fast thermal conversion rates, and short sintering times. For the preparation of ultrafine/nanocrystalline tungsten, the method offers significant advantages. The advantages of RSUHP are similar to those of SPS, but RSUHP applies pressures in the GPa range, which are much higher than the MPa pressures used in SPS, which greatly improves the density of sintered tungsten [38]. Zhou et al. [39, 40] prepared ultrafine-grain W by ultra-high pressure (6–9 GPa) assisted sintering. Zhou et al. [41] used the RSUHP method to sinter commercial pure W with particle sizes of 0.1 μm, 1 μm, and 10 μm, and the relevant performance test results are shown in Table 3.1. It can be found that the grains do not grow after sintering, which is because the sintering time of RSUHP is extremely short, not allowing enough time for grain growth. The relative density of pure W reached over 95%, and even up to 99% of the theoretical density, but the relative density and thermal conductivity decreased as the grain size decreased. Liu et al. [42] also used high-pressure technology to prepare a nearly fully dense sintered structure of W with a density of 99.5% under conditions of 1,400 °C/5.5 GPa. The grain size remained almost the same, and the material's fracture toughness was improved by about 50% compared to the traditional methods.

Table 3.1: Properties of tungsten powders with different particle sizes after ultra-high-pressure electrical sintering [41].

Sample	W02	W10	W100
Grain size (μm)	~0.2	~1	~10
Relative density (%)	95.25	97.93	99.05
Microhardness (HV)	1,107.5	811.5	710.5
Bending strength (MPa)	658.3	532.6	322.9
Thermal conductivity (W/m K)	105	128	140

(3) Hot isostatic pressing sintering

Hot isostatic pressing (HIP) is a process that uses inert gases, such as nitrogen or argon, as pressure-transmitting media. This approach involves placing the sintered compact into a sealed container and applying uniform pressure in all directions at temperatures ranging from 900 to 2,000 °C and pressures from 100 to 200 MPa to conduct the densification sintering treatment. As a thermomechanical processing method integrating simultaneous high-temperature and high-pressure effects, HIP produces materials with exceptional density uniformity, making it an advanced manufacturing process and a method for preparing and processing materials to enhance material performance [43–45].

HIP technology, developed by the American Battelle Institute in 1955 for the fabrication of nuclear reactor materials, demonstrates unique advantages in producing materials and components with high processing difficulty and stringent quality requirements. Kurishita et al. [46] used the HIP method to mechanically alloy W and TiC powders under controlled atmospheres (argon or hydrogen) to obtain highly distorted nanoscale powders. Subsequently, the powders were heated in a vacuum to remove the protective gas, followed by HIP sintering to prepare UFG W doped with TiC materials. After HIP processing, the W doped with additives exhibited fine grains, but the samples did not show significant ductility [47]. Upon subsequent thermal processing, recrystallization occurred, resulting in a significant increase in grain size. However, the material exhibited fracture strength at room temperature far greater than its yield strength, indicating excellent room-temperature ductility. Moreover, the increase in grain size did not decrease the material's fracture strength; instead, there was a substantial increase in fracture strength.

(4) Microwave sintering

Microwave sintering (MS) is a technique where microwave radiation in a specific frequency range interacts with the fine structure of materials, generating heat through dielectric loss. This process heats the entire material to the sintering temperature, enabling densification of the material. MS differs significantly from conventional sintering in terms of heating behavior and temperature gradients. It does not rely on conduction or convection for heat transfer but instead utilizes polarization relaxation to instantaneously convert microwave energy into heat, enabling rapid and efficient heating of the entire material [48]. As a result, the temperature distribution within the sintered material is uniform, internal thermal stresses are reduced, and heating rates can reach several tens to hundreds of degrees Celsius per minute. This shortens heating and holding times, prevents significant grain growth, and results in materials with fine and uniform grain structures after sintering. This approach offers great advantages for the preparation of fine-grained or nanocrystalline materials, overcoming drawbacks such as high surface temperatures and substantial energy loss associated with conventional heating methods. The technique effectively promotes the densification of materials and accelerates chemical reactions [49].

Metal powder compacts typically have particle sizes in the micron or nanometer range, which are comparable to the penetration depth of microwaves in metals. Their large surface area and high reactivity make them more amenable to uniform and comprehensive microwave heating compared with bulk metals, thereby facilitating densification through sintering. Liu et al. [14] successfully prepared fine-grained, high-density oxide dispersion-strengthened tungsten materials by MS at 1,500 °C for 30 min. For comparison, pure tungsten was sintered under the same conditions, resulting in densities of 96.9% for W, 96.8% for W-1wt% Y_2O_3, and 95% for W-1wt% La_2O_3.

3.3.3 Powder activation pretreatment

Powder activation pretreatment is generally achieved through the following methods: (1) altering the surface state of the powder to effectively improve the surface atomic activity and diffusion properties of the powder; (2) changing the contact interface properties of the powder particles, thereby significantly enhancing the atomic diffusion pathways. Mechanical activation, the addition of trace amounts of activating elements, or a combination of both, can significantly improve the sintering ability of the powder.

By using a high-energy ball mill for mechanical activation, the metal or alloy powders loaded into it undergo intense collisions and impacts with the grinding balls, causing continuous cold welding and fracturing among the powder particles, thereby accelerating the diffusion of internal atoms. Prabhu et al. [50] used high-energy ball milling to activate tungsten powder. The surface morphology of the activated tungsten powder underwent significant changes, transforming from a smooth cubic surface to a lamellar structure, as shown in Figure 3.8. The powder developed many new stepped surfaces, increasing the specific surface area (energy) and placing it in a high-

Figure 3.8: SEM images of raw tungsten powder (left) and activated tungsten powder (right) [50].

energy state. Under the same sintering conditions, the sintered body made from activated tungsten powder is more densified.

Adding trace amounts of activated elements can improve the sintering effect [51–53]. Since the solubility of these activating elements in tungsten is very low, they segregate at the contact interfaces of the powder particles, forming a new "activation layer." This, in turn, increases the atomic diffusion rate during the sintering process. Experiments conducted by Johnson, German [54, 55], and Boonyongmaneerat [56] revealed that transition elements such as Fe, Ni, Co, and Pd can enhance the sintering activity of W powders, resulting in a lower sintering temperature. The reason is that these added elements are insoluble in W, causing them to accumulate in the intermediate regions between W particles, providing fast diffusion channels and thus reducing the sintering activation energy [57]. Johnson et al. [54] also studied the application of transition elements such as Ni, Fe, Pd, and Co in tungsten-copper, and found that Fe and Co play the most obvious role in the sintering activation performance, mainly because their solubility in Cu is not significant, and during sintering, Fe and Co form highly stable interphases with W, creating an interface layer with strong diffusion properties and accelerating the sintering rate of solid-phase W particles. German et al. studied the effects of Ni, Co, Cu, Fe, Pd, and Pt on low-temperature sintering of W, and found that Pd was the best sintering activator for W, followed by Ni, Pt, and Co, as shown in Figure 3.9. Gupta et al. [58] investigated the effects of doping with different nickel contents under various sintering temperatures and found that the density of the pure tungsten sintered body changed very little, whereas the density of tungsten containing 0.5–1% nickel reached 89% under the condition of sintering at 1,400 °C for 2 h.

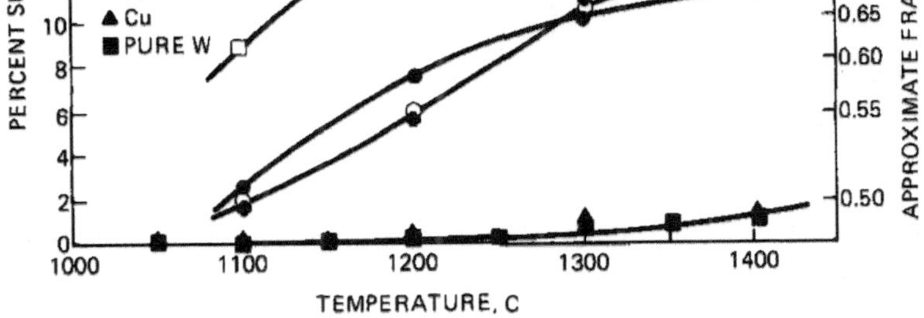

Figure 3.9: Influence of different activating elements on the sintering of W [57].

Combining the addition of activating elements with mechanical activation can also produce significant effects. Genc et al. [59] conducted mechanical activation and liquid-phase sintering with 1% Ni doping on the original W-2wt% TiC composite powder. The results showed that after sintering at 1,400 °C for 1 h, the density of the W-2wt% TiC sintered body reached 94.2% (Figure 3.10). However, the addition of activating elements and impurities brought in by ball milling is detrimental to the irradiation resistance of PFM.

3.4 Forming and processing of tungsten materials

The billets obtained from powder metallurgy sintering possess a certain degree of plasticity, allowing for direct pressure processing. Through repeated subsequent processing, the sintered body can even achieve full density. Multiple instances of plastic deformation during material forming can improve the microstructure, potentially enhancing the thermal, electrical, and mechanical properties. Thermal processing can further increase the density of tungsten materials, and the plastic deformation can result in ultrafine grains and allow for the control of tungsten grain boundary types. Altering the segregation of impurity elements at the grain boundaries and increasing the number of tungsten grain boundaries can absorb radiation-induced point defects, leading to improved mechanical properties and enhanced irradiation resistance of the material [60].

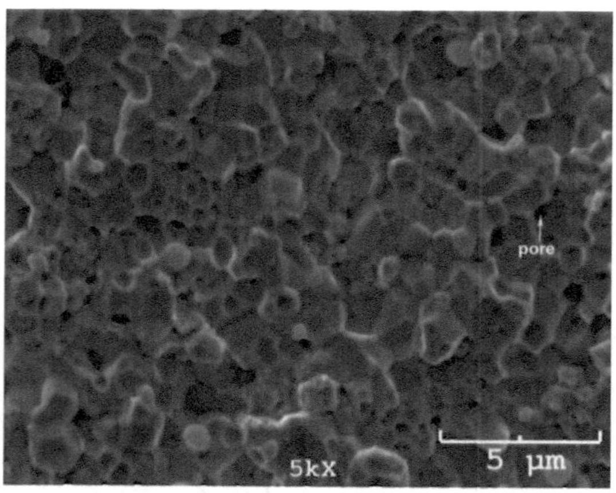

Figure 3.10: Fracture morphology of W2TiC+1Ni sintered sample after 12 h of mechanical ball milling [59].

Based on the requirements for the shape and microstructure of the material, various tungsten material processing methods have been developed industrially, such as rolling, forging, rotary forging, drawing, ECAP, and high-pressure torsion (HPT). The recrystallization temperature of pure tungsten is generally 1,100–1,300 °C, so processing above 1,300 °C is considered hot working, and below 1,100 °C is classified as cold working. Additionally, the DBTT is a critical point for tungsten forming. Below the DBTT, tungsten exhibits brittleness and cannot be formed; only above the DBTT can it be processed, and the DBTT decreases with increasing deformation. Currently, rolling and forging processes are widely used in production to obtain ultra-fine grains. Moreover, severe plastic deformation can be used to achieve submicron and nanoscale grains, which significantly enhance the material's strength while maintaining good ductility and toughness.

3.4.1 Rolling processing

Rolling processing is a type of plastic deformation in which the rolled piece passes between rotating rolls with grooves. During the process, the material is subjected to frictional forces as well as compressive forces, resulting in a reduction in cross-sectional dimensions and an increase in length. Compared with rotary forging, rolling is a more traditional process that allows for greater deformation and is used for the forming of plates and rods. Rolling increases the long aspect ratio of tungsten particles, making them fibrous. This process can be performed at a constant temperature, which significantly reduces the occurrence of defects and improves the density of tungsten materials [61]. A common rolling method is longitudinal rolling, where the metal workpiece passes between two oppositely rotating rolls, causing plastic deformation of the workpiece's plane.

Rolling is an effective method for producing large tungsten blocks suitable for fusion first-wall engineering applications, particularly for obtaining plate materials. Commonly, powder metallurgy sintered billets are used as raw materials, which are then heated in a high-temperature hydrogen furnace before being subjected to rolling processing. Reiser et al. [62, 63] used cold rolling technology to roll multiple layers of tungsten foil into thin tungsten sheets, which reached nanometer size in the thickness direction, and these thin tungsten sheets exhibit room-temperature ductility and have a tensile strength of up to 2 GPa. Krsjakt et al. [64] studied the mechanical properties of rolled commercial thick tungsten plates and found that the DBTT is around 200 °C, and the plates possess excellent bending resistance.

For sintered tungsten billets, hot rolling is generally used due to their relatively low initial density (~90%) and significant brittleness. The billets typically have a thickness greater than 20 mm, and the rolling heating temperature is approximately 1,000–1,550 °C, with a holding time of about 10–40 min. Rolling deformation is carried out under heating conditions, with multiple cycles of rolling and intermediate annealing stages, ultimately rolling the material to the desired dimensions and ensuring the rolled material meets performance requirements. The intermediate annealing heat treatments eliminate stress within tungsten grains, allowing for complete dislocation recovery, and induce recrystallization in polycrystalline tungsten, thereby enhancing the ductility of the deformed material. The hot rolling process involves significant deformation, which is particularly advantageous for improving the ductility and strength of brittle tungsten materials. The process finds primary application in deforming tungsten materials into thin sheets and foils [65]. Xie et al. [66] prepared W-Zr-Y_2O_3 plate billets through ball milling mixed powders, cold isostatic pressing at 200 MPa, and hot press sintering. The final tungsten material plates were obtained after four passes of hot rolling. Figure 3.11 shows the schematic diagram of the dynamic recrystallization of W-Zr-Y_2O_3 plates during hot rolling. The mechanical properties of the material were further improved by rolling, and the DBTT was significantly reduced to 150 °C.

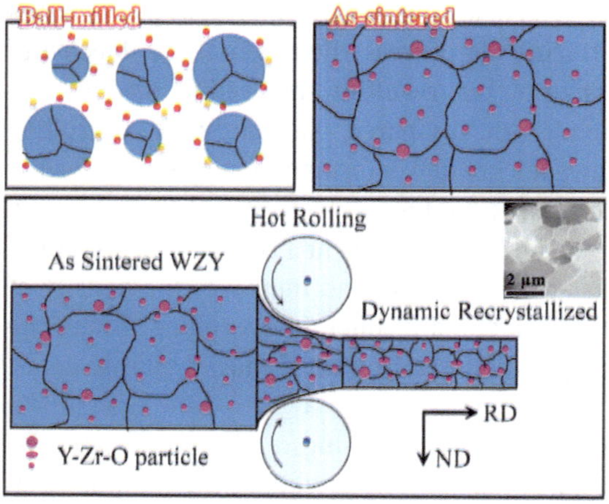

Figure 3.11: Schematic diagram of rolling for mechanically milled W-Zr-Y_2O_3 material [66].

The deformation rate of the first pass of pure tungsten (>99%) plates and rod rolling is generally 20–30%. During the rolling process, the billet is subjected to uniaxial compressive stress and bidirectional tensile stress, which can easily cause the initiation and propagation of cracks in brittle materials. In the cross-section perpendicular to the movement direction of the rolled plates and rods, the stress and strain distribution is uneven, leading to uneven deformation and microstructure distribution. As a result, the macroscopic mechanical properties of the rolled plates and rods decrease. By precisely measuring the deformation extent, deformation speed, and rolling temperature of each step, controlling the generation of dynamic recrystallization, grain growth, and even the refinement of tungsten grains could be made possible.

3.4.2 Rotary forging

Rotary forging is a forming process for long-axis rolled pieces using a rotary forging machine. It involves high-frequency radial reciprocating motion to strike the workpiece, which rotates and moves axially. Under the hammer's impact, the workpiece undergoes radial compression and length-extension deformation. Rotary forging deformation leads to the formation of cellular structures and longitudinally deformed grains formed by high-density dislocations in the grains of the material, thereby improving the mechanical properties of the material [67].

Rotary forging is a method that can be used to enhance the performance of tungsten materials. The strength of tungsten billets processed through the rotary forging process can be greatly improved. Hot rotary forging is often used for processing refractory metals like tungsten and molybdenum, as the process effectively eliminates the porosity of sintered billets, increases billet density, and obtains grains with special structures (rod-shaped). Habainy et al. [68] studied the effects of rolling and forging on the microstructure and mechanical properties of pure tungsten materials, finding that both processes can improve the strength and toughness of tungsten and reduce the DBTT. Due to the distinct advantages of both rolling and rotary forging, these two processes are effectively combined to process tungsten materials. Yan Qingzhi et al. [69] prepared pure W and W-1.0La_2O_3 using a method of rotary forging followed by rolling. The research team found that the room-temperature flexural strength of pure tungsten reached 2,180 MPa, an improvement of 34% compared with rolling alone, and it also lowered the DBTT of W-1.0La_2O_3. Xie [60] obtained high-density W-1.0 wt% Y_2O_3, W-0.2 wt%Zr-1.0 wt% Y_2O_3, and W-0.51.0 wt% ZrC through powder metallurgy combined with high-temperature swaging, all of which were nearly fully dense (99.5%). After swagging, the DBTT of all three tungsten materials dropped to 200 °C, and their performance was superior to materials prepared by SPS.

3.4.3 Hydrostatic extrusion

The process of inducing plastic deformation in a rod through high-pressure liquid is termed hydrostatic extrusion. During hydrostatic extrusion, the material of the workpiece undergoes deformation under extremely high triaxial compressive stress, causing inherent microcracks within the internal structure to close continuously throughout the extrusion process. This ultimately reduces internal defects and strengthens the material through deformation [70]. Hydrostatic extrusion can achieve a large deformation extent of 60–80% in a single pass, and further deformation can even exceed 90%. After cold hydrostatic extrusion, tungsten alloys often require annealing to relieve internal stresses and improve the crystalline state, thereby enhancing the overall performance of the alloy [71]. Hot hydrostatic extrusion not only provides excellent strengthening effects and higher elongation but also results in the best overall performance. Additionally, the extrusion force is reduced, which prolongs the life of the die [72]. Hot hydrostatic extrusion combines forming and strengthening processes, allowing simultaneous strengthening of powder materials and densification of sintered materials during the forming operation.

Compared with forging, hydrostatic extrusion significantly increases the deformation capability, achieving 60–80% deformation in a single extrusion while maintaining good uniformity of material deformation. After hydrostatic extrusion, the strength of tungsten materials is greatly improved. Due to the application of hydrostatic pressure, this process avoids uneven stress between the surface and the core and overcomes significant anisotropy caused by notable grain elongation. It is considered the most effective deformation-strengthening process for tungsten materials currently. Lu et al. [73] conducted HIP on tungsten-copper alloy materials and found that the density of the billets increased from about 96% before treatment to over 99.5% after treatment, achieving near-full density. The flexural strength was also significantly improved.

3.4.4 Equal-channel angular pressing

ECAP is a widely studied technique for applying pure shear plastic deformation to bulk materials without changing the cross-sectional shape of the material. The principle is illustrated in Figure 3.12(a): a round rod sample is pressed through a channel with equal cross-sectional areas and intersecting axes that form an angle. During passage through the shear plane at the intersection, the sample undergoes intense plastic deformation while maintaining its shape. By repeating this process multiple times, a high cumulative shear strain is achieved, ultimately refining the grains [74].

At present, the following four process routes have been developed in the ECAP process: (1) Route A: After each pass of extrusion, the sample is not rotated and directly undergoes the next pass of extrusion. (2) Route Ba: After each pass of extrusion, the sample is rotated 90°, and the next pass of extrusion is performed, with the rotation direction alternating. (3) Route Bc: After each extrusion, the sample is rotated 90°, and the next pass of extrusion is performed, with the rotation direction remaining the same. (4) Route C: After each extrusion, the sample is rotated 180° before the next pass of extrusion. These four process routes have significant impacts on the final microstructure and properties of materials. It is generally believed that materials processed through Route Bc exhibit superior properties, and such materials tend to achieve higher efficiency in grain refinement [75].

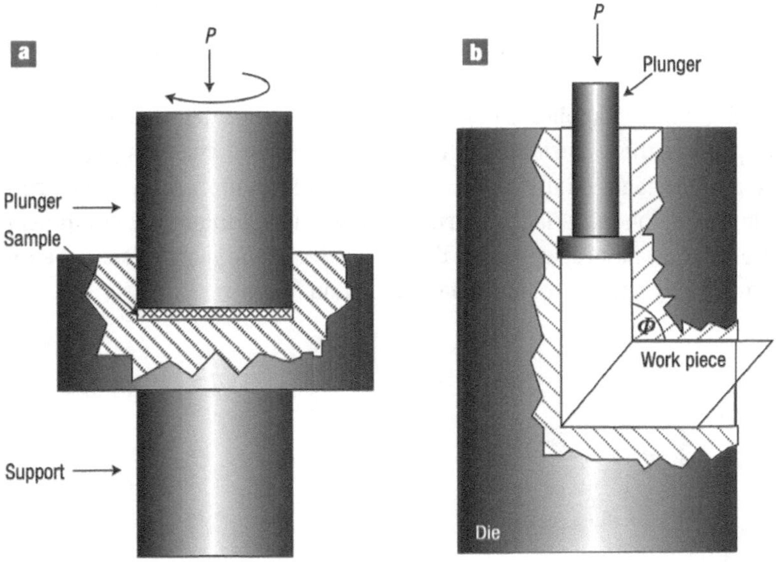

Figure 3.12: Schematic diagrams of (a) high-pressure torsion (HPT) and (b) equal-channel angular pressing (ECAP) [74].

3.4.5 High-pressure torsion

HPT is a severe plastic deformation method [76], which, compared with other severe plastic deformation methods, most closely adheres to the principles of severe plastic deformation processing and has the strongest grain refinement capability. Its working principle is shown in Figure 3.12(b). The sample is placed in a fixed concave die position and subjected to GPa-level high axial compressive stress. Simultaneously, the friction between the plunger and the sample induces shear deformation in the sample, leading to a continuous reduction in grain size to ultrafine or even nanocrystalline levels. The immense quasi-static pressure ensures that the sample maintains its original size and shape despite the large strain. HPT not only enhances the material through processing but also improves its ductility more effectively than ECAP. HPT produces a higher proportion of high-angle grain boundaries, facilitating the formation of more stable and finer nanograins.

3.4.6 Multidirectional forging

Multidirectional forging is a type of free forging process, with its operational principle illustrated in Figure 3.13. In this process, the material undergoes deformation under applied forces from the *X*, *Y*, and *Z* dimensions, forming a complete deformation cycle. The material is first deformed by applying force along the *X*-axis. After the *X*-axis deformation is complete, the material is rotated 90°along the *X*-axis, and force is applied along the *Y*-axis to induce deformation. Next, the material is rotated 90° along the *Y*-axis, and force is applied along the *Z*-axis for further deformation. During the deformation process, the material is continuously compressed and elongated as the direction of the applied load axially varies. Repeated deformation refines the grain structure and improves the material properties. Using existing industrial equipment, large dense materials can be prepared, and the properties of the materials can be improved, making it directly applied to industrial production [77].

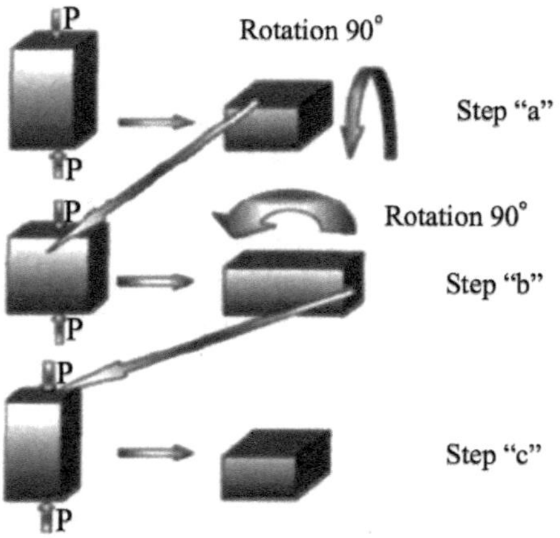

Figure 3.13: Schematic diagram of the multidirectional forging process [75].

With the change of the axial direction of the applied load, the orientation of the deformation bands in multidirectional forging will change, causing them to intersect within the grains. At the intersections of these deformation bands, there are significant dislocation pile-ups, resulting in high dislocation density. The dislocations entangle with each other to form cell structures. As the deformation continues, these cellular structures transform into subgrains (which have independent slip systems) and subsequently into new grains with low-angle or high-angle grain boundaries. These new grains can form not only at the grain boundaries but also extensively within the grains, promoting grain refinement [78]. Multidirectional forging with severe plastic deformation not only significantly enhances the mechanical properties of the material but also ensures relatively uniform deformation and mechanical properties in all directions due to the change in the axial variation of the applied load. This uniformity helps avoid the anisotropy that often occurs in other conventional forming processes, such as extrusion and rolling.

3.5 Summary

To meet the requirements of fusion reactor applications, it is necessary to improve the brittleness and radiation resistance of tungsten, and the development of UFG or nanocrystalline tungsten is a promising direction. Several material processing techniques can be employed to produce UFG/nanocrystalline tungsten, and specific processing routes have been explored for tungsten in fusion reactor applications. The commonly used methods for preparing tungsten alloy powders and composite materials include wet chemical methods and MM methods. The purity and particle size control of tungsten powder play a decisive role in preparing high-performance tungsten materials with fine-grained, fully dense sintered structures. Wet chemical methods provide precise control over powder composition, high purity, and uniformity, making them suitable for large-scale industrial production. Particularly in the preparation of tungsten-based materials reinforced with second-phase dispersion, nanoscale-sized, uniformly dispersed particles can be achieved.

To maintain the final tungsten material in nano or ultrafine grain size, it is crucial to employ effective sintering techniques after producing the tungsten powder to limit grain growth. The key is to control the sintering temperature and time, aiming for densification at low temperatures. Relying on plastic deformation can independently alter material phase transformation behavior, significantly refine microstructure, and unlock material performance potential. The effectiveness of grain refinement far surpasses traditional heat treatment techniques, making it a technology independent of heat treatment that can modify material phase, structure, and properties. Therefore, employing suitable and effective preparation and processing techniques holds the promise of enhancing the overall performance of tungsten-based materials, facilitating the production of high-performance tungsten-based first-wall materials.

References

[1] F. Jinglian, H. Boyun, Z. Chuanfu, Q. Xuanhui. Densification and grain growth of nano-tungsten alloy powder sintered under atmospheric pressure [J]. Journal of Central South University of Technology, 2001, 32(4): 391–393.

[2] R.G. Chaudhuri, S. Paria. Core/shell nanoparticles: Classes, properties, synthesis mechanisms, characterization, and applications [J]. Chemical Reviews, 2012, 112: 2373–2433.

[3] R. Malewar, K.S. Kumar, B.S. Murty, B. Sarma, S.K. Pabi. On sinterability of nanostructured W produced by high-energy ball milling [J]. Materials Research Society, 2007, 22(5): 1200–1206.

[4] R. Sarkar, P. Ghosal, M. Premkumar, A.K. Singh, K. Muraleedharan, A. Chakraborti, T.P. Bagchi, B. Sarma. Characterisation and sintering studies of mechanically milled nano tungsten powder [J]. Powder Metallurgy, 2008, 51(2): 166–170.

[5] C. Linsmeier, M. Rieth, J. Aktaa, T. Chikada, A. Hoffmann, J. Hoffmann, A. Houben, H. Kurishita, X. Jin, M. Li, A. Litnovsky, S. Matsuo, A. von Müller, V. Nikolic, T. Palacios, R. Pippan, D. Qu, J. Reiser,

J. Riesch, T. Shikama, R. Stieglitz, T. Weber, S. Wurster, J.-H. You, Z. Zhou. Development of advanced high heat flux and plasma-facing materials [J]. Nuclear Fusion, 2017, 57: 092007.

[6] L. Dai, Q. Chen, S. Lin, L. Ouyang. Study on the main promotion factors of powder refinement in high energy ball milling [J]. Materials Review, 2009, 23(11): 59–61.

[7] L. Ding, D.P. Xiang, Y.Y. Li, C. Li, J.B. Li. Effects of sintering temperature on fine-grained tungsten heavy alloy produced by high energy ball milling assisted spark plasma sintering [J]. International Journal of Refractory Metals and Hard Materials, 2012, 33: 65–69.

[8] M.A. Monge, M.A. Auger, T. Leguey, Y. Ortega, L. Bolzoni, E. Gordo, R. Pareja. Characterization of novel W alloys produced by HIP [J]. Journal of Nuclear Materials, 2009, 386–388: 613–617.

[9] T. Ryu, S. Hwang, Y.J. Choi, H.Y. Sohn. The sintering behavior of nanosized tungsten powder prepared by a plasma process [J]. International Journal of Refractory Metals and Hard Materials, 2009, 27 (4): 701–704.

[10] X. Xi, Z. Nie, J. Yang, X. Fu, W. Wang, T. Zuo. Preparation and characterization of Ce-W composite nanopowder [J]. Materials Science and Engineering: A, 2005, 394(1–2): 360–365.

[11] L. Wang, Y. Zhang, M. Muhammed. Synthesis of nanophase oxalate precursors of YBaCuO superconductor by coprecipitation in microemulsions [J]. Journal of Materials Chemistry, 1995, 5: 309–314.

[12] M. Wang, M. Muhammed. Novel synthesis of Al_{13}-cluster based alumina materials [J]. Nanostructured Materials, 1999, 11(8): 1219–1229.

[13] N. Lin, Y. He, C. Wu, Q. Zhang, J. Zhou, Z. Zhao. Fabrication of tungsten carbide-vanadium carbide core-shell structure powders and their application as an inhibitor for the sintering of cemented carbides [J]. Scripta Materialia, 2012, 67: 826–829.

[14] R. Liu, X.P. Wang, T. Hao, C.S. Liu, Q.F. Fang. Characterization of ODS-tungsten microwave-sintered from sol-gel prepared nano-powders [J]. Journal of Nuclear Materials, 2014, 450(1–3): 69–74.

[15] R.K. Barik, A. Bera, A.K. Tanwar, I.K. Baek, S.H. Min, O.J. Kwon, W.S. Lee, G.-S. Park. A novel approach to synthesis of scandia-doped tungsten nano-particles for high-current-density cathode applications [J]. International Journal of Refractory Metals and Hard Materials, 2013, 38: 60–66.

[16] S. Wahlberg, M.A. Yar, M.O. Abuelnage, H.G. Salem, M. Johnsson, M. Muhammed. Fabrication of nanostructured W-Y_2O_3 materials by chemical methods [J]. Journal of Materials Chemistry, 2012, 22: 12622–12628.

[17] M.A. Yar, S. Wahlberg, H. Bergqvist, H.G. Salem, M. Johnsson, M. Muhammed. Chemically produced nanostructured ODS-lanthanum oxide-tungsten composites sintered by spark plasma [J]. Journal of Nuclear Materials, 2011, 408: 129–135.

[18] M. Xia, Q. Yan, L. Xu, H. Guo, L. Zhu, C. Ge. Bulk tungsten with uniformly dispersed La2O3 nanoparticles sintered from co-precipitated La2O3/W nanoparticles [J]. Journal of Nuclear Materials, 2013, 434: 85–89.

[19] M. Xia, Q. Yan, L. Xu, L. Zhu, H. Guo, C. Ge. Synthesis of TiC/W core-shell nanoparticles by precipitate-coating process [J]. Journal of Nuclear Materials, 2012, 430: 216–220.

[20] G. Liu, G.J. Zhang, F. Jiang, X.D. Ding, Y.J. Sun, J. Sun, Y. Ma. Nanostructured high-strength molybdenum alloys with unprecedented tensile ductility [J]. Nature Materials, 2013, 12: 344–350.

[21] F. Jinglian, H. Boyun, Z. Chuanfu, Q. Xuanhui, W. Denglong. Preparation and sintering technology of nano-tungsten alloy powder [J]. Tungsten Carbide, 2001, 18(4): 225–231.

[22] Z.Z. Fang, H. Wang. Densification and grain growth during sintering of nanosized particles [J]. International Materials Reviews, 2008, 53(6): 326–352.

[23] Y. Kim, K.H. Lee, E.P. Kim, D.I. Cheong, S.H. Hong. Fabrication of high temperature oxides dispersion strengthened tungsten composites by spark plasma sintering process [J]. International Journal of Refractory Metals and Hard Materials, 2009, 27(5): 842–846.

[24] Z. Zhou, G. Pintsuk, J. Linke, T. Hirai, M. Rödig, Y. Ma, C. Ge. Transient high heat load tests on pure ultra-fine grained tungsten fabricated by resistance sintering under ultra-high pressure [J]. Fusion Engineering and Design, 2010, 85(1): 115–121.

[25] H. Kurishita, S. Matsuo, H. Arakawa, T. Sakamoto, S. Kobayashi, K. Nakai, T. Takida, M. Kato, M. Kawai, N. Yoshida. Development of re-crystallized W–1.1%TiC with enhanced room-temperature ductility and radiation performance [J]. Journal of Nuclear Materials, 2010, 398(1–3): 87–92.

[26] R. Liu, Y. Zhou, T. Hao, T. Zhang, X.P. Wang, C.S. Liu, Q.F. Fang Microwave synthesis and properties of fine-grained oxides dispersion strengthened tungsten [J]. Journal of Nuclear Materials, 2012, 424(1): 171–175.

[27] M. Omori. Sintering, consolidation, reaction and crystal growth by the spark plasma system (SPS) [J]. Materials Science and Engineering: A, 2000, 287(2): 183–188.

[28] E. Olevsky, L. Froyen. Constitutive modeling of spark-plasma sintering of conductive materials [J]. Scripta Materialia, 2006, 55(12): 1175–1178.

[29] V. Viswanathan, T. Laha, K. Balani, A. Agarwal, S. Seal. Challenges and advances in nanocomposite processing techniques [J]. Materials Science and Engineering: R: Reports, 2006, 54(5–6): 121–285.

[30] F. Monteverde. Ultra-high temperature HfB2-SiC ceramics consolidated by hot-pressing and spark plasma sintering [J]. Journal of Alloys and Compounds, 2007, 428(1–2): 197–205.

[31] X. Song, X. Liu, J. Zhang. Neck formation and self-adjusting mechanism of neck growth of conducting powders in spark plasma sintering [J]. Journal of the American Ceramic Society, 2006,89 (2): 494–500.

[32] M. Pellizzari, M. Zadra, A. Fedrizzi. Development of a hybrid tool steel produced by spark plasma sintering [J]. Materials and Manufacturing Processes, 2009, 24(7–8): 873–878.

[33] S. Ran, L. Gao. Spark plasma sintering of nanocrystalline niobium nitride powders [J]. Journal of the American Ceramic Society, 2008, 91(2):599–602.

[34] G. Molenat, M. Thomas, J. Galy, A. Couret. Application of spark plasma sintering to titanium aluminide alloys [J]. Advanced Engineering Materials, 2007, 9(8): 667–669.

[35] G. Maizza, S. Grasso, Y. Sakka, T. Noda, O. Ohashi. Relation between microstructure, properties and spark plasma sintering (SPS) parameters of pure ultrafine WC powder [J]. Science and Technology of Advanced Materials, 2007, 8(7–8): 644–654.

[36] Y. Kim, K.H. Lee, E. Kim, K. Cheong, S.H. Hong. Fabrication of high temperature oxides dispersion strengthened tungsten composites by spark plasma sintering process [J]. International Journal of Refractory Metals and Hard Materials, 2009, 27(5): 842–846.

[37] T. Jun, Z. Zhangjian, Q. Dandan, M. Yao, L. Ming. Preparation of ultra-fine-grained W-TiC composites by sintering of discharge plasma [J]. Journal of Rare Metal Materials and Engineering, 2011, 40(11): 1990–1993.

[38] Z. Zhou, Y. Ma, J. Du, J. Linke. Fabrication and characterization of ultra-fine grained tungsten by resistance sintering under ultra-high pressure [J]. Materials Science and Engineering: A, 2009, 505 (1–2): 131–135.

[39] Z.J. Zhou, Y.S. Kwon. Fabrication of W-Cu composite by resistance sintering under ultra-high pressure [J]. Journal of Materials Processing Technology, 2005, 168: 107–111.

[40] Z.J. Zhou, J. Du, S.X. Song, Z.H. Zhong, C.C. Ge. Microstructural characterization of W/Cu functionally graded materials produced by a one-step resistance sintering method [J]. Journal of Alloys and Compounds, 2007, 428: 146–150.

[41] Z. Zhou, G. Pintsuk, J. Linke, T. Hirai, M. Rödig, Y. Ma, C. Ge. Transient high heat load tests on pure ultra-fine grained tungsten fabricated by resistance sintering under ultra-high pressure [J]. Fusion Engineering and Design, 2010, 85(1): 115–121.

[42] P. Liu, F. Peng, F. Liu, H. Wang, C. Xu, Q. Wang, X. Zhou, W. Yin, S. Yin, Y. Li, D. He. High-pressure preparation of bulk tungsten material with near-full densification and high fracture toughness [J]. International Journal of Refractory Metals and Hard Materials, 2014, 42: 47–50.

[43] T. Shikama, T. Tanabe, M. Fujitsuka, A. Ono, T. Baba, H. Shinno. Thermal diffusivity of boron-carbon-titanium compounds as plasma facing materials [J]. Journal of Nuclear Materials, 1992, 196–198: 1140–1143.

[44] H.T. Larker, R. Lundberg. Near net shape production of monolithic and composite high temperature ceramics by hot isostatic pressing (HIP) [J]. Journal of the European Ceramic Society, 1999, 19(13–14): 2367–2373.

[45] M.T. Kim, S.Y. Chang, J.B. Won. Effect of HIP process on the micro-structural evolution of a nickel-based superalloy [J]. Materials Science and Engineering: A, 2006, 441(1–2): 126–134.

[46] H. Kurishita, S. Matsuo, H. Arakawa, T. Sakamoto, S. Kobayashi, K. Nakai, T. Takida, M. Kato, M. Kawai, N. Yoshida. Development of re-crystallized W–1.1%TiC with enhanced room-temperature ductility and radiation performance [J]. Journal of Nuclear Materials, 2010, 398(1–3): 87–92.

[47] Y. Ishijima, H. Kurishita, K. Yubuta, H. Arakawa, M. Hasegawa, Y. Hiraoka, T. Takida, K. Takebe. Current status of ductile tungsten alloy development by mechanical alloying [J]. Journal of Nuclear Materials, 2004, 329–333: 775–779.

[48] H. Xiangdong, L. Jianbao, X. Zhipeng, H. Yong. Interaction between microwaves and inorganic non-metallic media [J]. Journal of Inorganic Materials, 1998(13): 282–290.

[49] M.A. Janney, H.D. Kimrey, W.R. Allen, J.O. Kiggans. Enhenced diffusion in sapphire during microwave heating, Journal of Materials Science, 1997(32): 1374–1355.

[50] G. Prabhu, A. Chakraborty, B. Sarma. Microwave sintering of tungsten [J]. International Journal of Refractory Metals and Hard Materials, 2009, 27(5): 545–548.

[51] N.M. Hwang, Y.J. Park, D.Y. Kim, D.Y. Yoon. Activated sintering of nickel-doped tungsten: Approach by grain boundary structural transition [J]. Scripta Materialia, 2000, 42: 421–425.

[52] J.I. Toth, N.A. Lockington. Activation sintering of tungsten metals by theplatinum-group [J]. Journal of the Less Common Metals, 1967, 12:353–365.

[53] C.W. Corti. Sintering aids in powder metallurgy [J]. Platinum Metals Review, 1986, 30(4):184–195.

[54] J.L. Johnson, R.M. German. Solid-state contribution to densification during liquid phase sintering [J]. Metallurgical and Materials Transactions B, 1996, 27(6):901–909.

[55] J.L. Johnson, R.M. German. Phase equilibria effects on the enhanced liquid phase sintering of tungsten-copper [J]. Metallurgical Transactions A, 1993, 24(11): 2369–2377.

[56] Y. Boonyongmaneerat. Effects of low-content activators on low-temperature sintering of tungsten [J]. Journal of Materials Processing Technology, 2008, 209(8):4084–4087.

[57] R.M. German, Z.A. Munir. Enhanced low-temperature sintering tungsten [J]. Metallurgical Transactions A, 1976, 7(12):1873–1877.

[58] V.K. Gupta, D. Yoon, H.M. Meyer III, J. Luo. Thin intergranular films and solid-state activated sintering in nickel-doped tungsten [J]. Acta Materialia, 2007, 55: 3131–3142.

[59] A. Genc, S. Coskun, M.L. Öveçoğlu. Microstructural characterizations of Ni activated sintered W-2wt %TiC composites produced via mechanical alloying [J]. Journal of Alloys and Compounds, 2010, 497 (1–2): 80–89.

[60] X. Zhuoming. Preparation and properties of carbide/oxide diffusion-strengthened tungsten-based alloys [D]. Ph.D., Hefei: University of Science and Technology of China. 2017.

[61] Y.Y. Wilde. Research progress on deformation strengthening process of W-Ni-Fe high-density tungsten alloy [J]. Powder Metallurgy Technology, 2004, 22(5): 303–307.

[62] J. Reiser, M. Rieth, B. Dafferner, A. Hoffmann. Tungsten foil laminate for structural divertor applications-Basics and outlook [J]. Journal of Nuclear Materials, 2012, 423(1): 1–8.

[63] J. Reiser, M. Rieth, A. Möslang, B. Dafferner, A. Hoffmann, X. Yi, D.E.J. Armstrong. Tungsten foil laminate for structural divertor applications-Tensile test properties of tungsten foil [J]. Journal of Nuclear Materials, 2013, 434(1): 357–366.

[64] V. Krsjak, S.H. Wei, S. Antusch, Y. Dai. Mechanical properties of tungsten in the transition temperature range [J]. Journal of Nuclear Materials, 2014, 450(1–3): 81–87.

[65] F. Jinglian, H. Boyun. Study on new deformation strengthening process and fracture mechanism of tungsten alloy [J]. Tungsten Carbide, 2007, 24(2): 119–123.
[66] Z.M. Xie, R. Liu, T. Zhang, Q.F. Fang, C.S. Liu, X. Liu, G.N. Luo. Achieving high strength/ductility in bulk W-Zr-Y_2O_3 alloy plate with hybrid microstructure [J]. Materials and Design, 2016, 107: 144–152.
[67] W. Yu Yang. Effect of deformation strengthening on microstructure and properties of 93W-4.9Ni-2.1Fe alloy [J]. Materials Science and Technology, 2005, 13(4): 442–444.
[68] J. Habainy, S. Iyengar, Y. Lee, Y. Dai. Fatigue behavior of rolled and forged tungsten at 25°, 280° and 480 °C [J]. Journal of Nuclear Materials, 2015, 465: 438–447.
[69] Q.Z. Yan, X.F. Zhang, Z.J. Zhou, W.P. Shen, Y.C. Zhang, S.M. Wang, L. Xu, C. Ge. Status of R&D on plasma facing materials in China [J]. Journal of Nuclear Materials, 2013, 442(1): S190–S197.
[70] L. Zhiqiang, Y. Guoyi, Z. Hongmei, W. Changyu, Z. Yao. Structure and defect analysis of hydrostatic extrusion of high-density tungsten alloy [J]. Ordnance Materials Science and Engineering, 2004, 27(4): 40–43.
[71] W. Guangda, Y. Haibing, L. Guirong, X. Ning. Research progress on deformation processing of high specific gravity alloys [J]. Powder Metallurgy Technology, 2014, 32(3): 221–225
[72] W. Hu Lianxi. Advances in materials in China, 2011, 30(7): 48–54.
[73] L.Y.U. Daming, T. Anqing, M. Keqiang, L. Xianye. Hot isostatic pressing of tungsten copper materials [J]. Electrotechnical Materials, 1990, 1: 42–45.
[74] R. Valiev. Nanostructuring of metals by severe plastic deformation for advanced properties [J]. Nature Materials, 2004, 3: 511–516.
[75] Z. Xin, G. Yuwei, N. Yun, J. Tianfu. Large plastic deformation technology for the preparation of bulk nano/ultrafine grain materials [J]. Materials Reports, 2003, 17(12): 5–8
[76] R.Z. Valiev, A.V. Korznikov, R.R. Mulyukov. Structure and properties of ultrafine-grained materials produced by severe plastic deformation [J]. Materials Science and Engineering: A, 1993, 168(2): 141–148.
[77] G. Qiang, Y. Hongge, C. Zhenhua, Z. Hui. Research progress of multi-directional forging technology [J]. Materials Reports, 2007, 21(2): 106–108.
[78] O. Sitdikov, T. Sakai, A. Goloborodko, H. Miura, R. Kaibyshev. Effect of pass strain on grain refinement in 7475 Al alloy during hot multidirectional forging [J]. Materials Transactions, 2004, 45: 2232–2238.

Chapter 4
Strengthening and toughening of tungsten-matrix materials for fusion reactors

4.1 Introduction

The high ductile-to-brittle transition temperature (DBTT) of pure tungsten leads to low-temperature embrittlement. Conversely, the low recrystallization temperature of it will give rise to high-temperature embrittlement. Additionally, pure tungsten is susceptible to irradiation embrittlement and hardening [1–3]. Therefore, the operating window temperature of tungsten in divertor components is significantly limited. Therefore, enhancing the toughness and radiation resistance of tungsten materials is crucial for the development and application of new tungsten-matrix PFMs. At present, four viable approaches are being considered: (1) Alloying and solid solution strengthening of tungsten, the main elements of which are rhenium (Re), iridium (Ir), tantalum (Ta), vanadium (V), molybdenum (Mo), niobium (Nb), titanium (Ti), and so on. Among them, the addition of Re and Ir can have a strong toughening effect on tungsten, but they are expensive and have the problem of irradiation embrittlement; alloying tantalum, vanadium, molybdenum, niobium with tungsten will cause the DBTT of tungsten to rise and become more brittle; whether titanium has a toughening effect on tungsten is still under investigation. (2) Dispersion strengthening of oxide/carbide, which involves adding particles such as Y_2O_3, La_2O_3, TiC, and other particles to pure tungsten to form tungsten-matrix materials with high strength and excellent radiation resistance. (3) Superplastic deformation such as high-pressure torsion and equal-channel angular pressing to refine the grain structure. With the increase of deformation, the fracture toughness of tungsten increases and DBTT decreases.

4.2 Brittleness of tungsten

Refractory materials such as tungsten (W) and molybdenum exhibit inherent brittleness due to their body-centered cubic (BCC) structure, less slip systems, and sensitivity to grain boundary impurities. These characteristics pose significant challenges for the application of tungsten material in future fusion devices. BCC metals exhibit numerous distinctive mechanical characteristics, such as obvious ductile-to-brittle transition, sharply decreasing strength with increasing temperature, and the slip that does not follow the critical resolved shear stress (CRSS, Schmidt) law, and so on [4].

 | https://doi.org/10.1515/9783111556987-004

4.2.1 Low-temperature brittleness

Tungsten, with its BCC structure, is prone to brittleness at room temperature due to the low bonding strength between its grains that cracks can easily form and propagate along these grain boundaries. DBTT is an important characteristic parameter of the ductility and brittleness of tungsten and tungsten-based materials, which is related to their microstructure, stress-strain state, dislocations, impurities, and so on. The DBTT of unalloyed tungsten is approximately 373 K to 673 K [5], while the DBTT of most tungsten alloys (W-Cu, W-Ni-Fe) is below room temperature. Tungsten experiences plastic deformation and exhibits ductility above the DBTT. In contrast, when deformed below DBTT, tungsten tends to develop brittle fracture. Tungsten's DBTT decreases through plastic deformation and increases through annealing. After hot rolling, the DBTT of tungsten is approximately 373 K to 473 K, which, following recrystallization treatment, increases to around 633 K. Tungsten features extremely low solubility for interstitial impurities at room temperature and in fact, tungsten alloy exhibits multiphase characteristics, with the presence of supersaturated impurities. The impact of impurity content and distribution on DBTT is of particular significance. The presence of a small quantity of O, C, N, S, and P in the alloy results in the weakening of the grain boundaries due to the formation of precipitates on the grain boundaries. This phenomenon leads to an increase in the alloy's DBTT. However, the formation of some compounds can strengthen the grain boundaries, increase the area and resistance of transgranular cleavage fracture, and thus improve the strength of the alloy [6].

4.2.2 Recrystallization brittleness

When tungsten is processed above its recrystallization temperature, grain growth occurs, which leads to an increase in grain size and a relative decrease in grain boundary area. Consequently, impurities become more concentrated at the grain boundaries, which reduce the material's hardness and strength. For instance, the strength of tungsten at 1,000 °C is only about 20% to 40% of its strength at room temperature. The recrystallization temperature of tungsten is primarily dependent on the plastic deformation, as well as the addition of alloying elements and the dispersion strengthening phase, with a recrystallization temperature range of approximately 1,423 K to 1,623 K. In nuclear fusion reactors, the service temperature of plasma-facing materials often exceeds this temperature range, so increasing the recrystallization temperature of tungsten materials emerges as a vital strategy to optimise their utilization. After recrystallization, the DBTT of tungsten increases [7–9], which not only reduces the high-temperature strength, but also increases its brittleness at low temperatures. It is viable to improve the mechanical properties of tungsten by controlling the extent of recrystallization and grain growth.

4.2.3 Neutron irradiation effects

When tungsten is subjected to neutron irradiation in a nuclear fusion environment, the introduction of various microscopic defects, such as atom displacement damage, will degrade the mechanical properties of material and increase the DBTT [10]. Even at low neutron flux levels, the resulting irradiation-induced embrittlement can be quite severe. In addition, irradiation-induced swelling, sedimentation, and segregation will lead to changes in the composition and properties of tungsten materials. The irradiation activation of tungsten can be categorized into short-term and long-term effects. The short-term issues primarily arise from the presence of stable tungsten isotopes, and the decay heat of those isotopes requires active cooling for several weeks after reactor shutdown to dissipate. Also, the high activity of the transmutation product 186Re limits the service life of tungsten. The long-term activation problem is mainly due to impurity elements in tungsten. To meet the standard of 100 mSv/h for recyclable materials, neutron-irradiated tungsten needs to be stored for 50 years before it can be reused [11]. In the DEMO reactor, the elemental transmutation of tungsten is related to neutron wall loading and neutron energy spectrum. With an average neutron loading of 2 MW/m^2 on the first wall, after 5 full-power years of service, pure tungsten would be composed of approximately 94%

W, 3.8% Re, 1.4% Os, and 0.8% Ta, as shown in Figure 4.1(a) [12]. In addition to transmutation products, neutron irradiation can also produce large number of microscopic defects such as dislocations and voids in tungsten. Even at low irradiation damage levels, there is a significant increase in the retention of fuel particles within the tungsten material. Furthermore, neutron irradiation reduces the thermal conductivity of tungsten, with this effect being more pronounced at low temperatures (below 750 °C) (Figure 4.1(b)). This reduction is associated with recoverable irradiation defects [13].

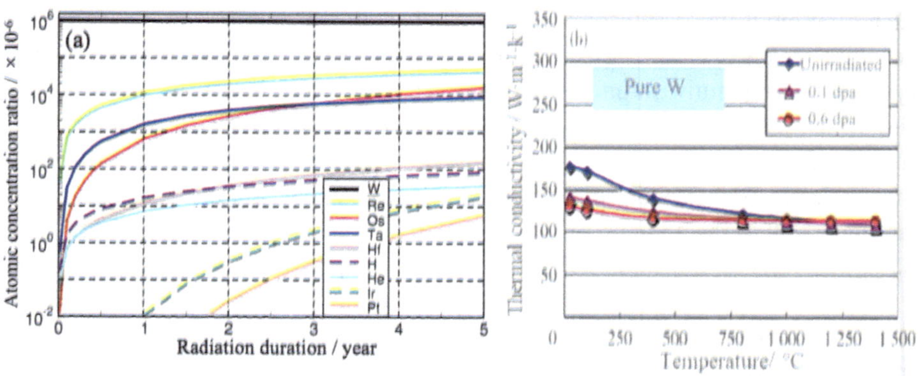

Figure 4.1: (a) Elemental transmutation [20] and (b) thermal conductivity change [21] of pure tungsten under neutron irradiation.

4.3 The intrinsic cause of tungsten brittleness

To fundamentally address the issue of brittleness, it is essential to understand the intrinsic causes of tungsten's brittleness. A significant number of researchers have engaged in comprehensive investigations into this critical issue from both theoretical frameworks and empirical studies.

4.3.1 Chemical bonds

Plastic deformation of metal materials is intimately linked to the properties of metal bonds. For the transition group elements VI B, Pauling previously postulated that the electrons outside the nucleus were all d^4s^2, an explanation that is difficult to reconcile with the significant physical differences in the physical properties of Cr, Mo, and W elements in this group. Xie et al. [14] made significant progress by employing the "new valence bond theory" they had developed to study the valence bond structure of many pure metals, including these three elements. The findings for the Group VI B elements are as follows: $[Ar](3d_c)3.32(3d_n)2.26(4s_c)0.25(4s_f)0.17$, Mo: $[Kr](4d_n)0.12(4d_c)3.88(5s_c)0.68(5s_f).1.32$, W: $[Xe](5d_c)5.16(6s_c)0.25(6s_f)0.59$. Here the subscripts c, f, n, and m represent covalent electrons, near-free electrons, nonbonding electrons, and magnetic electrons, respectively.

The results indicate that tungsten possesses the largest proportion of covalent components. This finding provides a qualitative explanation for tungsten's higher DBTT compared to molybdenum.

4.3.2 Dislocation mobility

It is widely accepted that the plastic deformation of metals occurs through the actuation of the Frank-Read source. Although the models for screw and edge dislocations are relatively simple and easy to understand, mixed dislocations are often observed in real materials. The plastic deformation of the BCC metal depends significantly on the mobility of the most challenging type of dislocation, that is, the screw dislocation. Computer simulations have revealed that the core structure of the screw dislocation in tungsten metal adopts a symmetrical non-planar structure, as depicted in the differential displacement diagram (Figure 4.2) [15]. This dislocation, which expands non-coplanar expansion at the core, can slip only if partial dislocation is compressed back into coplanar expansion. This requirement results in a higher CRSS for screw dislocation compared to edge dislocation. Notably, the Peierls stress for the screw dislocation obtained from the simulation results exceeds the actual measured value. Groger [16] suggested that this discrepancy arises because the simulations do not account for edge component, a phenomenon present in real materials. Under high temperature thermal activation, the "kink pair" effect of screw dislocations will play a significant role. In this way, the high temperature reduces the CRSS value of dislocation starting; and high temperatures activate a large number of dislocations and thus increase the plasticity of the material (the slip direction of BCC metal is found to be < 111 >, while the slip surface is very uncertain, generally {110} at low temperatures, and the slip on {112} and {123} surfaces can be activated at high temperatures). Therefore, the plastic deformation and properties of tungsten are related to the dislocation mobility and mixing ratio. And in fact, temperature exerts a profound influence, which makes the performance of tungsten under high temperature irradiation more complex.

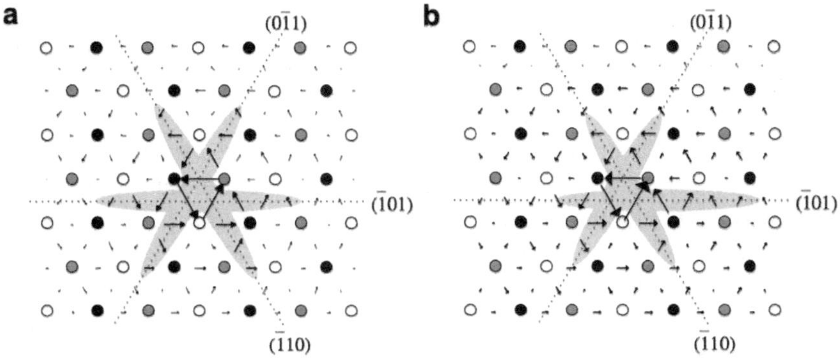

Figure 4.2: The core structure of the tungsten 1/2 < 111 > screw dislocation simulated by the BOP method and the response of the screw dislocation under elastic stress [23].

4.3.3 Internal defects in actual materials

Tungsten materials in a powder-sintered state typically fracture under tension of around 500 MPa without showing significant plastic (or yield) deformation [17]. Due to the fact that tungsten materials is difficult to be fully densified by ordinary powder sintering, and the presence of cracks in the materials results in stress concentration and fracture before yielding. Pressureless sintering, typically performed in the industry at temperatures above 2,000 °C [18], tends to cause the growth of tungsten grains, thus intensifying the brittleness tendency.

In recent years, many researchers have tried to obtain fine-grained tungsten materials by using fine powders and strengthening sintering (such as spark plasma sintering and ultra-high voltage electrification sintering). However, the implementation of the process is frequently more challenging due to the narrow kinetic window during the sintering process (the temperature range in which densification can be carried out while grain growth is suppressed). In view of this sintering limitation, the resulting material may appear to be "undersintered" [19] because the sintering time is short and the sintering temperature is low, resulting in insufficient diffusion between the sintered particles. The addition of low melting point components (commonly known as nickel for tungsten) can activate sintering and promote densification. However, the same process causes grain growth. Ni is a highly activated element and is unsuitable for use in fusion PFMs. To a certain extent, plastic processing can improve the density of W sintered body and close micro cracks and holes. At the same time, some defects of refractory materials in the sintered state will also be elongated under the action of plastic processing (such as rolling), which is one of the reasons why the strength of the rolled W plate in the rolling direction is significantly higher than that in the normal direction. Furthermore, the inherent brittleness of tungsten renders it susceptible to cracking and other defects during plastic processing.

4.3.4 Grain boundary characteristics

In a broad sense, grain boundaries act as a type of defect in the material, distinct from cracks. They play a significant role in the material's fracture behavior, particularly in the brittle regime. Simulation of short-term and high-energy flux density ELM using high-energy electron gun of tungsten materials with low matrix temperature, revealed that different commercial materials always tend to crack along grain boundaries [20]. Kurishita et al. [21] suggested that the grain boundaries of tungsten are the source of cracks in polycrystalline tungsten materials, despite theoretical simulations showing high strength at these boundaries [22]. The powder is susceptible to contamination with impurities during the actual production process, a consequence of the inherent properties of powder metallurgy. Moreover, during sintering process, large number of impurity elements (such as C, O, and N) are often introduced into the sintered products [23] in the absence of effective protective measures (such as powder pre-degassing and full reduction). Due to their low solubility in tungsten lattice, these impurities are easy to segregate at grain boundaries, weaken grain boundary bonding, and reduce grain boundary strength. For the hot isostatic pressing (HIP) preparation method, the importance of degassing needs to be emphasized again. In the conventional sintering process, the heat is transferred from the outside to the inside of the embryo body. In the early stage of the densification process, due to the poor thermal conductivity of the embryo body, closed pores often appear on the outer surface and a large amount of gas in the embryo body cannot overflow, resulting in a porous structure, which affects the quality of the sintered body. SEM observations show that sintered pores are often located at the intersection of multiple grains, and such pores are sharp and polygonal, which can easily induce stress concentration [24].

As mentioned above, low grain boundary strength of tungsten has long been attributed to impurities. However, related studies have found that the polycrystals prepared by high-pressure torsion of single crystals purified in the high-purity region of the initial material also exhibit significant intergranular fracture [23]. This finding suggests that attributing intergranular fracture to impurity-induced embrittlement is an inaccurate assumption; however, it does not refute the notion that impurity segregation exerts an influence on grain boundary strength.

Tungsten products are often manufactured through rolling processes. During the rolling process, the grains are elongated along the rolling direction, the rolling surface is compressed transversely, and finally an elliptical disk-shaped grain structure is formed. Experimental studies have shown that the relationship of tensile strength is: rolling direction > rolling transverse > rolling normal, and the fracture toughness K_{IC} also shows the same trend [25]. From the grain shape, the maximum effective size of grain boundary can be obtained at the normal rolling direction. This variation demonstrates the influence of grain boundaries on mechanical properties.

4.3.5 Cleavage fracture

For tungsten single crystal and tungsten plate with strong texture ({100} < 110 > plate texture will appear during rolling), brittle cleavage fracture should be taken into consideration. It is found that the cleavage plane of bcc metals is typically {100} plate. Generally, in such strongly textured rolled plate, there are two such surfaces with an angle of 45° between the normal and the rolling direction. (Therefore, this specific type of embrittlement in the rolled plate is also commonly called 45° angular embrittlement [26].) Cottrell believes that this embrittlement is actually related to the intersection of edge dislocations [27]: a/2[-111](0-11) + a/2[1-11](011)→a[001](010), and that this dislocation reaction is energy-reducing, but dislocations with this Burger vector are difficult to move. Moreover, the computer simulations show that the underside of the semi-atomic plane of this edge dislocation generates large tensile stresses. This immovable dislocation is liable to result in dislocation congestion. Therefore, such a dislocation reaction finally causes a source of cracks.

4.4 Ways to solve tungsten's brittleness

In view of the brittleness of tungsten materials, many previous studies have been focused on the preparation of ultrafine grained tungsten by alloying elements addition, second phase doping and severe plastic deformation (SPD).

4.4.1 Addition of alloying elements

The toughness of tungsten can only be improved by tungsten-rhenium alloying. The addition of rhenium changes the core structure of screw dislocations from symmetrical to asymmetrical (Figure 4.3), which reduces the Peierls stress (σ_p) (from 2.49 GPa at 0% Re to 2.09 GPa at 12% Re, and further to 1.84 GPa at 25% Re) [28]. The change of the symmetry of the dislocation core changes the slip system, and the symmetrical core structure slips on the {110} surface. The asymmetrical core slides in a zigzag pattern on these surfaces, so that the overall slip surface becomes {112}, and the number of slip surfaces increases from 6 to 12, thus improving the deformation coordination ability of the material.

Common tungsten-rhenium alloys contain between 10% and 26% rhenium, and a small amount (~5%) of rhenium has been found to significantly increase the initial recrystallization temperature of tungsten. However, the nuclear transmutation reaction of tungsten to form a brittle phase of rhenium and the requirement of low activation as a fusion material, the large-scale application of tungsten-rhenium alloys is limited due to the rare and expensive nature of rhenium [29]. There are very few metal elements that can form solid solution with tungsten at room temperature, and even fewer can form brittle mesophase. Only tantalum, vanadium, niobium, and molybdenum are completely miscible with tungsten, while others, such as titanium and tungsten-rhenium, and are only partially miscible with a maximum solubility of 12% and 27%, respectively, forming an intermediate phase [30, 31]. Given the fact that niobium and molybdenum are subject to nuclear transmutation, resulting in the formation of radiation-activated isotopes characterised by a prolonged half-life, so those two elements cannot be used in fusion materials, leaving only tantalum, vanadium, and titanium to be considered. However, the alloying of tantalum and vanadium does not reduce the DBTT or Charpy impact energy of tungsten materials, and the fracture toughness decreases with the increase of Ta content [32]. Titanium, which is of a low melting point, can promote densification with grain growth. And there are no other metal elements which could improve the plasticity of tungsten by alloying except tungsten-rhenium.

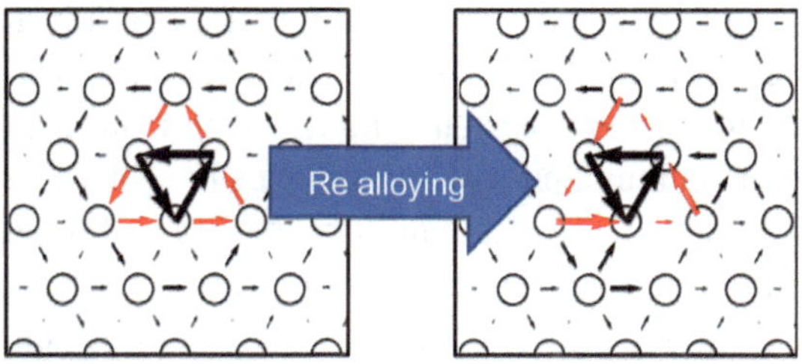

Figure 4.3: Rhenium alloying leads to the transformation of the dislocation core structure [38].

4.4.2 Second phase doping

An effective method to improve the mechanical properties of tungsten is to uniformly disperse nanoparticles, such as carbides or oxides, to strengthen the grain and grain boundaries of tungsten [33–35]. On the one hand, nanoparticles can pin dislocations and grain boundaries, prevent grain growth, and improve the high-temperature strength of tungsten. On the other hand, the refined grain can greatly increase the grain boundary area and reduce the average concentration of impurities at the grain boundary, thereby weakening the embrittlement of impurities on the grain boundary and reducing the DBTT of tungsten. At the same time, the high-density crystal/phase boundary acts as a defect trap to absorb and disperse irradiated defects, inhibit the formation of voids and hydrogen/helium (H/He) bubbles, and thus improving the anti-irradiation performance and plasma etching ability of tungsten [36, 37]. Using mechanical alloying or wet chemical methods, a small amount of nanoscale oxides could be uniformly diffused into the tungsten matrix, forming oxide dispersion strengthened tungsten (ODS-W). The dispersion of ultrafine oxide is used to hinder the movement of grain boundaries and dislocations, so that the strength at room temperature and mechanical properties could be improved at high temperature. Although oxides have the best oxidation and ablation resistance, the melting point of oxides is still relatively low when washed by high-temperature plasma, and the thermal stability of carbides is higher. In the context of high-temperature sintering, gathering and growing carbides present a significant challenge; consequently, carbides maintain their nanostructure. Despite the fact that a proportion of the research has recently been directed toward carbides with higher melting points, the carbon-containing components are disadvantageous when subjected to neutron irradiation in fusion reactors.

4.4.3 Severe plastic deformation

SPD refers to a processing method in which a large amount of shear strain is generated in bulk materials to make the grains ultrafine, and can be used to prepare bulk ultrafine/nanocrystalline materials with high density and no impurities [38]. Ultrafine/nanocrystalline tungsten can be obtained by SPD, to improve the low-temperature brittleness of tungsten. SPD-treated tungsten exhibits exceptional high strength and toughness [39, 40], a great combination that is not typically achievable with traditional treatment method [41]. After SPD treatment of coarse-grained tungsten, large numbers of nonequilibrium large-angle grain boundaries are generated, and the impurities are redistributed along these grain boundaries through diffusion, which reduces the average concentration of impurity atoms at the grain boundaries, thus improving the ductility of the material. It is also believed that after SPD treatment, the material produces nonequilibrium large-angle grain boundaries, resulting in both high strength and high toughness [42]. Ultrafine equiaxed grains with large angular grain boundaries hinder the movement of dislocations, thereby increasing the strength of the material. Such grains may contribute to the development of other deformation mechanisms (e.g., grain boundary sliding and/or grain rotation), which improves the ductility of the material. The DBTT of dispersion strengthened tungsten can be reduced by appropriate plastic processing; however, machining will inevitably introduce anisotropy into the structure and properties of the material.

4.5 Summary

Tungsten and tungsten-based materials are of great importance for future engineering applications as plasma-facing materials. However, they are faced with significant challenges, including difficulties in processing, high DBTT, and low recrystallization temperatures. These issues need to be fully understood and solved by comprehensive processing methods. At present, tungsten properties can be improved through the composition and structure design of tungsten-based materials, including alloying, dispersion strengthening, and SPD modification. It is difficult to achieve the ideal toughening effect by a single method, and the comprehensive application of multiple toughening methods (such as diffusion-strengthened tungsten-rhenium alloy and dispersion-strengthened tungsten alloy with appropriate machining) can obtain tungsten-based materials with excellent comprehensive properties. It is of great significance to explore the deformation, strengthening mechanism and damage behavior of tungsten and tungsten alloys to promote their application.

References

[1] V. Philipps. Tungsten as material for plasma-facing components in fusion devices [J]. Journal of Nuclear Materials, 2011, 415(1): S2–S9.

[2] G. Qilu, X. Zhensheng, H. Huanqing. Application and development trend of high-performance refractory materials in cutting-edge fields [J]. Powder Metallurgy Industry, 2000, 10(1): 7–13.

[3] W. Dingchun, X. Yaoqin. New progress in research of tungsten and tungsten alloy in China [J]. China Tungsten Industry, 2001, 16(5–6): 91–95.

[4] N.A.H. Wan. Mobility of screw dislocations in BCC crystals: A review on modeling methods [J]. Progress in Natural Science, 2000, 10: 721–729.

[5] I. Smid, M. Akiba, G. Vieider, L. Plöchl. Development of tungsten armor and bonding to copper for plasma-interactive components [J]. Journal of Nuclear Materials, 1998, 258–263: 160–172.

[6] H. Kurishita, S. Kobayashi, K. Nakai, H. Arakawa, S. Matsuo, T. Takida, K. Takebe, M. Kawai. Current status of ultra-fine grained W-TiC development for use in irradiation environments [J]. Physica Scripta, 2007, T128: 76–80.

[7] R. Matera, G. Federici, The ITER Joint Central Team. Design requirements for plasma facing materials in ITER [J]. Journal of Nuclear Materials, 1996, 233–237: 17–25.

[8] M. Fujitsuka, I. Mutoh, T. Tanabe, T. Shikama. High heat load test on tungsten and tungsten containing alloys [J]. Journal of Nuclear Materials, 1996, 233–237: 638–644.

[9] K. Nakamura, S. Suzuki, K. Satoh, M. Araki, K. Yokoyama, M. Dairaku, M. Akiba. Erosion of CFCs and W at high temperature under high heat loads [J]. Journal of Nuclear Materials, 1994, 212–215: 1201–1205.

[10] J.M. Steichen. Tensile properties of neutron irradiated TZM and tungsten [J]. Journal of Nuclear Materials, 1976, 60(1): 13–19.

[11] L. Feng, L. Guangnan, L. Qiang, W. Wanjing. Research on the application of tungsten in nuclear fusion reactor [J]. China Tungsten, 2017, 32(2): 41–48.

[12] M.R. Gilbert, J.C. Sublet. Neutron-induced transmutation effects in W and W-alloys in a fusion environment [J]. Nuclear Fusion, 2011(51): 043005.

[13] J. Linke. Plasma facing components and plasma materials interactions [R]. 16th International Conference on Fusion Reactor Materials, Beijing, China, 2013.
[14] Z. Yingjiu, X. Youqing, Z. Xiaodong, C. Shiding. Determination of electronic structure and physical properties of tungsten metal [J]. Rare Metals, 1994, 3: 203.
[15] R. Gröger, A.G. Bailey, V. Vitek. Multiscale modeling of plastic deformation of molybdenum and tungsten: I. Atomistic studies of the core structure and glide of 1/2 screw dislocations at 0 K [J]. Acta Materialia, 2008, 56(19): 5401–5411.
[16] R. Gröger, V. Vitek. Explanation of the discrepancy between the measured and atomistically calculated yield stresses in body-centered cubic metals [J]. Philosophical Magazine Letters, 2007, 87(2): 113–120.
[17] Q. Wei, L.J. Kecskes. Effect of low-temperature rolling on the tensile behavior of commercially pure tungsten [J]. Materials Science and Engineering A, 2008, 491(1–2): 62–69.
[18] W. Fazhan, T. Lixia, F. Pengfa. Tungsten materials and their processing [M]. Beijing: Metallurgical Industry Press, 2008.
[19] S. Wahlberg, M.A. Yar, M.O. Abuelnaga, H.G. Salem, M. Johnsson, M. Muhammed. Fabrication of nanostructured W-Y_2O_3 materials by chemical methods [J]. Journal of Materials Chemistry, 2012, 22: 12622.
[20] H. Kurishita. Recent progress in toughening nanostructured W alloys through domestic and international research collaboration [R]. Charleston, SC, USA, October 17–21, 2011.
[21] H. Kurishita, Y. Amano, S. Kobayashi, K. Nakai, H. Arakawa, Y. Hiraoka, T. Takida, K. Takebe, H. Matsui. Development of ultra-fine grained W-TiC and their mechanical properties for fusion applications [J]. Journal of Nuclear Materials, 2007, 367–370: 1453–1457.
[22] H. Zhou, S. Jin, Y. Zhang, G. Lu. Effects of hydrogen on a tungsten grain boundary: A first-principles computational tensile test [J]. Progress in Natural Science: Materials International, 2011, 21(3): 240–245.
[23] B. Gludovatz, S. Wurster, T. Weingärtner, A. Hoffmann, R. Pippan. Influence of impurities on the fracture behaviour of tungsten [J]. Philosophical Magazine, 2011, 91(22): 3006–3020.
[24] S.X. Zhao, F. Liu, S.G. Qin, J.P. Song, G.-N. Luo. Preliminary results of W fiber reinforced W (Wf/W) composites fabricated with powder metallurgy [J]. Fusion Science and Technology, 2013, 64(2): 225–229.
[25] B. Gludovatz, S. Wurster, A. Hoffmann, R. Pippan. Fracture toughness of polycrystalline tungsten alloys [J]. International Journal of Refractory Metals and Hard Materials, 2010, 28(6): 674–678.
[26] J. Neges, B. Ortner, G. Leichtfried, H.P. Stüwe. On the 45° embrittlement of tungsten sheets [J]. Materials Science and Engineering: A, 1995, 196(1–2): 129–133.
[27] Y. Yongning. Principles of metallurgy [M]. Metallurgical Industry Press, 2000.
[28] L. Romaner, C. Ambrosch-Draxl, R. Pippan. Effect of rhenium on the dislocation core structure in tungsten [J]. Physical Review Letters, 2010, 104(19): 195–503.
[29] L. El-Guebaly, R. Kurtz, M. Rieth, H. Kurishita, A. Robinson, Aries Team. W-based alloys for advanced divertor designs: Options and environmental impact of state-of-the-art alloys [J]. Fusion Science and Technology, 2011, 60: 185–189.
[30] S. Wurster, N. Baluc, M. Battabyal, T. Crosby, J. Du, C. Garcia-Rosales, A. Hasegawa, A. Hoffmann, A. Kimura, H. Kurishita, R.J. Kurta, H. Li, S. Nosh, J. Reiser, J. Riesch, M. Rieth, W. Setyawan, M. Walter, J.H. You, R. Pippan. Recent progress in R&D on tungsten alloys for divertor structural and plasmafacing materials [J]. Journal of Nuclear Materials, 2013, 442(1–3): S181–S189.
[31] J.W. Coenen, V. Philipps, S. Brezinsek, G. Pintsuk, I. Uytdenhouwen, M. Wirtz, A. Kreter, K. Sugiyama, H. Kurishita, Y. Torikai, Y. Ueda, U. Samm, TEXTOR-Team. Melt-layer ejection and material changes of three different tungsten materials under high heat-flux conditions in the tokamak edge plasma of textor [J]. Nuclear Fusion, 2011, 51: 113020.

[32] S. Wurster, B. Gludovatz, A. Hoffmann, R. Pippan. Fracture behaviour of tungsten-vanadium and tungsten-tantalum alloys and composites [J]. Journal of Nuclear Materials, 2011, 413(3): 166–176.
[33] S. Wurster, N. Baluc, M. Battabyal. Recent progress in R&D on tungsten alloys for divertor structural and plasma facing materials [J]. Journal of Nuclear Materials, 2013, 442: S181–S189.
[34] Z.M. Xie, R. Liu, S. Miao, X.D. Yang, T. Zhang, Q.F. Fang, X.P. Wang, C.S. Liu, Y.Y. Lian, X. Liu, G.N. Luo. High thermal shock resistance of the hot rolled and swaged bulk W-ZrC alloys [J]. Journal of Nuclear Materials, 2016, 469: 209–216.
[35] I. Wesemann, W. Spielmann, P. Heel, A. Hoffmann. Fracture strength and microstructure of ODS tungsten alloys [J]. International Journal of Refractory Metals and Hard Materials, 2010, 28(6): 687–691.
[36] X. Li, W. Liu, Y. Xu, C.S. Liu, B.C. Pan, Y. Liang, Q.F. Fang, J.L. Chen, G.-N. Luo, G.H. Lu, Z. Wang. Radiation resistance of nano-crystalline iron: Coupling of the fundamental segregation process and the annihilation of interstitials and vacancies near the grain boundaries [J]. Acta Materialia, 2016, 109: 115–127.
[37] X. Li, Y. Xu, C.S. Liu, Q.F. Fang, B.C. Pan, Z. Wang. Energetic and kinetic behaviors of small vacancy clusters near a symmetric Σ5 (310)/[001] tilt grain boundary in bcc Fe [J]. Journal of Nuclear Materials, 2013, 440: 250–256.
[38] R.Z. Valiev, R.K. Islamgaliev, I.V. Alexandrov. Bulk nanostructured materials from severe plastic deformation [J]. Progress in Materials Science, 2000, 45: 103–189.
[39] Q. Wei, T. Jiao, K.T. Ramesh, E. Ma, L.J. Kecskes, L. Magness, R. Dowding, V.U. Kazykhanov, R.Z. Valiev. Mechanical behavior and dynamic failure of high-strength ultrafine grained tungsten under uniaxial compression [J]. Acta Materialia, 2006, 54(1): 77–87.
[40] Q. Wei, H.T. Zhang, B.E. Schuster, K.T. Ramesh, R.Z. Valiev, L.J. Kecskes, R.J. Dowding, L. Magness, K. Cho. Microstructure and mechanical properties of super-strong nanocrystalline tungsten processed by high-pressure torsion [J]. Acta Materialia, 2006, 54(15): 4079–4089.
[41] R. Valiev. Nanostructuring of metals by severe plastic deformation for advanced properties [J]. Nature Materials, 2004, 3: 511–516.
[42] R.Z. Valiev, I.V. Alexandrov, Y.T. Zhu, T.C. Lowe. Paradox of strength and ductility in metals processed by severe plastic deformation [J]. Journal of Materials Research, 2002, 17(1): 5–8.

Chapter 5
Radiation damage of tungsten-matrix materials in fusion reactors

5.1 Introduction

The first wall materials facing the plasma are subjected to high H/He plasma flux (10^{20}–10^{24} m^{-2} s^{-1}), high heat flux (10–20 MW/m^2) [1] density, and high-energy neutron irradiation (14 MeV) [2, 3]. The damage dose is 10^4 times higher than that experienced by the cladding material of nuclear power plant components. Additionally, their transmutation reactions with neutrons [4, 5] produce a large amount of impurity gases (H, He, etc.), and produce transmutation elements (Re, Os, etc.). Some data show that the contents of the transmutation elements, Re and Os, in tungsten can reach 3.8 at% and 1.38 at%, respectively, after 5 years of service in the future fusion reactor power plant environment [6]. This change will significantly alter the composition of tungsten materials, thereby affecting their service properties. Under the high heat and high flow irradiation field, the properties of plasma-facing materials (PFMs) are of great importance, as such properties are directly related to the safe and reliable operation of fusion reactors. Every advancement in this technology is intimately connected to improvements and enhancements in the materials' irradiation performance.

The changes in the macroscopic properties of materials, such as irradiation swelling, irradiation growth, irradiation hardening, and irradiation embrittlement, caused by lattice defects generated from collisions between neutrons/radiation particles and lattice atoms, as well as by transmutation elements produced in nuclear reactions, are referred to as irradiation effects. These effects, which lead to material weakening or performance degradation, are known as irradiation damage. In the tokamak irradiation environment, PFMs are continuously subjected to high-flux plasma irradiation. The intense interaction between the plasma and materials results in irradiation damage, significantly affecting the microstructure, properties, and tritium cycling efficiency of the materials, thereby affecting the service performance of the divertor and even the operation stability of the tokamak device. Therefore, to study the radiation damage process of tungsten materials under fusion service conditions is of great significance [7, 8]. Moreover, high-dose neutron and ion irradiation at high temperatures is liable to cause serious problems such as irradiation swelling, growth, hardening, creep and fracture, and hydrogen and helium embrittlement, so the development of a new type of irradiation-resistant material that balances good thermal conductivity and irradiation resistance performance has become both urgent and necessary.

Currently, there is a lack of 14 MeV high-flux neutron facilities to conduct irradiation experiments, which means researchers must rely on the theory and models of

 | https://doi.org/10.1515/9783111556987-005

irradiation effects to accurately design irradiation simulation experiments. These models help in processing data to assess the feasibility of materials for potential applications. Because charged particles can achieve the amount of damage equivalent to years or even decades of neutron irradiation in a relatively short amount of time, such particles are used as a valuable tool for simulating neutron irradiation damage. In the existing high-flux neutron experimental reactors, even if an experimental setup with a neutron flux of up to 10^{15} n/cm^2 has been found, the irradiation experiment will take 3 years to achieve the required neutron dose value. By contrast, conducting the experiment in a facility with a typical neutron flux of 10^{14} n/cm^2 would require 30 years, a timeline that renders practical evaluation difficult [9]. The high damage rate of heavy ions and the high flux attainable via electron irradiation enable irradiation simulation studies of fusion reactor materials, and this approach has been proven to be an effective means of shortening irradiation timescales by several orders of magnitude.

5.2 Irradiation damage behavior of tungsten materials

5.2.1 Damage and retention of tungsten materials irradiated with hydrogen isotope

Under irradiation by low-energy high-flux deuterium plasma, tungsten experiences significant irradiation damage. Both single-crystal and polycrystalline tungsten develop surface blistering when exposed to hydrogen plasma, and this blistering may occur even under low-energy particle irradiation conditions [10]. The formation of hydrogen blisters reduces the bonding strength at grain boundaries and increases material brittleness, which severely impacts the service performance of the material and tritium retention, thus hindering the long-term stable operation of reactors. The main reasons are as follows: firstly, the defect itself acts as a trap for the incident particles, which increases the retention of fuel components inside the material; secondly, severe defects formed by irradiation can create rapid pathways for element loss, causing valuable nuclear fuel to quickly escape through these channels. To address this issue, extensive research has been conducted on the retention and accumulation of hydrogen isotopes at tungsten defects. It is difficult to study the behavior of tritium in materials due to the high cost of tritium treatment and its radiation hazard to humans. Since deuterium and tritium are isotopes of hydrogen atoms with the same electronic structure and similar chemical properties, using hydrogen or deuterium to simulate tritium in research is feasible and safe.

Hydrogen, due to its strong mobility and fast diffusion rate, can be trapped by defects in tungsten both intrinsic and irradiation-induced, including vacancies, impurities, dislocations, and grain boundaries, which combines with the defects to form defect clusters or hydrogen bubbles. It is found that the blistering shape on the tungsten surface varies under the irradiation of deuterium plasma at different temperatures with low energy and high flux [11]. As shown in Figure 5.1, high injection doses lead to the formation of two types of bubbles on the surface of the material: (1) large bubbles with tens of microns in size; (2) small bubbles with a few microns in size. At the same time, bubble bursting occurs, which can take place in three ways: (1) edge rupture with a small tail-like appearance; (2) partial rupture at the bubble top; and (3) complete bubble explosion. As irradiation temperature increases, the tungsten surface melts and cracks [12] (Figure 5.2). The deuterium retention and blistering phenomena are significantly influenced by factors such as plasma temperature, flux size, the surface morphology, and internal structure of the tungsten. Jia et al. [13] discovered that the grain orientation of tungsten affects the formation of deuterium bubbles, with notable blistering occurring in the [111] orientation, while the [001] orientation suppresses deuterium bubble formation.

Liu et al. [14] proposed the concept of optimal charge density isosurface inside the vacancy to reveal the specific formation process of hydrogen bubbles. When a certain number of hydrogen atoms inside a single vacancy are reached, the hydrogen atoms in the center will combine and form a hydrogen molecule, as shown in Figure 5.3. First, the hydrogen atoms are captured by the vacancies and sequentially occupy the inner surface of the vacancies; when the number of hydrogen atoms in the vacancy reaches a critical density, there will be no extra isosurface inside the vacancy to accommodate more hydrogen atoms and reach a saturation state; and the hydrogen atoms that are subsequently added have no choice but to form a hydrogen molecule in the middle of the vacancy, which results in the hydrogen bubble. The growth of the hydrogen bubble consists of two stages: (1) The accumulation of numerous hydrogen atoms inside the bubble generates an internal pressure at the GPa level [15], but the size of the hydrogen bubble remains unchanged; (2) when the internal pressure of the bubble reaches the threshold, the hydrogen bubble ring pushes outward against the surrounding interstitial atoms, causing the bubble to expand. The growth of the hydrogen bubble is related to the pressure inside the hydrogen bubble and the outward thrust of the gap ring. When the pressure inside the bubble reaches the limit, the hydrogen bubble will stop growing. At present, there is no consensus on the maximum number of hydrogen atoms a single vacancy in tungsten can accommodate. While Heinola et al. [16] suggested at most 5, Johnson and Carter [17] proposed at least 6, Liu et al. [14] estimated the number to be 10, You et al. [18] and Guerrero et al. [19] put it at 12, and Ohsawa et al. [20] concluded the number to be 14.

Calculations show that the threshold energy required for hydrogen ions to produce displacement damage in tungsten is 2 keV [21]. Hu et al. [22] carried out studies on the formation and growth process of hydrogen bubbles in pure tungsten at different experimental temperatures (room temperature, 350 °C, 500 °C, 600 °C, and 800 °C) using a hydrogen ion beam (H^+) of 6.67 keV, as shown in Figure 5.4. Compared with the unirradiated sample, when the H^+ injection amount is 2.25×10^{21} m^{-2} at room temperature, many dislocation loops with a size of 2–6 nm are generated on the surface

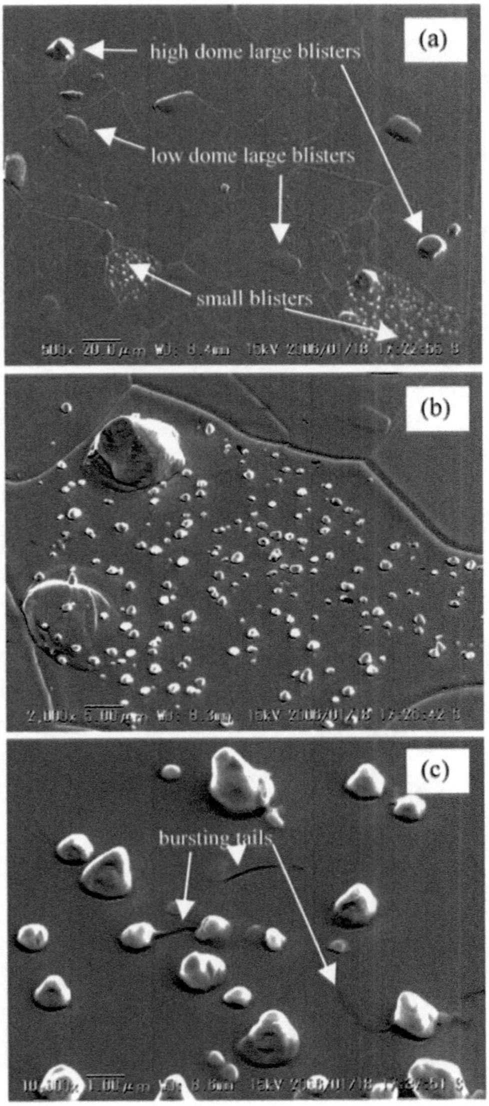

Figure 5.1: SEM image of blistering on the surface of pure tungsten after deuterium plasma irradiation. The deuterium ions have an energy of 38 eV, a dose of 1.0×10^{26} D/m^2, and a surface temperature of 520 K [11].

of the material, but no blistering occurs. The mobility of hydrogen atoms at room temperature is poor, with a diffusion coefficient of only $D_H = 4.1 \times 10^{-7}$ exp(−0.39 eV/kT) m^2/s [23], and the movement of vacancies can only occur above 200 °C [18]. When the temperature reaches 350 °C, a few hydrogen bubbles become apparent in tungsten, at which point the vacancies within the material begin to move and the mobility of

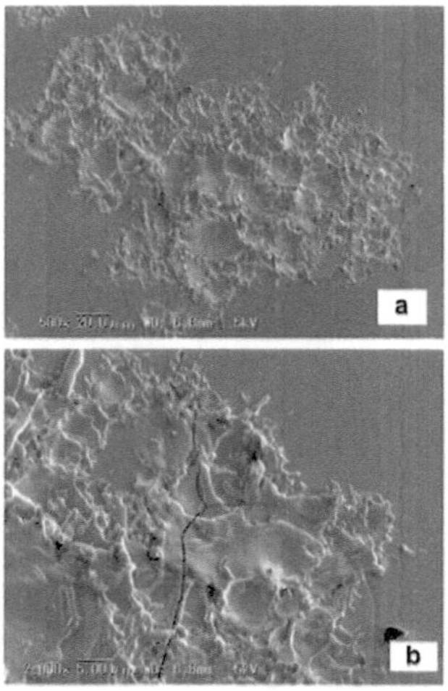

Figure 5.2: SEM image of melting and cracking on the surface of pure tungsten after deuterium plasma irradiation. The deuterium ions have an energy of 38 eV, a dose of 1.0×10^{26} D/m^2, and a surface temperature of 618 K [12].

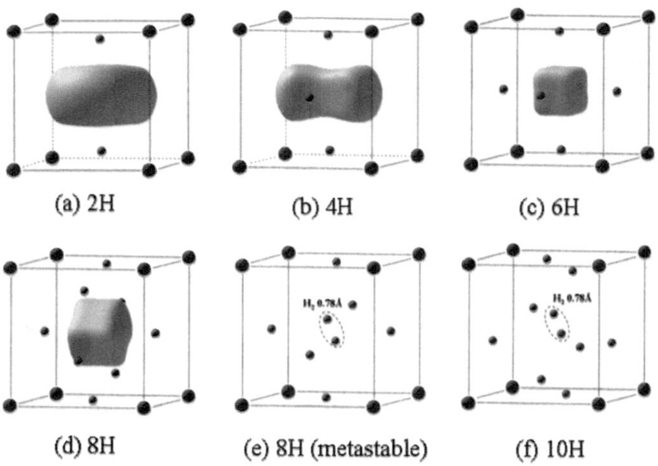

Figure 5.3: Atomic structure diagram of different numbers of hydrogen atoms trapped by a single vacancy [14].

hydrogen atoms increases. At 600 °C, the average size of the hydrogen bubble is 2.4 nm, and the density reaches its peak. Because the temperature increases further, the mobility of hydrogen bubbles increases, allowing them to combine into larger bubbles. However, when the temperature rises too high, some hydrogen bubbles migrate to the surface and disappear, leading to a sudden decrease in the number of hydrogen bubbles. At 800 °C, the hydrogen bubbles begin to aggregate, and the density decreases dramatically, but their size significantly increases. Shu et al. [24] studied the behavior of hydrogen irradiation in W materials at low energy (38 eV) and high injection doses (~10^{27} D^{+}/m^{2}). The research shows that hydrogen bubbles are formed in tungsten even when the energy of the injected particles is far below that required to cause dislocation damage.

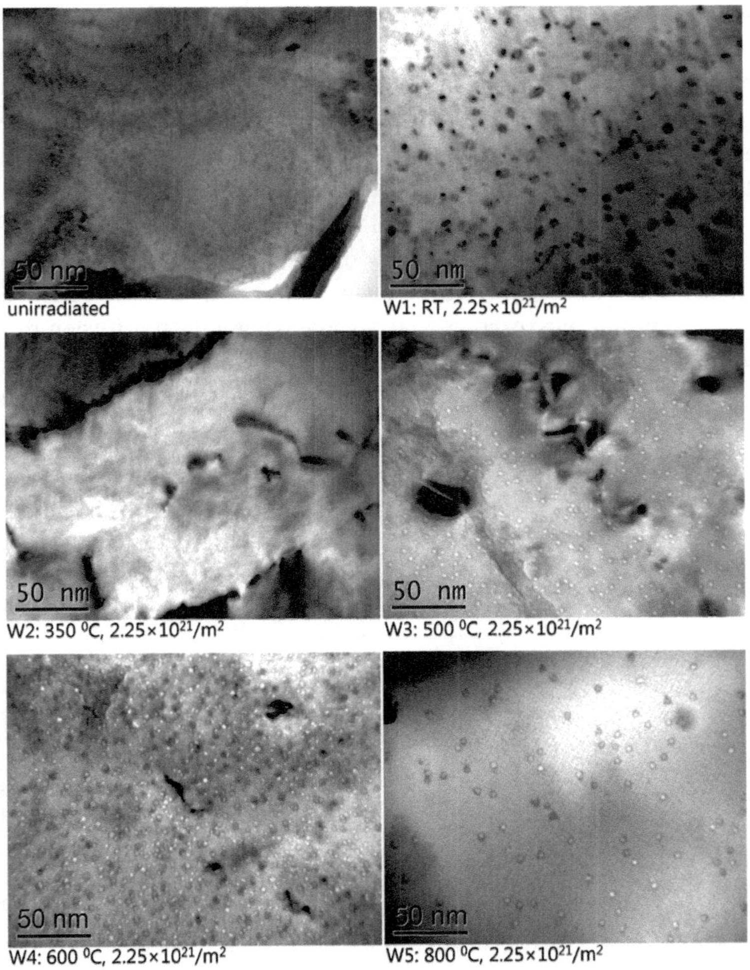

Figure 5.4: Surface morphology of tungsten with the same hydrogen ion beam injection dose and different experimental temperatures [22].

The isotopic behavior of hydrogen in materials is crucial to ensure the safety and economics of fusion reactors [25, 26]. In most cases, majority of hydrogen isotopes injected into the tungsten materials will reform into gas molecules and diffuse away from the surface. Only a part of the injected hydrogen isotopes can diffuse deeper and then be captured by various types of defects in the materials (vacancies, dislocations, impurities, grain boundaries, etc.). Figure 5.5 illustrates a schematic diagram of hydrogen retention in polycrystalline tungsten [27]. The results of the thermal desorption spectroscopy (TDS) experiment contain multiple peaks, reflecting the dynamic behavior of different temperatures when desorption occurs. According to the number and intensity of peaks in the thermal desorption curve, the adsorption state and distribution law of the surface gas can be analyzed; the kinetic parameters of desorption can be obtained from the thermal desorption curve, including the desorption activation energy, the chemical form of the desorption gas, the desorption rate, the desorption series, and the speed constant that refers to the pre-exponential factor; and the amount of desorption gas (holdup capacity) can also be calculated by analyzing the thermal desorption spectrum. Figure 5.6 demonstrates the variation of TDS profile with irradiation temperature and dose for a typical ITER-W (International Thermonuclear Experiment Reactor-grade tungsten) under D^+ bombardment [28]. Notably, the deuterium retention values measured by TDS are always larger than those obtained by ion beam analysis and nuclear reaction analysis, implying a deep diffusion distance of hydrogen in tungsten [27]. As shown in Figure 5.7, the amount of deuterium retention decreases at high temperatures, as the desorption rate is higher than the capture rate at high temperatures [28].

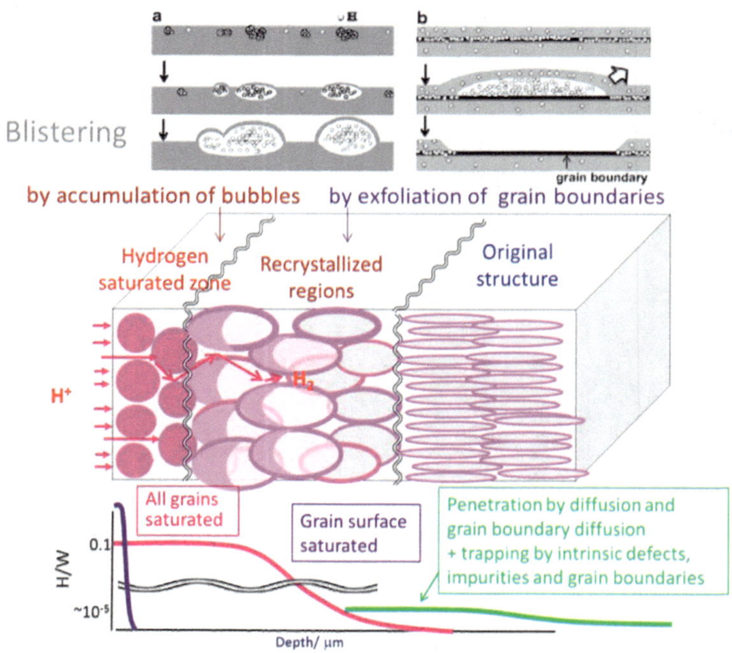

Figure 5.5: Schematic diagram of polycrystalline tungsten hydrogen retention under high H ion flux [27].

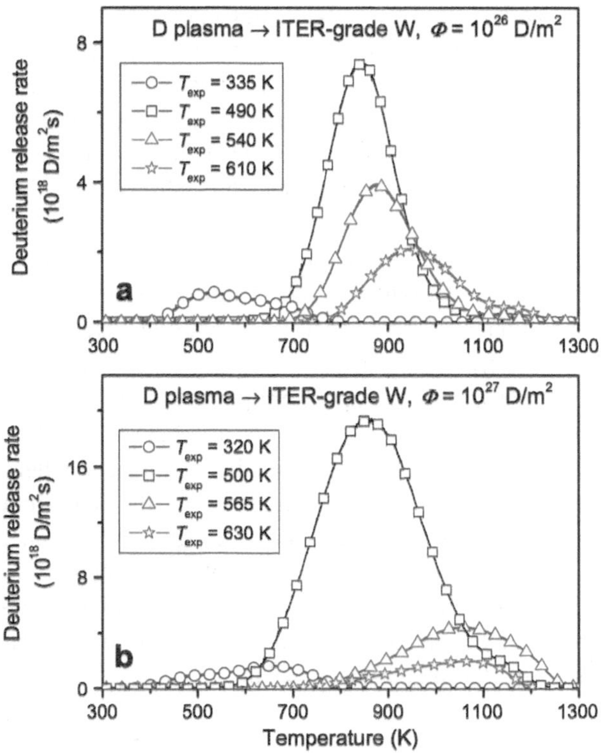

Figure 5.6: TDS curves of ITER-grade tungsten with temperature and irradiation dose after 38 eV deuterium plasma implantation [28].

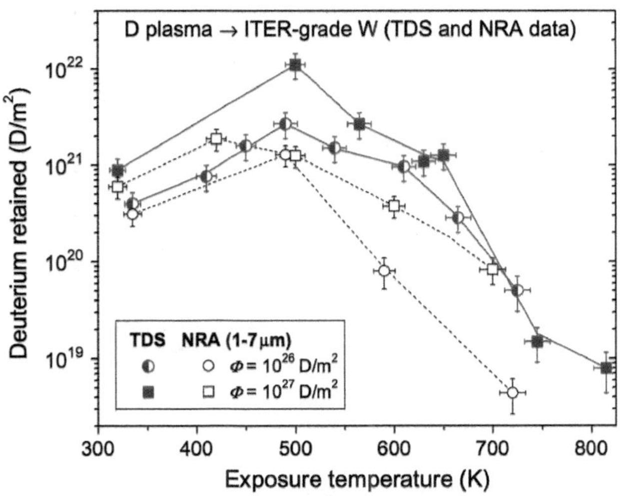

Figure 5.7: Variation of deuterium retention with temperature under deuterium plasma bombardment with an energy of 38 eV and a flux of 10^{22} $D^+/m^2 \cdot s$ [28].

Deuterium retention in tungsten could be affected by a series of factors such as the microstructure of the material, the machining process, and the treatment process [29]. Tungsten materials can be obtained by different preparation methods such as powder metallurgy, which inevitably brings in various impurities during preparation and processing, as well as different types of lattice defects. These defects trap the injected hydrogen isotopes, affecting the deuterium retention of the material. The hydrogen retention of tungsten materials with the same purity (99.96%) varies by several times due to different crystal structures [30]. The low-temperature desorption peak in the TDS spectrum of pure tungsten results mainly from the release of deuterium from the intrinsic defects (grain boundaries and dislocations) of the material. The dislocation exhibits the lowest trap energy among all the defect types of tungsten, so the low-temperature peak is attributed to the dislocation, whereas the high-temperature peaks (or shoulders) in the TDS spectrum are mainly contributed by defects (vacancies and holes) caused by ion irradiation or by the release of deuterium from locations in the material where there is originally a high desorption energy [31]. As shown in Figure 5.8, the deuterium retention increases with the increasing irradiation dose, which is roughly proportional to the square root of the retention and the irradiation time, indicating that deuterium retention is a diffusion-dominated process [32]. Recently, the retention is found to follow a simple square root relationship and the deuterium retention does not have a saturation trend, even when the dose of deuterium plasma is 10^{28} m^{-2} [33].

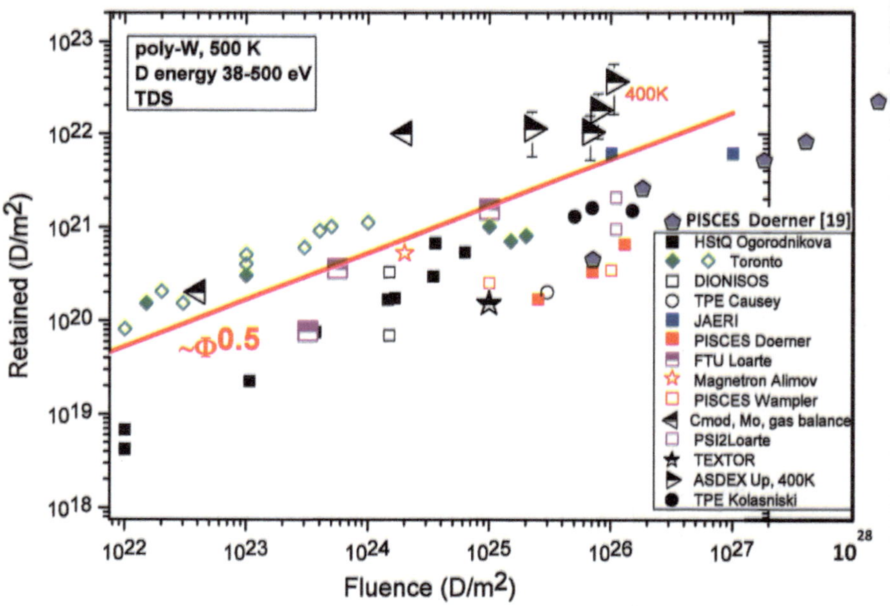

Figure 5.8: Variation curve of polycrystalline tungsten deuterium retention with irradiation dose [32].

Currently, most findings at home and abroad focus on pure tungsten materials, while studies on the effects of dispersed phase particles on tungsten irradiation remain limited due to their late start. Kurishita et al. [34] prepared toughened, fine-grained, and recrystallized W-1.1% TiC and found that the W-TiC sample showed almost no bubbles and voids under deuterium plasma irradiation (Figure 5.9). The deuterium retention amount was two orders of magnitude smaller than that of pure tungsten. Zhang Tao et al. [35] found that the bubble size on the surface of carbide-doped tungsten was significantly smaller than that of pure tungsten, with the bubble size on the surface of W-0.5%HfC and W-0.5%TiC about 1 μm, while the bubble size on the surface of W-0.5%ZrC reached the nanometer level, as shown in Figure 5.10. Zhao et al. [36] discovered that Y_2O_3-doped particles that could inhibit the formation of deuterium bubbles, and the size and density of deuterium bubbles formed on the surface were significantly reduced (Figure 5.11), and there was no significant increase in deuterium retention.

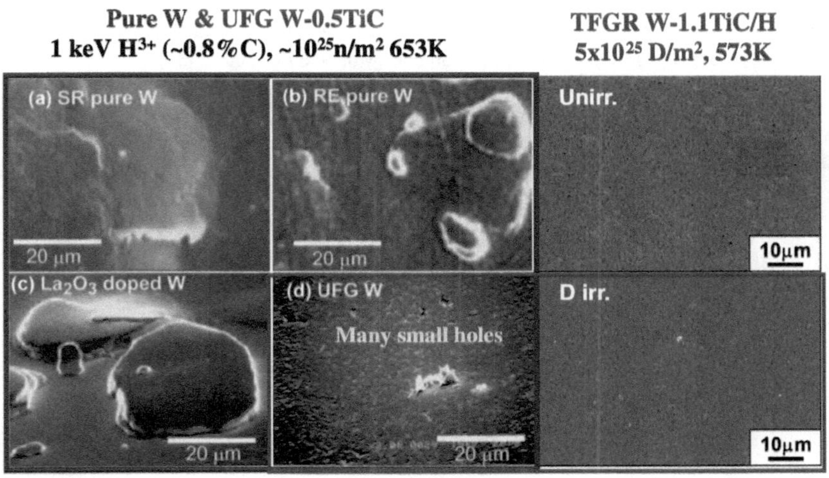

Figure 5.9: Surface morphology of different W materials irradiated with hydrogen isotopes of the PISCES-A device [34].

Oya et al. [37] measured the deuterium retention of pure W, W-TiC, and W-TaC after irradiation with JAEA and HiFIT devices (Figure 5.13). The results indicate the deuterium retention of carbide-doped tungsten at high temperatures is significantly higher than that of pure tungsten, which may be due to the formation of TiD_2 and TaD from Ti, Ta, and D. Zibrov et al. [38] investigated the deuterium retention of W, W-1.1TiC, and W-3.3TaC after different doses of deuterium ion implantation at an energy of 30 eV and a temperature of 300 K, and found the same results (Figure 5.14). The differences in deuterium retention between TiC and TaC-doped tungsten were mainly attributed to the distinct stoichiometric ratios of the carbides. Zhao et al. [39] also found that the retention of oxide-doped tungsten was also higher than that of pure tungsten, which may be the consequence of the high desorption energy introduced by Y_2O_3 doping (Figure 5.15).

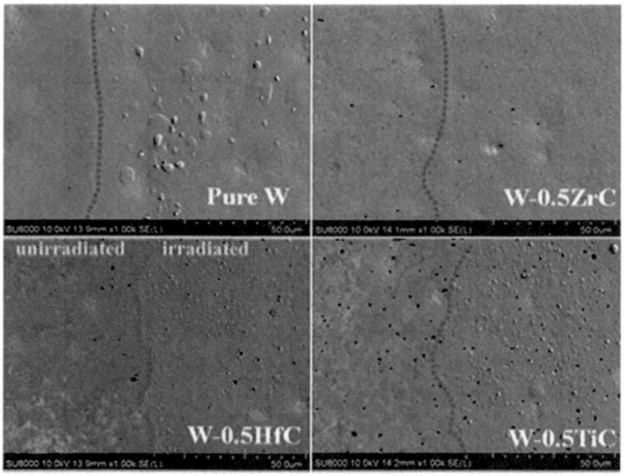

Figure 5.10: Surface morphology of different tungsten-based materials after irradiation at a fluence of 5×10^{21} ions/($m^2 \cdot s$), a dose of 7.02×10^{25} ions/m^2, and an energy of 90 eV D^+ 200 °C [35].

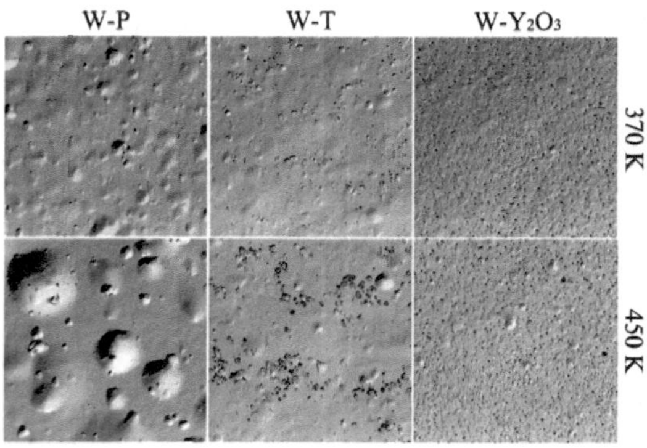

Figure 5.11: Surface morphology of pure tungsten and W-Y_2O_3 irradiated at a dose of 6×10^{24} ions/m^2 and an energy of 38 eV D^+ 450 K [36].

5.2.2 Damage behavior of tungsten materials under helium ion irradiation

The low solubility of helium in crystals makes it prone to aggregation at defects such as dislocations, vacancies, grain boundaries, and interfaces between particles and the material matrix, resulting in the formation of helium bubbles. In tungsten materials, the formation of lattice defects like vacancies, dislocations, and impurities under bombardment by energy-carrying helium ions makes tungsten an unsaturated trap for capturing helium. As helium continues to accumulate, migrate, and diffuse, small defects gradually aggregate to form nano-sized defects, that is, helium bubbles. Consonant with the hydrogen dissolution and diffusion mechanisms, helium atoms tend to

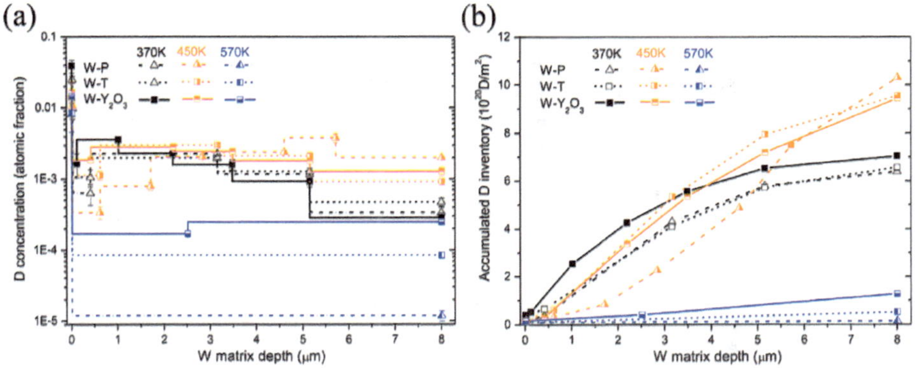

Figure 5.12: (a) Variation of deuterium concentration with depth for W-Y_2O_3 and two types of pure tungsten measured by nuclear reaction analysis; (b) variation curve of deuterium retention with depth [36].

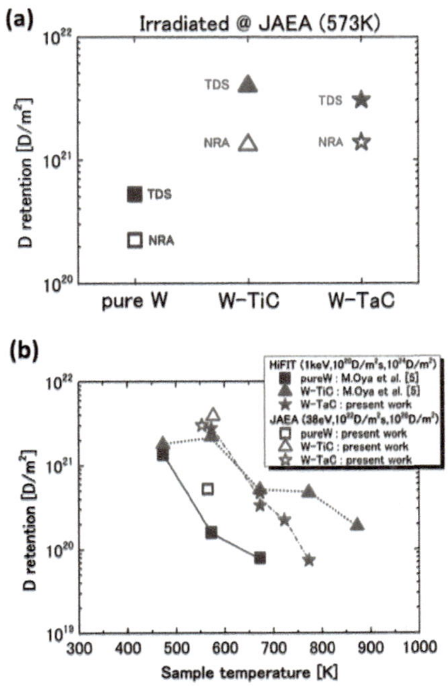

Figure 5.13: Total deuterium retention of pure W, W-TiC, and W-TaC after irradiation on (a) JAEA and (b) HiFIT [37].

be trapped in the tetrahedral gap in the material, and helium diffusion is mainly related to the transition between the two adjacent tetrahedral gap structures [40]. Compared with the low binding energy of two hydrogen atoms at the interstitial position (less than 0.1 eV), which is difficult to spontaneously aggregate to form stable clusters at the interstitial position, helium atoms are more likely to self-aggregate into clusters at the tetrahedral interstitial position due to their higher binding energy (about 1 eV) and can aggregate to form monolayer clusters on the (110) crystal plane [41, 42]. In

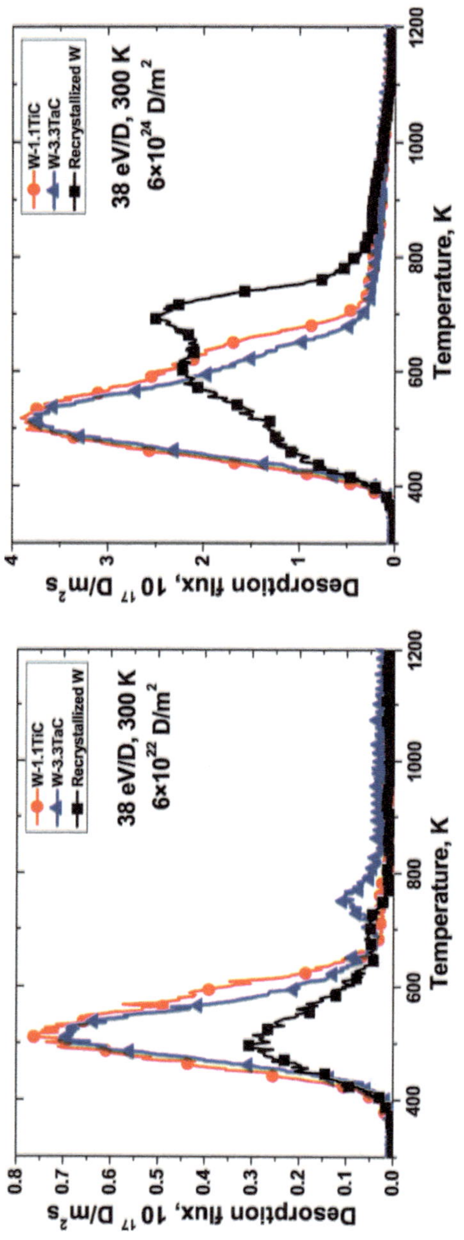

Figure 5.14: TDS spectra of pure W, W-TiC, and W-TaC with 30 eV energy and 300 K temperature with different doses of deuterium ion implantation [38].

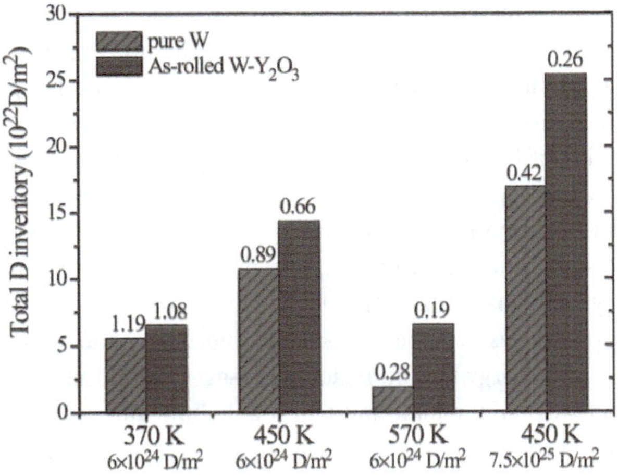

Figure 5.15: The total deuterium retention of pure W and W-Y_2O_3 after deuterium ion implantation at different temperatures and doses [39].

general, gap positions and vacancies in materials are regarded as the main cause for the formation of bubbles by trapping helium atoms. When the energy of helium ions is insufficient to cause displacement damage in tungsten, interstitial dislocation loops can still be observed. This phenomenon is related to the formation of He platelets. (see Figure 5.16). You et al. [43] explain the substantial helium accumulation observed in experiments.

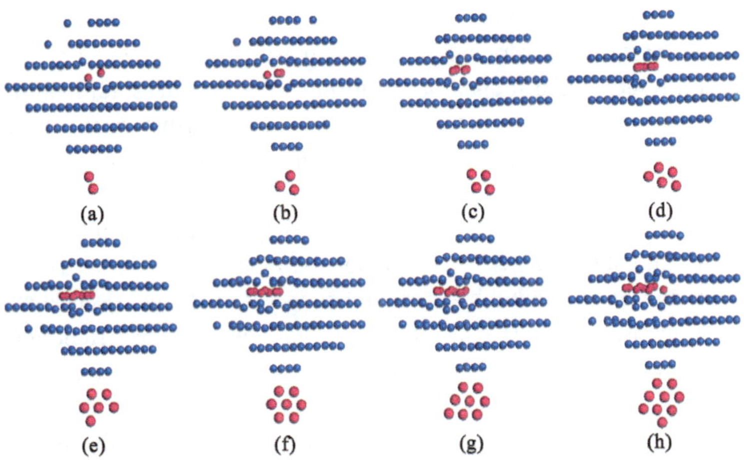

Figure 5.16: Schematic diagram of the layered structure of interstitial helium atoms formed by the aggregation of helium atoms at the crystal plane (110) in tungsten [43].

The ITER divertor site is exposed to He ion irradiation at a flux of 10^{22}–10^{24} m^{-2} s^{-1}. He ions, mainly produced by the D-T fusion reaction with an energy of 3.5 MeV, are used for plasma self-sustained heating to maintain plasma combustion. After self-sustaining heating, their energy is reduced to 10–10^6 eV, and they are guided to the divertor by the magnetic field to prevent the dilution of D-T plasma density caused by excessive He ion concentration. Given the extremely low solubility of He in metals, helium atoms tend to accumulate and form He bubbles in fusion reactor materials. Compared to hydrogen irradiation, helium irradiation causes more severe damage to materials, resulting in the formation of nanofilamentary structures (fuzz). These structures can significantly affect the physical and chemical properties of materials, such as thermal conductivity and mechanical properties. Under low-energy helium irradiation (several hundred eV), the impurity atoms, vacancies, and pores of the materials will trap helium atoms, forming helium bubbles and interstitial dislocation loops. The size, density, and distribution of nanoscale bubbles and dislocation loops are related to the energy of helium irradiation, the injection dose, and the tungsten surface temperature.

Baldwin and Doerner [44] used the PISCES-B linear plasma device to study the helium irradiation behavior of nine different tungsten materials relevant to fusion engineering. They found that helium irradiation can cause irradiation damage on the surface of tungsten materials, resulting in obvious "fuzz," as shown in Figure 5.17. Figure 5.18 illustrates the conditions for the formation of "fuzz" on the surface of tungsten materials: ion incident energy >20 eV, 1,000 K < T < 2,000 K [45–47]. Kajita et al. [47] summarized that the formation of "fuzz" was due to the formation, growth, and aggregation of subsurface helium bubbles. Ito et al. [48] further discovered that the formation of tungsten nanowires is related to the fusion growth and migration of these helium bubbles. By combining molecular dynamics and Monte Carlo method, three specific growth processes of nanofilamentary structures were hybrid simulated as follows: (1) helium atoms aggregated into helium bubbles and topped out to form convex surfaces; (2) helium atoms aggregated and ruptured to form a concave surface; and (3) further retention of helium atoms in tungsten exacerbates the degree of radiation damage of the above two types, as depicted in Figure 5.19.

The injection of He induces large number of such point defect formation in the material as vacancies and interstitial atoms, which hinder the movement of dislocations and lead to irradiation hardening. He bubbles accumulate in vacancies and dislocation loops, forming three-dimensional voids that increase in volume, causing irradiation swelling. Under irradiation, the number of interstitial atoms in tungsten will continue to increase and many dislocation loops will be formed. The attraction of the dislocation loop to the gap atoms is usually much higher than that of the vacancy because the migration energy of the gap atoms is lower than that of the vacancy atoms, and the dislocation loop always absorbs more interstitial atoms than the vacancy. The interstitial dislocation loop grows after absorbing the interstitial atoms, and the dislocation loop produced by helium injection is an interstitial dislocation loop. Watanabe et al. [49] investigated the surface morphology of tungsten material under the experimental conditions of helium

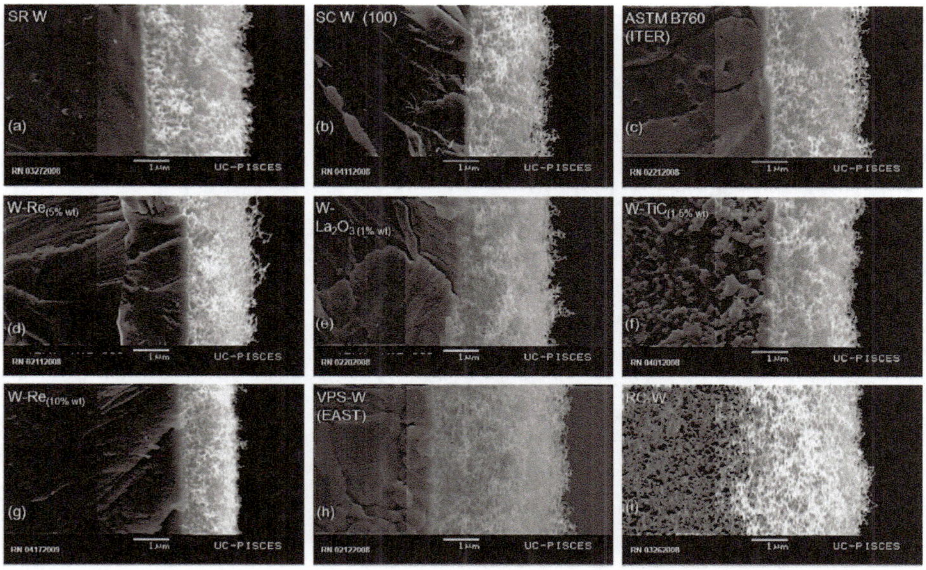

Figure 5.17: SEM cross sections of nine different tungsten materials related to fusion engineering after irradiation with helium ions at 1,120 K, 40 eV [44].

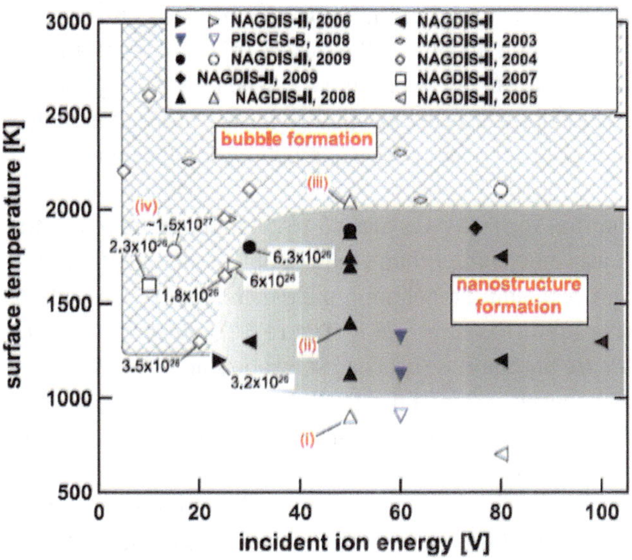

Figure 5.18: SEM cross sections of nine different tungsten materials after irradiation with helium ions at 1,120 K, 40 eV [44].

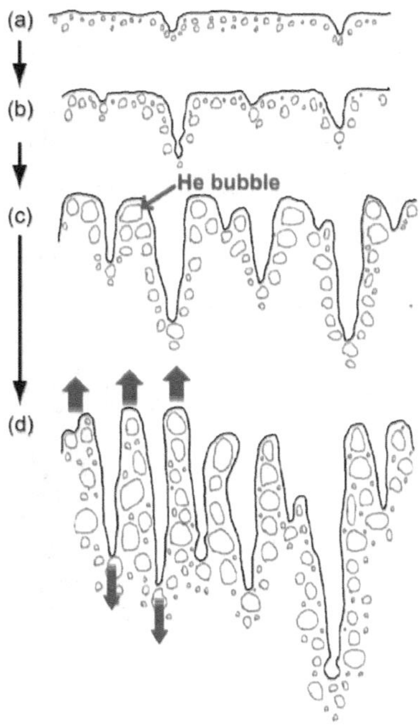

Figure 5.19: Schematic diagram of the formation process of "fuzz" [47].

ion energy of 8 keV and helium injection dose from 1.0×10^{18} He^+/m^2 to 2.0×10^{21} He^+/m^2, and found that interstitial loops appeared initially on the surface of the material, and the density and size increased with the increase of injection dose. When the density reaches saturation, part of the interstitial loops is stacked together to form a dislocation tangle. When the injected dose continues to increase, noticeable helium bubbles emerge, as shown in Figure 5.20.

Miyamoto et al. [50] found that for 60 eV He plasma irradiation, the depth of the helium bubble layer (>30 nm) was much larger than the ion implantation depth by several nm due to the diffusion process, and the helium bubble increased in size with increasing temperature, reaching up to 10 nm at 973 K (Figure 5.21). Regardless of the sample temperature, the 60 eV He plasma dose reached saturation at 5×10^{23} m^{-2}, and the helium bubble density was larger than that of the 3 keV He^+ saturation dose 10^{22} m^{-2} (Figure 5.22). El-Atwani et al. [51] found that nanocrystalline W (NCW) exhibited better resistance to He^+ irradiation than ultrafine-grained W, as shown in Figure 5.23.

Tungsten is regularly of a very low deuterium retention, while tungsten could be irradiated not only by deuterium ions, but also by high-flux helium plasma under fusion service conditions. Therefore, the irradiation efficiency of helium ions should be taken into consideration when evaluating tungsten as a fusion material, and it is of importance to study how helium ion irradiation impacts deuterium retention in tungsten materials [52, 53]. Most studies have shown that pre-irradiation of helium ions increases the deuterium retention in tungsten, although the position of the TDS

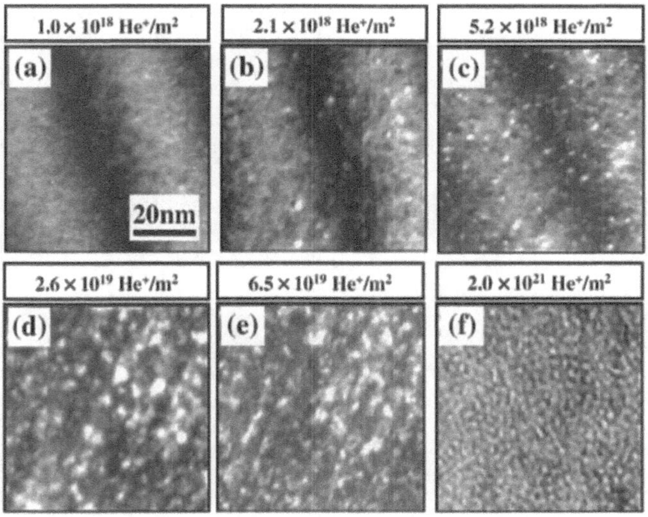

Figure 5.20: TEM plot of W surface morphology after irradiation with 300 K, 8 keV, and 2.0×10^{21} He^+/m^2 helium ion beam [49].

peak does not always shift toward high temperatures, as illustrated in Figure 5.24. For low-dose injections, the TDS peak shifts toward high temperatures or a new high-temperature desorption peak appears (Figure 5.24a) [52], and for high-dose injections, the TDS peaks shift in the opposite direction (Figure 5.24b) [54].

5.2.3 Damage behavior of tungsten materials under neutron irradiation

In most studies, the experimental conditions were controlled below the injection dose of 1.5 dpa (displacements per atom) and the temperature range of 400–800 °C. It was found that the defect formation process in the material is closely related to both irradiation dose and temperature [55]. Hasegawa et al. [56, 57] have extensively studied the effects of irradiation dose and temperature on neutron irradiation of tungsten materials. They identified the types of damage defects in tungsten under different irradiation devices, different irradiation damage degrees, and different temperatures. As shown in Figure 5.25, the primary defects produced by neutron irradiation are dislocation loops and voids. At low dpa levels, dislocation loops can be clearly observed, and their density is much lower than that of voids, with a ratio between 1/10 and 1/100. Under neutron irradiation in High Flux Isotope Reactor (HFIR) device, the dislocation loop disappears, and a small amount of voids and large number of acicular precipitates are generated as the damage approaches 1 dpa and the temperature increases from 500 to 800 °C, which are composed of ReW or Re_3W phases. Figure 5.26 illustrates the evolution process and mechanism of microstructure in tungsten under

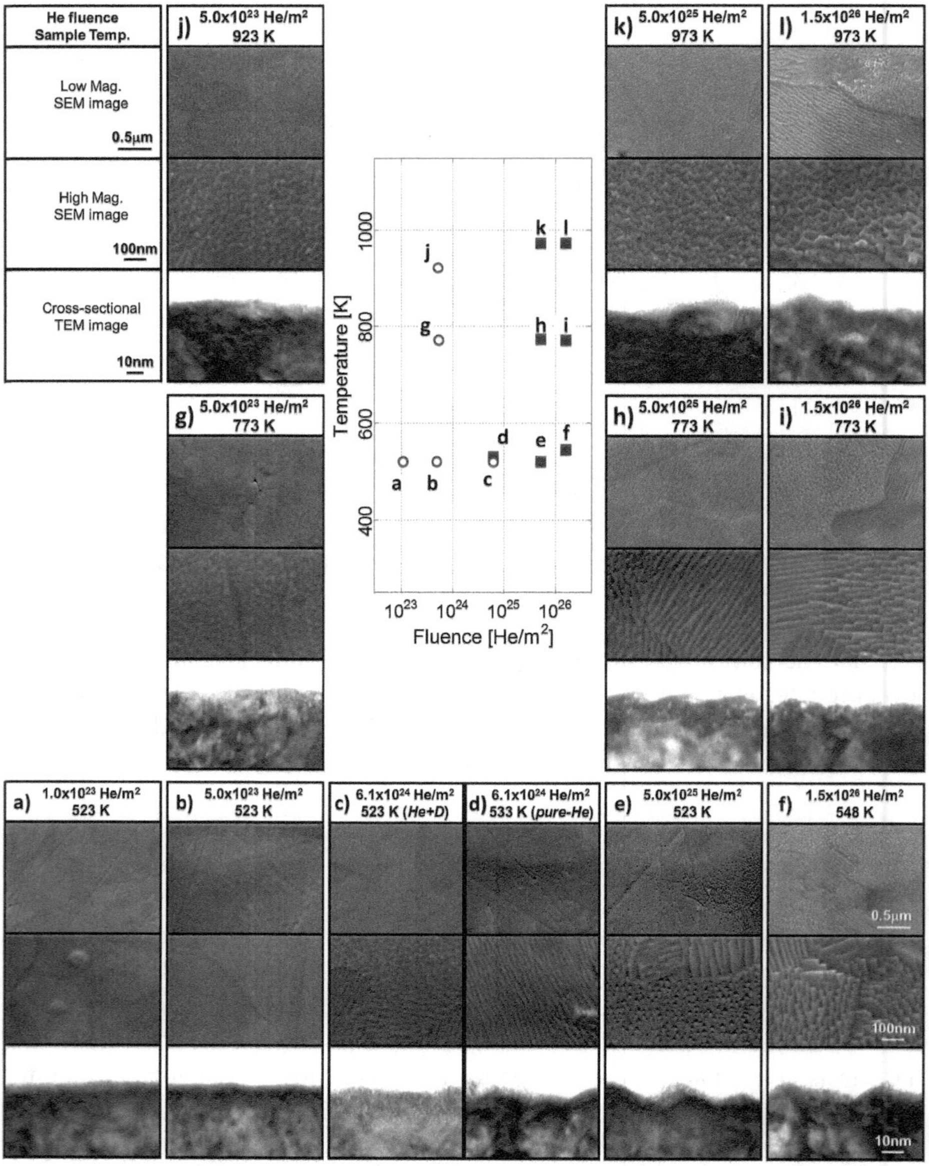

Figure 5.21: Microstructure evolution of tungsten under 60 eV He plasma irradiation of the PISCES-A device [50].

neutron irradiation between 400 and 800 °C [58]. With the increase of irradiation dose, dislocation loops and voids are generated sequentially in tungsten. After that, the dislocation loops disappeared, the number and size of voids decreased, accompanied by the formation of an acicular precipitate, and the content of Re in the material increased. The irradiation hardening of materials caused by neutron irradiation is

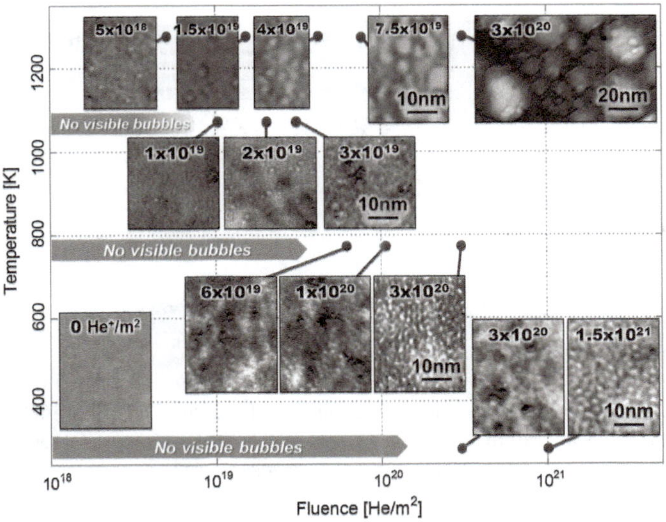

Figure 5.22: Microstructure evolution of tungsten irradiated with 3 keV He^+ at different temperatures [50].

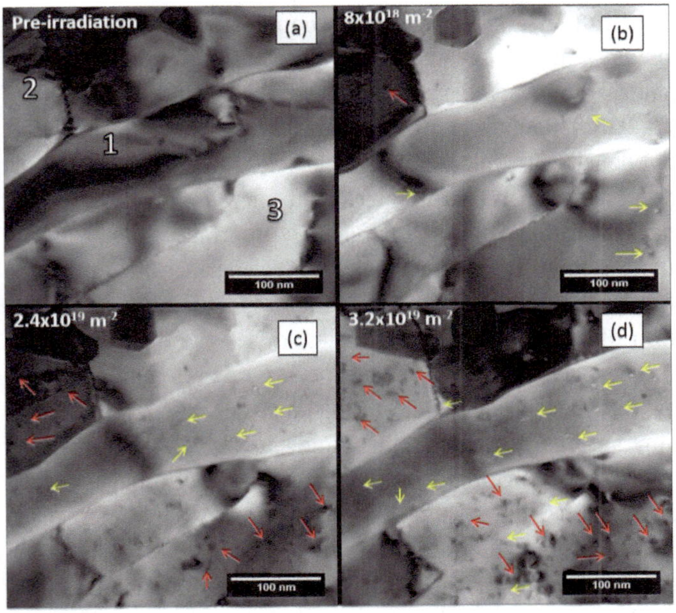

Figure 5.23: TEM image of microstructure of tungsten irradiated at 950 °C, 2 keV He^+: 1 for nanocrystalline W; 2 and 3 for ultrafine-grained W [51].

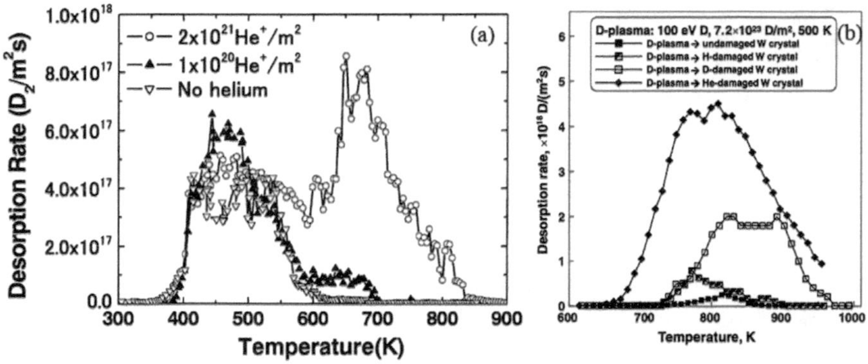

Figure 5.24: Effect of He^+ pre-irradiation on tungsten-deuterium retention: (a) TDS plots of 1.0×10^{21} D_2/m^2 injected after 1.0×10^{20} He^+/m^2 and 2.0×10^{21} He^+/m^2 pre-irradiation at room temperature energy of 8 keV [55]; (b) TDS plots of 7.2×10^{23} D_2/m^2 injected after 350 K room temperature energy of 12 keV and dose of 0.5×10^{22} He^+/m^2 pre-TDS map of 7.2×10^{23} D_2/m^2 injected after irradiation [54].

also related to the degree of irradiation damage, temperature, and Re content in the materials, and is mainly brought about by dislocation loops, voids, and precipitates in the materials after irradiation damage. Fukuda et al. [59] and Hu et al. [60] drew a similar conclusion that when the degree of irradiation damage was low (<0.16 dpa), dislocation loops and voids are the main causes of irradiation hardening. However, the second phase precipitate of intermetallic precipitate would result in hardening of the material when the irradiation damage degree exceeds 0.6 dpa.

By incorporating high-density precipitated phase particles, dispersing oxide particles, increasing dislocation density, and refining grains to increase the specific surface area, the preferential adsorption effect of high-density interfaces (such as grain boundaries, phase boundaries, interlayer boundaries, and free surfaces) is utilized to increase the recombination probability of vacancy and interstitial atoms. This process reduces the concentration of supersaturated vacancies, raises the dose threshold required to trigger irradiation swelling, and decreases the swelling rate under irradiation, ultimately improving the material's irradiation resistance.

Most researchers adopt various approaches, including alloying, grain refinement, and diffusion strengthening, to enhance the irradiation resistance of tungsten-based PFMs. Fukuda et al. [61, 62] conducted a systematic investigation into the microstructural changes induced by neutron irradiation in La- and K-doped W materials. The irradiation experiment was carried out on the fast reactor Joyo, with irradiation temperatures of 531–756 °C and neutron irradiation doses of 0.42–0.47 dpa in order to be nearly as the actual operating conditions of ITER (200–1,000 °C, 0.7 dpa). Figure 5.27 presents the micromorphology of pure W, La-doped W, and K-doped W after neutron irradiation, indicating the formation of numerous defects, such as voids, black spots, and dislocation loops, in all three materials.

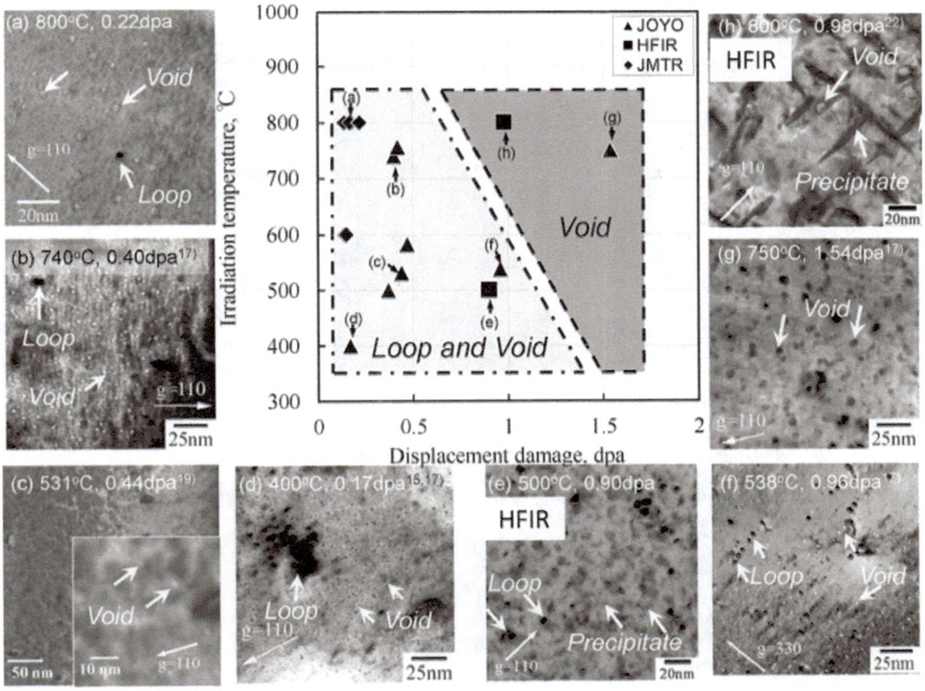

Figure 5.25: TEM diagram of microstructure defects of pure tungsten under different irradiation conditions [56].

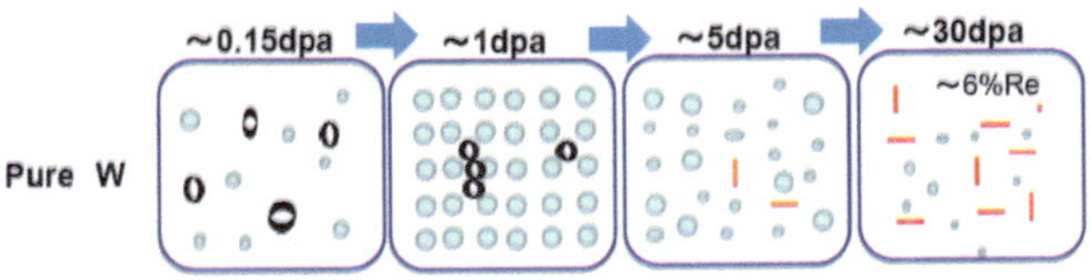

Figure 5.26: Schematic diagram of microstructure evolution in tungsten under fast neutron irradiation (circles represent voids, the black rings represent dislocation loops, and the red bars represent needle-like Re phase precipitates [58]).

Fukuda et al. [62] also investigated the effects of neutron irradiation at 1,029 K and 0.42 dpa on the microstructure and properties of ultrafine W-0.5TiC/H. The irradiation damage exhibited by several materials was found to be similar, as shown in Figure 5.28. Kurishita et al. [63] investigated the effect of neutron irradiation on the microstructure and properties of ultrafine W-0.5TiC composites (performed on a JMTR (Japan Materials Test Reactor)) at an irradiation temperature of 600 °C and a neutron irradiation dose of 2×10^{24} n/m^2. The microstructure morphology of pure W, W-0.5TiC-H2, and W-0.5TiC-Ar after neutron irradiation is shown in Figure 5.29. Defects such as small black spots or type I dislocation loops appear in the grains of the three materials after neutron irradi-

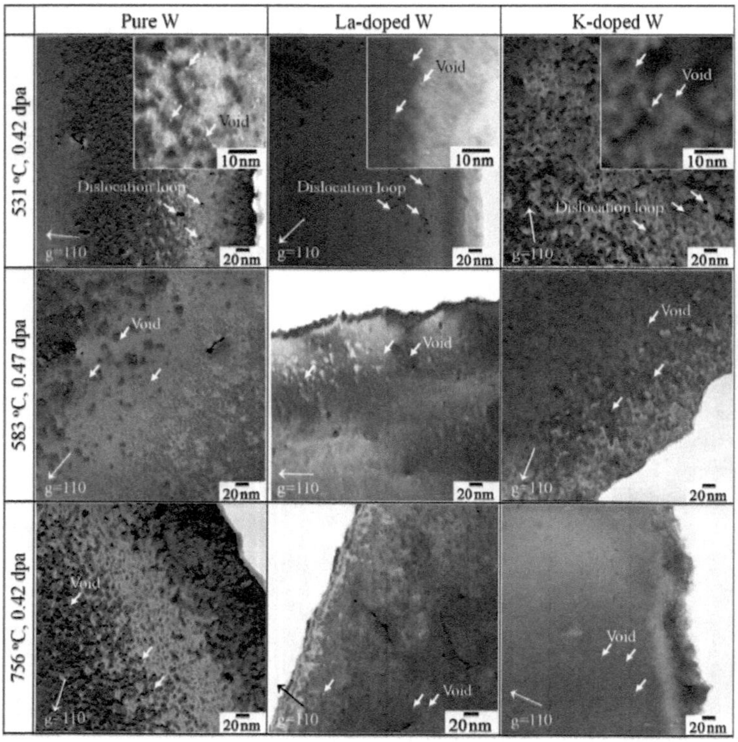

Figure 5.27: Microstructure of pure tungsten, La-doped tungsten, and K-doped tungsten after neutron irradiation [61].

ation. However, the number and density of defects in W-TiC are significantly lower than those of pure tungsten. As shown in Figure 5.30, the irradiation hardening values (△HV) of several materials show that the ultrafine W-TiC composites have stronger irradiation hardening resistance than other reinforced phase materials.

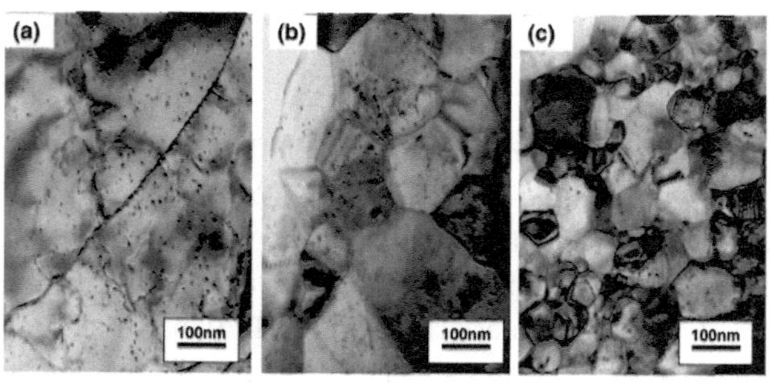

Figure 5.28: TEM bright-field phases of pure tungsten, W-0.5TiC-H, and W-0.5TiC-Ar irradiated at 873 K, 2×10^{24} n/m^2 neutrons on a JMTR [62].

Figure 5.29: Formation of voids in heat-treated stress-relieving samples after neutron irradiation at 1,029 K, 0.42 dpa on Joyo reactor: (a) pure tungsten; (b) La-doped tungsten; (c) K-doped tungsten; and (d) ultrafine-grained W-0.5 TiC/H [63].

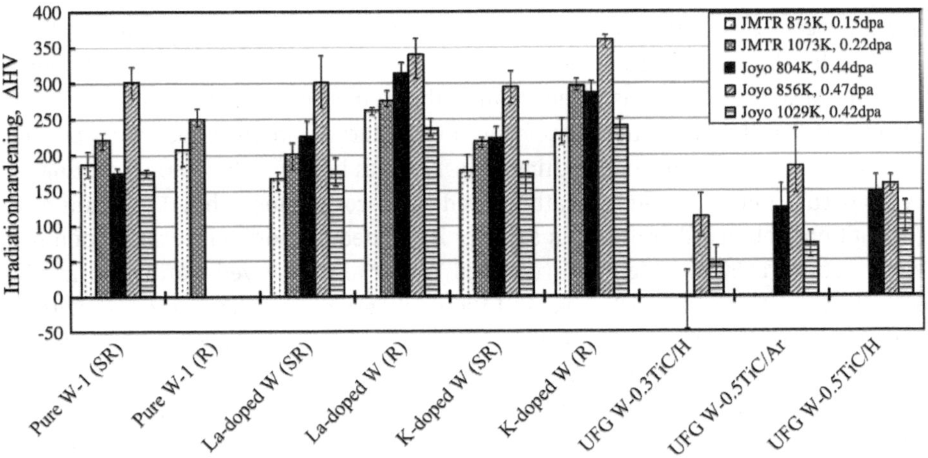

Figure 5.30: Irradiation hardening of samples after neutron irradiation [63].

5.2.4 Damage behavior of tungsten materials under heavy ion irradiation

The deuterium-tritium fusion reaction in a nuclear fusion reactor releases neutrons with an energy of 14 MeV, causing severe irradiation damages to materials. The radiation damage weakens material properties, and as irradiation accumulates to a certain dose, the properties of the material significantly deteriorate. The deterioration includes a loss of ductility (low-temperature irradiation embrittlement and high-temperature helium embrittlement) and void swelling, eventually leading to material failure. The high damage rate of heavy ions and the high flux that can be achieved by electron irradiation enable the application of alternative irradiation simulation evaluation of fusion reactor materials [64, 65]. Since the 1960s and 1970s, heavy ion irradiation has been used to simulate neutron irradiation damage in fast breeder reactors [66]. With the rapid development of nuclear fusion reactor construction and research, ion irradiation has also been applied to simulate neutron damage in nuclear fusion reactors to reflect the degree of irradiation damage to structural materials in nuclear fusion reactors.

High-energy heavy ion irradiation often induces high-density off-position damage near the surface of materials, including dislocations, vacancies, and vacancy clusters [67]. Dislocation loops, as the most common microstructure in irradiated materials, along with irradiation defects such as voids, stacking fault tetrahedra, bubbles, and interstitial atomic clusters, play a crucial role in the degradation of material properties during irradiation. One of the most striking phenomena is the formation of "loop rafting" [68]. Brimhall and Mastel [69] found "rafting" in neutron-irradiated Mo, which occurs at temperatures above 400 °C, mainly due to dislocation loop slip and climbing. Similarly, Sikka and Moteff [70] reported "rafting" in neutron-irradiated W samples at 430 and 580 °C, and found that the dislocation loops in the raft had the same Burgers vector, that is, <111>. Zinkle and Singh [71] found in neutron-irradiated Fe samples that raft occurs when the dose exceeds 0.3 dpa and that the number of rafts in thick film samples is higher than in thin film samples. Barnes and Mazey [72] suggested that the dislocation loop forms a low-energy state by adjusting its position and orientation, thus forming "rafting." Most studies have concluded that "rafting" always occurs under high-temperature irradiation conditions, while El-Atwani et al. [73] not only observed "rafting" in samples irradiated at room temperature and high temperature (1,050 K) 3 MeV Cu^+ irradiated W, but also observed "rafting" at irradiation doses as low as 0.2–0.25 dpa, and the thickness of "rafting" increased with increasing doses. This is shown in Figure 5.31 [73]. This discovery aligns with the formation mechanism proposed by Wen et al. [74], who demonstrated through kinetic Monte Carlo simulations that "rafting" would result from the slip of dislocation loops and the rotation of Burger vectors from self-interstitial clusters, without the need for dislocation climb. Therefore, the formation of "rafting" could occur at room temperature.

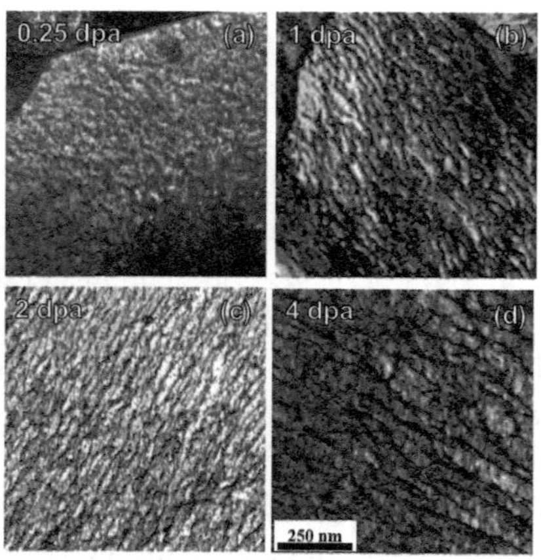

Figure 5.31: TEM diagram of W tissue irradiated with 3 MeV at room temperature with a dpa rate of 0.000167 dpa/s [73].

Figure 5.32 illustrates the dislocation distribution of tungsten irradiated at different temperatures of 6.4 MeV Fe^{3+} and 1 MeV He^+ [75]. The region between the irradiated surface and the depth where defect density rapidly changes, where the dislocation density is uniformly distributed, is called the front zone (FZ), while the high dislocation density zone (HDDZ) lies at a depth of approximately 1,800–2,000 nm. In this zone, the dislocation loop is prone to form acicular polymers along a fixed direction, which may be related to the specific orientation of the crystal. In general, at 300 °C, the damage zone can be divided into two parts: FZ and HDDZ, where the diffusion zone (DZ) can be ignored. At 500 °C, the HDDZ disappears and it is replaced by the DZ, which begins at a depth of ~2,000 nm, and the dislocation loop within the DZ forms "rafting." The low dislocation density zone (LDDZ) occurs at temperatures of 500 °C and higher, with a width of only ~100 nm, separating FZ and DZ. As the irradiation temperature increases, the front division line A shifts toward the surface, moving to 1.8 μm at 500 °C, 1.7 μm at 700 °C, and 1 μm at 1,000 °C. The position of the LDDZ demarcation line B remains unchanged at 2 μm, and the DZ demarcation line C moves deeper into the sample, indicating that the distribution of defects is affected by the irradiation temperature. In addition, the distribution of defects is also affected by grain boundaries, with grain boundaries functioning as sinks for irradiation defects, a denuded zone of dislocation loops will appear along the grain boundaries, as shown in Figure 5.33. As shown in Figure 5.33a, a denuded zone with a width of 20 nm is formed near the grain boundaries, and the density of the helium bubble at the grain boundaries is lower than that of the matrix (Figure 5.33b). El-Atwani et al. [73] also investigated the effect of grain boundary density on the "rafting" of nanocrystalline and ultrafine-grained tungsten samples under the same irradiation conditions. Higher grain boundary density promotes the absorption of interstitial atoms, contributing to defect annihilation and limiting raft formation. As shown in Figure 5.34a and b, NCW exhibits fewer rafts than coarse-grained W after 0.25 dpa Cu^+ irradiation at room temperature. At high temperatures, rafts appear in both grain sizes of W, and the rafts within the NCW are shorter in length and less dense, as shown in Figure 5.34c and d.

The tungsten material in the nuclear fusion reactor is subjected to high heat loads, with ion irradiation occurring at high temperatures ranging from 500 to 1,200 ° C. Therefore, studying the evolution behavior of irradiation defects in tungsten at elevated temperatures is of great significance. This involves observing how dislocation loops evolve in typical irradiated samples, determining the temperature at which they evolve, and assessing whether they could be completely removed by annealing [76–81]. Based on the study of fast neutron-irradiated tungsten (≥1 MeV) [82], the annealing process can be divided into five stages. Stage I (20–100 K): migration of free self-interstitial atoms (SIAs) and recombination of interstitial atom-Frenkel pairs; stage II (100–700 K): release of SIAs from dislocations and impurity atoms; stage III (700–920 K): vacancy migration; stage IV (920–1,270 K): migration of vacancy clusters and vacancy impurity complexes; stage V (1,270–1,800 K): thermal dissociation of vacancy and interstitial clusters. While the resistivity test helps to find the temperature range and activation energy during the recovery stages, without the direct observa-

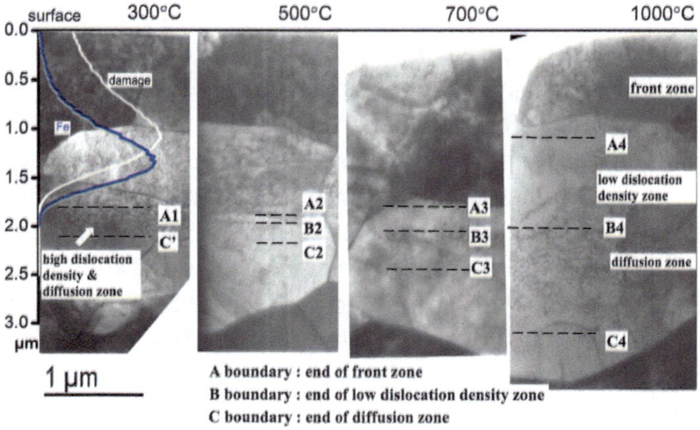

Figure 5.32: Overall distribution of dislocations after irradiation of W at different temperatures of 6.4 MeV Fe^{3+} and 1 MeV He^{+}, with the dashed lines representing the boundaries of the different regions. Boundary line A: the end of the front zone; boundary line B: the end of the low dislocation density zone; boundary line C: the end of the diffusion zone [75].

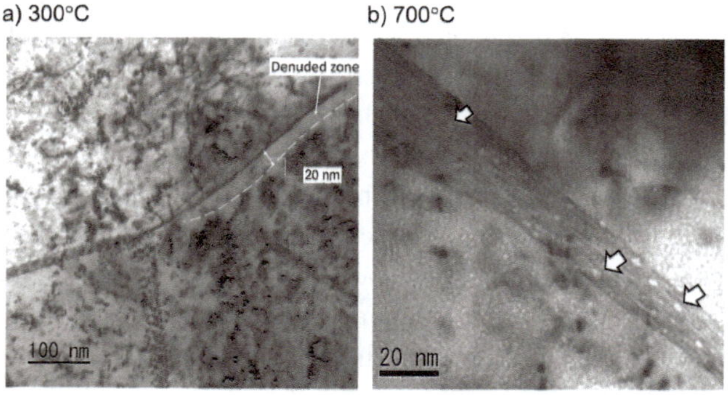

Figure 5.33: Effect of grain boundaries on (a) dislocations, a defect-free zone with a width of 20 nm; (b) effect of helium bubbles, which are smaller and larger in size than the matrix at grain boundaries [75].

tion of the defects, their intrinsic properties, or the evolution of these defects under various annealing conditions. Recently, TEM has been used to observe the recovery process of defects caused by irradiation of annealed materials at different temperatures. This not only makes possible the immediate quantified analysis of the irradiated defects but also helps with the observation of the recovery change process with temperature and time, which is helpful to obtain the recovery mechanism of irradiation defects.

Ferroni et al. [76] used TEM to observe the microstructure and defect evolution of tungsten samples during high-temperature annealing by self-ion irradiation (2 MeV W^{+}, 500 °C 10^{14} W^{+}/cm^{2}). Figure 5.35 indicates that the density of the dislocation loop decreases and the size increases (especially between 800 and 950 °C) as the annealing temperature

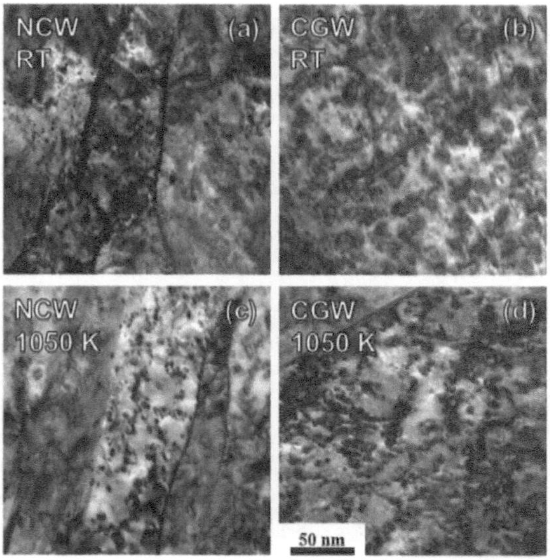

Figure 5.34: TEM diagram of coarse grain and nanocrystalline tungsten microstructure irradiated with 3 MeV, 0.25 dpa Cu^+ [73].

increases. At 950 and 1,100 °C, dislocation lines and dislocation nets appeared, and the gap dislocation loop annealed at 1,400 °C for 1 h was basically eliminated. Although TEM is a powerful tool for characterizing the evolution of interstitial dislocation loops and voids, it is limited to detecting defects larger than 1 nm (the resolution limit of TEM). Thus, the evolution mechanism for defects smaller than 1 nm, particularly for small vacancy clusters, annealing remains unclear during post-irradiation. As shown in Figure 5.36, voids are visible in TEM images at 800 °C, and the density decreases as the voids grow to 4 nm at 1,400 °C.

Liu et al. [82] used the Lanzhou heavy ion accelerator to study the effect of 122 MeV Ne^{7+} ions on the irradiation damage and deuterium retention of tungsten foil. The deuterium retention in tungsten increased significantly at an irradiation dose of 0.3 dpa, and the thermal desorption temperature range of deuterium became larger (730–1,173 K), which was the result of heavy ion irradiation defects. Watanabe et al. [83] investigated the effect of 2.4 MeV Cu^{2+} ion irradiation on deuterium retention in pure tungsten, and found that, even at room temperature, low doses of heavy ion irradiation could result in high-density dislocation loops and vacancy clusters in tungsten materials, and these irradiation defects will become strong deuterium traps and increase deuterium retention, as shown in Figure 5.37. Shimada et al. [84] examined the influence of 2.8 MeV Fe^{2+} ion irradiation on deuterium retention in pure tungsten and found that as the irradiation dose increased from 0.027 to 0.3 dpa, the intensity of the desorption peak increased, although the overall curve shape remained essentially unchanged. As the irradiation dose increased from 0.3 to 3 dpa, the inten-

Figure 5.35: TEM images of microstructure of 500 °C, 2 MeV, 1.5 dpa W^+ irradiated tungsten (a and b) annealed at different temperatures for 1 h (c) 800 °C; (d) 950 °C;(e) 1,100 °C;(f) 1,400 °C [76].

sity increased significantly, and the desorption peak moved to a higher temperature. This may be attributed to the high density of dislocation loops produced by heavy ion irradiation in tungsten, which became strong deuterium traps, resulting in an increase in deuterium retention. Tyburska et al. [85] studied the effect of W^{2+} pre-irradiation at 5.5 MeV with different dpa on the deuterium retention of tungsten, and the deuterium retention of tungsten at the irradiation dose of ~0.3 dpa under self-ion irradiation showed an obvious saturation trend, and the increase of the W^{2+} irradiation dose did not increase the deuterium retention. The result shown in Figure 5.38 is consistent with the findings of Ogorodnikova et al.'s [86] experiment. However, there are other studies suggesting that tungsten does not reach saturation at a deuterium retention level of 3 dpa, similar to findings at comparable doses [87] (Figure 5.39). The specific reasons for this discrepancy are still under hot discussion.

Hatano and coworkers [87] studied the effects of neutron irradiation and heavy ion irradiation (20 MeV tungsten ion self-irradiation) on deuterium retention in pure W with the HFIR at Oak Ridge National Laboratory (ORNL) in the United States. Both neutron irradiation and tungsten heavy ion self-irradiation significantly increase the retention of deuterium in the material. After deuterium plasma bombardment with a

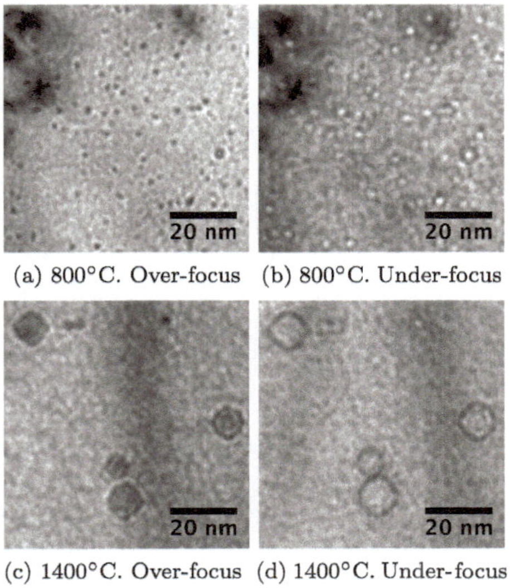

Figure 5.36: TEM diagram of voids after annealing at 800 °C (a and b) and 1,400 °C (c and d) by self-ion irradiation (2 MeV W^+, 500 °C 10^{14} W^+/cm^2).

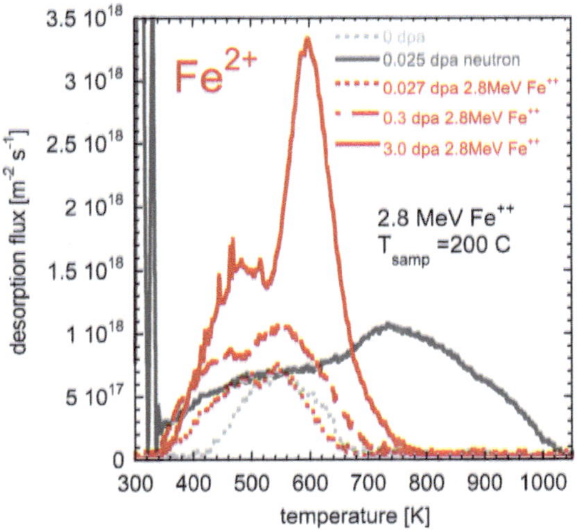

Figure 5.37: TDS spectra of tungsten injected with different doses of Fe^{2+} [86].

neutron irradiation dose of 0.3 dpa, a sample temperature of 473 K, and deuterium plasma bombardment, the deuterium concentration in tungsten reached up to 0.8 at% due to the irradiation defects and the defects are thermally stable. Oya and coworkers [88] compared the effects of heavy ion irradiation (2.8 MeV Fe^{2+}) and neutron irradiation on the deuterium desorption peak of tungsten. They found that the TDS curve of tungsten after Fe^{2+} irradiation exhibited two desorption peaks at 450 and 550 K. In

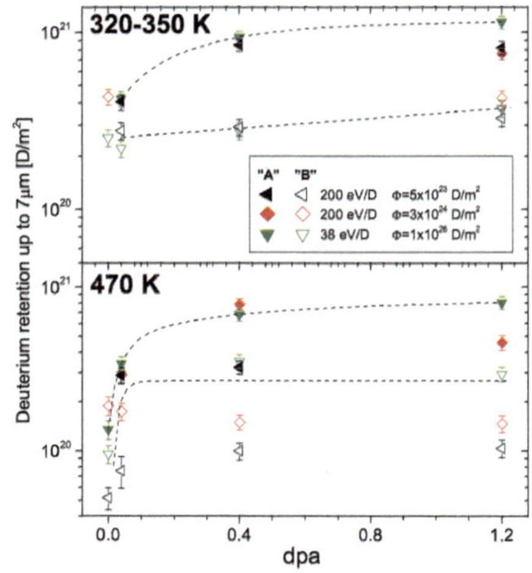

Figure 5.38: Deuterium retention of 5.5 MeV W^{2+} pre-irradiated tungsten with different dpa [87].

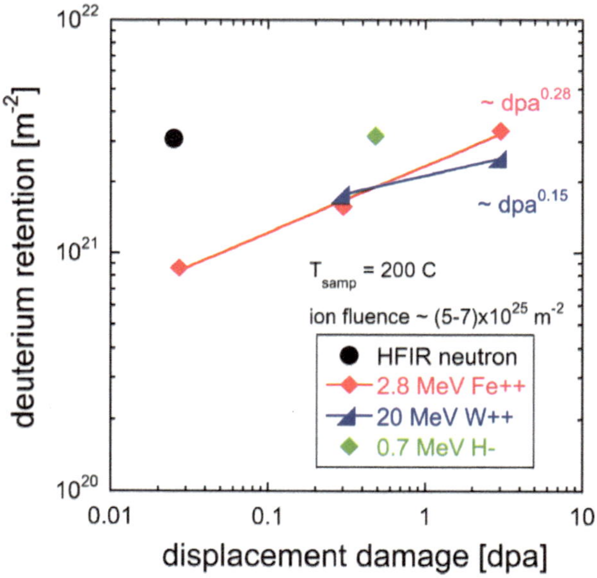

Figure 5.39: Deuterium retention of three different high-energy ion beams and neutron-irradiated tungsten after plasma implantation under similar conditions [88].

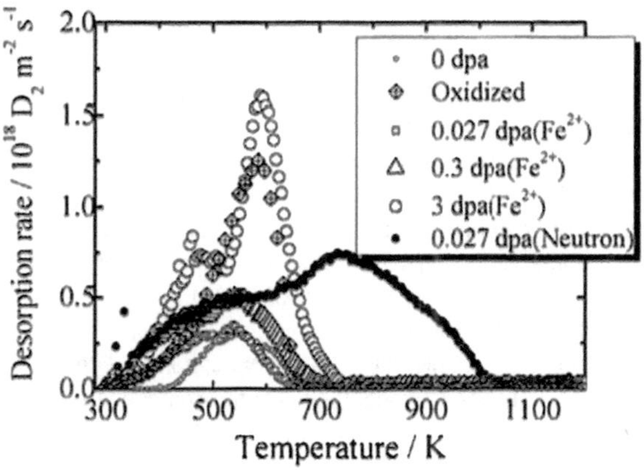

Figure 5.40: TDS spectra of W after 2.8 MeV Fe^{2+} irradiation and neutron irradiation [89].

contrast, neutron irradiation of tungsten produced a unique third high-temperature desorption peak at 750 K, suggesting that heavy ion irradiation does not fully replicate deuterium retention under neutron irradiation (Figure 5.40).

5.3 Summary

The ITER is designed to operate with the goal of achieving steady-state long-pulse deuterium-tritium fusion plasma discharges, largely dependent on PFMs. Typical PFMs in fusion reactors include pure tungsten and tungsten matrix materials. Beyond meeting the requirement of mechanics, high-temperature, and other properties, irradiation tests turn out to be crucial for the screening of application, with irradiation performance as the decisive factor of the materials selection.

The first wall material will withstand such three high environmental tests as the high H/He plasma flux (10^{20}–10^{24} m^{-2} s^{-1}), high heat flux (10–20 MW/m^2) density, and high-energy neutron irradiation (14 MeV). The damage dose in these conditions is 10^4 times greater than that of traditional nuclear power plant cladding. Changing the composition and microstructure of the tungsten material and adjusting its overall service performance determine what level of performance state the fusion reactor can achieve. The performance and variation of materials under irradiation field are of great importance, which directly associated with the safety and reliability of fusion reactors, and the progress of nuclear technology is indispensable for the improvement of irradiation performance and anti-irradiation ability of materials.

References

[1] C. Linsmeier, M. Rieth, J. Aktaa, T. Chikada, A. Hoffmann, J. Hoffmann, A. Houben, H. Kurishita, X. Jin, M. Li, A. Litnovsky, S. Matsuo, A. von Müller, V. Nikolic, T. Palacios, R. Pippan, D. Qu, J. Reiser, J. Riesch, T. Shikama, R. Stieglitz, T. Weber, S. Wurster, J.-H. You, Z. Zhou. Development of advanced high heat flux and plasma-facing materials [J]. Nuclear Fusion, 2017, 57: 092007.

[2] R.A. Pitts, S. Carpentier, F. Escourbiac, T. Hirai, V. Komarov, A.S. Kukushkin, S. Lisgo, A. Loarte, M. Merola, R. Mittear, A.R. Raffray, M. Shimada, P.C. Stangeby. Physics basis and design of the ITER plasma-facing components [J]. Journal of Nuclear Materials, 2011, 415(1): 957–964.

[3] A.R. Raffray, R. Nygren, D.G. Whyte, S. Abdel-Khalik, R. Doerner, F. Escourbiac, T. Evans, R.J. Goldston, D.T. Hoelzer, S. Konishi, P. Lorenzetto, M. Merola, R. Neu, P. Norajitra, R.A. Pitts, M. Rieth, M. Roedig, T. Rognlien, S. Suzuki, M.S. Tillack, C. Wong. High heat flux components-readiness to proceed from near term fusion systems to power plants [J]. Fusion Engineering and Design, 2010, 85(1): 93–108.

[4] J. Knaster, A. Moeslang, T. Muroga. Materials research for fusion [J]. Nature Physics, 2016, 12: 424–434.

[5] D.M. Duffy. Modeling plasma facing materials for fusion power [J]. Materials Today, 2009, 12(11): 38–44.

[6] M.R. Gilbert, J.C. Sublet. Neutron-induced transmutation effects in W and W-alloys in a fusion environment [J]. Nuclear Fusion, 2011, 51: 043005.

[7] V.K. Alimov, B. Tyburska-Püschel, Y. Hatano, J. Roth, K. Isobe, M. Matsuyama, T. Yamanishi. The effect of displacement damage on deuterium retention in ITER-grade tungsten exposed to low-energy, high-flux pure and helium-seeded deuterium plasmas [J]. Journal of Nuclear Materials, 2012, 420(1–3): 370–373.

[8] X. Yang, A. Hassanein. Molecular dynamics simulation of deuterium trapping and bubble formation in tungsten [J]. Journal of Nuclear Materials, 2013, 434(1–3): 1–6.

[9] Y. Jinnan. Irradiation effect of materials [M]. Beijing: Chemical Industry Press, 2007.

[10] W.M. Shu, A. Kawasuso, T. Yamanishi. Recent findings on blistering and deuterium retention in tungsten exposed to high-fluence deuterium plasma [J]. Journal of Nuclear Materials, 2009, 386–388: 356–359.

[11] W.M. Shu, E. Wakai, T. Yamanishi. Blister bursting and deuterium bursting release from tungsten exposed to high fluences of high flux and low energy deuterium plasma [J]. Nuclear Fusion, 2007, 47: 201–209.

[12] W.M. Shu, M. Nakamichi, V.K. Alimov, G.-N. Luo, K. Isobe, T. Yamanishi. Deuterium retention, blistering and local melting at tungsten exposed to high-fluence deuterium plasma [J]. Journal of Nuclear Materials, 2009, 390–391: 1017–1021.

[13] Y.Z. Jia, W. Liu, B. Xu, G.-N. Luo, S.L. Qu, T.W. Morgan, G. De Temmerman. Mechanism for orientation dependence of blisters on W surface exposed to D plasma at low temperature [J]. Journal of Nuclear Materials, 2016, 477: 165–171.

[14] Y.L. Liu, Y. Zhang, H.B. Zhou, G.H. Lu, F. Liu, G.N. Luo. Vacancy trapping mechanism for hydrogen bubble formation in metal [J]. Physical Review B, 2009, 79: 172103.

[15] A.V. Veen, H.A. Filius, J. De Vries, K.R. Bijkerk, G.J. Rozing, D. Segers. Hydrogen exchange with voids in tungsten observed with TDS and PA [J]. Journal of Nuclear Materials, 1988, 155–157: 1113–1117.

[16] K. Heinola, T. Ahlgren, K. Nordlund, J. Keinonen. Hydrogen interaction with point defects in tungsten [J]. Physical Review B, 2010, 82: 094102.

[17] D.F. Johnson, E.A. Carter. Hydrogen in tungsten: Absorption, diffusion, vacancy trapping, and decohesion [J]. Journal of Materials Research, 2010, 25(2): 315–327.

[18] Y.W. You, X.S. Kong, X.B. Wu, Y.C. Xu, Q.F. Fang, J.L. Chen, G.N. Luo, C.S. Liu, B.C. Pan, Z.G. Wang. Dissolving, trapping and detrapping mechanisms of hydrogen in bcc and fcc transition metals [J]. AIP Advances, 2013, 3: 012118.
[19] C. Guerrero, C. González, R. Iglesias, J.M. Perlado, R. González-Arrabal. First principles study of the behavior of hydrogen atoms in W monovacancy [J]. Journal of Materials Science, 2016, 51(3): 1445–1455.
[20] K. Ohsawa, K. Eguchi, H. Watanabe, M. Yamaguchi, M. Yagi. Configuration and binding energy of multiple hydrogen atoms trapped in monovacancy in bcc transition metals [J]. Physical Review B, 2012, 85: 094102.
[21] R. Sakamoto, T. Muroga, N. Yoshida. Microstructural evolution induced by low energy hydrogen ion irradiation in tungsten [J]. Journal of Nuclear Materials, 1995, 220–222: 819–822.
[22] W.H. Hu, F.F. Luo, Z.Y. Shen, L.P. Guo, Z.C. Zheng, Y.M. Wen, Y.Y. Ren. Hydrogen bubble formation and evolution in tungsten under different hydrogen irradiation conditions [J]. Fusion Engineering and Design, 2015, 90: 23–28.
[23] R. Frauenfelder. Solution and diffusion of hydrogen in tungsten [J]. Journal of Vacuum Science and Technology, 1969, 6: 388–397.
[24] W.M. Shu, E. Wakai, T. Yamanishi. Blister bursting and deuterium bursting release from tungsten exposed to high fluences of high flux and low energy deuterium plasma [J]. Nuclear Fusion, 2007, 47: 201–209.
[25] Y. Zayachuk, M.H.J. 't Hoen, P.A. Zeijlmans Van Emmichoven, I. Uytdenhouwen, G. Oost. Deuterium retention in tungsten and tungsten-tantalum alloys exposed to high-flux deuterium plasmas [J]. Nuclear Fusion, 2012, 52: 103021.
[26] B. Tyburska-Püschel, V.K. Alimov. On the reduction of deuterium retention in damaged Re-doped W [J]. Nuclear Fusion, 2013, 53: 123021.
[27] T. Tanabe. Review of hydrogen retention in tungsten [J]. Physica Scripta, 2014, 159: 014044.
[28] V.K. Alimov, B. Tyburska-Püschel, S. Lindig, Y. Hatano, M. Balden, J. Roth, K. Isobe, M. Matsuyama, T. Yamanishi. Temperature dependence of surface morphology and deuterium retention in polycrystalline ITER-grade tungsten exposed to low-energy, high-flux D plasma [J]. Journal of Nuclear Materials, 2012, 420(1–3): 519–524.
[29] K. Tokunaga, M.J. Baldwin, R.P. Doerner, N. Noda, Y. Kubota, N. Yoshida, T. Sogabe, T. Kato, B. Schedler. Blister formation and deuterium retention on tungsten exposed to low energy and high flux deuterium plasma [J]. Journal of Nuclear Materials, 2005, 337–339: 887–891.
[30] A.V. Golubeva, M. Mayer, J. Roth, V.A. Kurnaev, O.V. Ogorodnikova. Deuterium retention in rhenium-doped tungsten [J]. Journal of Nuclear Materials, 2007, 363–365: 893–897.
[31] O.V. Ogorodnikova, J. Roth, M. Mayer. Ion-driven deuterium retention in tungsten [J]. Journal of Applied Physics, 2008, 103(3): 034902.
[32] Y. Ueda, K. Schmid, M. Balden, J.W. Coenen, T. Loewenhoff, A. Ito, A. Hasegawa, C. Hardie, M. Porton, M. Gilbert. Baseline high heat flux and plasma facing materials for fusion [J]. Nuclear Fusion, 2017, 57: 092006.
[33] R.P. Doerner, M.J. Baldwin, T.C. Lynch, J.H. Yu. Retention in tungsten resulting from extremely high fluence plasma exposure [J]. Nuclear Materials and Energy, 2016, 9: 89–92.
[34] H. Kurishita, H. Arakawa, S. Matsuo, T. Sakamoto, S. Kobayashi, K. Nakai, G. Pintsuk, J. Linke, S. Tsurekawa, V. Yardley, K. Tokunaga, T. Takida, M. Katoh, A. Ikegaya, Y. Ueda, M. Kawai, N. Yoshida. Development of nanostructured tungsten based materials resistant to recrystallization and/or radiation induced embrittlement [J]. Materials Transactions, 2013, 54: 456–465.
[35] Z. Tao, Y. Wei, X. Zhuoming, M. Shu, Y. Junfeng, W. Xianping, F. Qianfeng, L. Changsong. Research progress on carbide/oxide dispersion strengthened tungsten-based materials [J]. Acta Metallurgica Sinica, 2018, 54: 831–843.

[36] M. Zhao, W. Jacob, A. Manhard, L. Gao, M. Balden, U. von Toussaint, Z. Zhou. Deuterium implantation into Y_2O_3-doped and pure tungsten: Deuterium retention and blistering behavior [J]. Journal of Nuclear Materials, 2017, 487: 75–83.
[37] M. Oya, H.T. Lee, Y. Ohtsuka, Y. Ueda, H. Kurishita, M. Oyaidzu, T. Yamanishi. Deuterium retention in various toughened, fine-grained recrystallized tungsten materials under different irradiation conditions [J]. Physica Scripta, 2014, 159: 014048.
[38] M. Zibrov, M. Mayer, E. Markina, K. Sugiyama, M. Betzenbichler, H. Kurishita, Y. Gasparyan, O.V. Ogorodnikova, A. Manhard, A. Pisarev. Deuterium retention in TiC and TaC doped tungsten under low-energy ion irradiation [J]. Physica Scripta, 2014, T159: 014050.
[39] M. Zhao, W. Jacob, L. Gao, A. Manhard, T. Dürbeck, Z. Zhou. Deuterium retention behavior of pure and Y_2O_3-doped tungsten investigated by nuclear reaction analysis and thermal desorption spectroscopy [J]. Nuclear Materials and Energy, 2018, 15: 32–42.
[40] C.S. Becquart, C. Domain. A density functional theory assessment of the clustering behaviour of He and H in tungsten [J]. Journal of Nuclear Materials, 2009, 386–388: 109–111.
[41] C.S. Becquart, C. Domain. Migration energy of He in W revisited by ab initio calculations [J]. Physical Review Letters, 2006, 97: 196402.
[42] K.O.E. Henriksson, K. Nordlund, A. Krasheninnikov, J. Keinonen. Difference in formation of hydrogen and helium clusters in tungsten [J]. Applied Physics Letters, 2005, 87: 163113.
[43] Y.W. You, D.D. Li, X.S. Kong, X.B. Wu, C.S. Liu, Q.F. Fang, B.C. Pan, J.L. Chen, G.N. Luo. Clustering of H and He, and their effects on vacancy evolution in tungsten in a fusion environment [J]. Nuclear Fusion, 2014, 54: 103007.
[44] M.J. Baldwin, R.P. Doerner. Formation of helium induced nanostructure 'fuzz' on various tungsten grades [J]. Journal of Nuclear Materials, 2010, 404: 165–173.
[45] Y. Ueda, H. Peng, H. Lee, N. Ohno, S. Kajita, N. Yoshida, R. Doerner, G. De Temmerman, V. Alimov, G. Wright. Helium effects on tungsten surface morphology and deuterium retention [J]. Journal of Nuclear Materials, 2013, 442(1): S267–S272.
[46] E. Gao, W. Nadvornick, R. Doerner, N.M. Ghoniem. The influence of low-energy helium plasma on bubble formation in micro-engineered tungsten [J]. Journal of Nuclear Materials, 2018, 501: 319–328.
[47] S. Kajita, W. Sakaguchi, N. Ohno, N. Yoshida, T. Saeki. Formation process of tungsten nanostructure by the exposure to helium plasma under fusion relevant plasma conditions [J]. Nuclear Fusion, 2009, 49: 095005.
[48] A.M. Ito, A. Takayama, Y. Oda, T. Tamura, R. Kobayashi, T. Hattori, S. Ogata, N. Ohno, S. Kajita, M. Yajima, Y. Noiri, Y. Yoshimoto, S. Saito, S. Takamura, T. Murashima, M. Miyamoto, H. Nakamura. Molecular dynamics and Monte Carlo hybrid simulation for fuzzy tungsten nanostructure formation [J]. Nuclear Fusion, 2015, 55: 073013.
[49] Y. Watanabe, H. Iwakiri, N. Yoshida, K. Morishita, A. Kohyama. Formation of interstitial loops in tungsten under helium ion irradiation: Rate theory modeling and experiment [J]. Nuclear Instruments and Methods in Physics Research B, 2007, 255: 32–36.
[50] M. Miyamoto, S. Mikami, H. Nagashima, N. Iijima, D. Nishijima, R.P. Doerner, N. Yoshida, H. Watanabe, Y. Ueda, A. Sagara. Systematic investigation of the formation behavior of helium bubbles in tungsten [J]. Journal of Nuclear Materials, 2015, 463: 333–336.
[51] O. El-Atwani, J.A. Hinks, G. Greaves, S. Gonderman, T. Qiu, M. Efe, J.P. Allain. In-situ TEM observation of the response of ultrafine- and nanocrystalline-grained tungsten to extreme irradiation environments [J]. Scientific Report, 2014, 4: 4716.
[52] H. Iwakiri, K. Morishita, N. Yoshida. Effects of helium bombardment on the deuterium behavior in tungsten [J]. Journal of Nuclear Materials, 2002, 307–311: 135–138.

[53] Y. Nobuta, Y. Hatano, M. Matsuyama, S. Abe, Y. Yamauchi, T. Hino. Helium irradiation effects on tritium retention and long-term tritium release properties in polycrystalline tungsten [J]. Journal of Nuclear Materials, 2015, 463: 993–996.
[54] I.I. Arkhipov, S.L. Kanashenko, V.M. Sharapov, R.K. Zalavutdinov, A.E. Gorodetsky. Deuterium trapping in ion-damaged tungsten single crystal [J]. Journal of Nuclear Materials, 2007, 363–365: 1168–1172.
[55] M. Klimenkov, U. Jäntsch, M. Rieth, H.C. Schneider, D.E.J. Armstrong, J. Gibson, S.G. Roberts. Effect of neutron irradiation on the microstructure of tungsten [J]. Nuclear Materials and Energy, 2016, 9: 480–483.
[56] A. Hasegawa, M. Fukuda, S. Nogami, K. Yabuuchi. Neutron irradiation effects on tungsten materials [J]. Fusion Engineering and Design, 2014, 89: 1568–1572.
[57] A. Hasegawa, M. Fukuda, T. Tanno, S. Nogami. Neutron irradiation behavior of tungsten [J]. Materials Transactions, 2013, 54(4): 466–471.
[58] J. Marian, C.S. Becquart, C. Domain, S.L. Dudarev, M.R. Gilbert, R.J. Kurtz, D.R. Mason, K. Nordlund, A.E. Sand, L.L. Snead, T. Suzudo, B.D. Wirth. Recent advances in modeling and simulation of the exposure and response of tungsten to fusion energy conditions [J]. Nuclear Fusion, 2017, 57: 092008.
[59] M. Fukuda, N.A.P.K. Kumar, T. Koyanagi, L.M. Garrison, L.L. Snead, Y. Katoh, A. Hasegawa. Neutron energy spectrum influence on irradiation hardening and microstructural development of tungsten [J]. Journal of Nuclear Materials, 2016, 479: 249–254.
[60] X.X. Hu, T. Koyanagi, M. Fukuda, N.A.P.K. Kumar, L.L. Snead, B.D. Wirth, Y. Katoh. Irradiation hardening of pure tungsten exposed to neutron irradiation [J]. Journal of Nuclear Materials, 2016, 480: 235–243.
[61] M. Fukuda, A. Hasegawa, S. Nogami, K. Yabuuchi. Microstructure development of dispersion-strengthened tungsten due to neutron irradiation [J]. Journal of Nuclear Materials, 2014, 449(1–3): 213–218.
[62] M. Fukuda, A. Hasegawa, T. Tanno, S. Nogami, H. Kurishita. Property change of advanced tungsten alloys due to neutron irradiation [J]. Journal of Nuclear Materials, 2013, 442(1–3): S273–S276.
[63] H. Kurishita, S. Kobayashi, K. Nakai, T. Ogawa, A. Hasegawa, K. Abe, H. Arakawa, S. Matsuo, T. Takida, K. Takebe, M. Kawai, N. Yoshida. Development of ultra-fine grained W-(0.25–0.8) wt%TiC and its superior resistance to neutron and 3 MeV He-ion irradiations [J]. Journal of Nuclear Materials, 2008, 377(1): 34–40.
[64] S. Zhu, Y. Xu, Z. Wang, Y. Zheng, D. Zhou, E. Du, D. Yuan, M. Fukuda, M. Mihara, K. Matsuta, T. Minamisono. Positron annihilation lifetime spectroscopy on heavy ion irradiated stainless steels and tungsten [J]. Journal of Nuclear Materials, 2005, 343(1–3): 330–332.
[65] S. Zhu, Y. Zheng, P. Ahmat, Y. Xu, D. Zhou, Z. Wang, E. Du, D. Yuan, Y. Zuo, Y. Ruan, X. Duan. Temperature and dose dependences of radiation damage in modified stainless steel [J]. Journal of Nuclear Materials, 2005, 343(1–3): 325–329.
[66] C. Abromeit. Aspect of simulation of neutron damage by ion irradiation [J]. Journal of Nuclear Materials, 1994, 216: 78–96.
[67] T. Schwarz-Selinger. Deuterium retention in MeV self-implanted tungsten: Influence of damaging dose rate. Nuclear Materials and Energy, 2017, 12: 683–688.
[68] O. El-Atwani, E. Esquivel, M. Efe, E. Aydogan, Y.Q. Wang, E. Martinez, S.A. Maloy. Loop and void damage during heavy ion irradiation on nanocrystalline and coarse grained tungsten: Microstructure, effect of dpa rate, temperature, and grain size [J]. Acta Materialia, 2018, 149: 206–219.
[69] J. Brimhall, B. Mastel. Neutron irradiated molybdenum-relationship of microstructure to irradiation temperature [J]. Radiation Effects, 1970, 3(2): 203–215.
[70] V. Sikka, J. Moteff. "Rafting" in neutron irradiated tungsten [J]. Journal of Nuclear Materials, 1973, 46(2): 217–219.

[71] S.J. Zinkle, B.N. Singh. Microstructure of neutron-irradiated iron before and after tensile deformation [J]. Journal of Nuclear Materials, 2006, 351(1): 269–284.
[72] R. Barnes, D. Mazey. Stress-generated prismatic dislocation loops in quenched copper [J]. Acta Materialia, 1963, 11(4): 281–286.
[73] O. El-Atwani, E. Aydogan, E. Esquivel, M. Efe, Y.Q. Wang, S.A. Maloy. Detailed transmission electron microscopy study on the mechanism of dislocation loop rafting in tungsten [J]. Acta Materialia, 2018, 147: 277–283.
[74] M. Wen, N.M. Ghoniem, B.N. Singh. Dislocation decoration and raft formation in irradiated materials [J]. Philosophical Magazine, 2005, 85(22): 2561–2580.
[75] Z. Zhang, K. Yabuuchi, A. Kimura. Defect distribution in ion-irradiated pure tungsten at different temperatures [J]. Journal of Nuclear Materials, 2016, 480: 207–215.
[76] F. Ferroni, X. Yi, K. Arakawa, S.P. Fitzgerald, P.D. Edmondson, S.G. Roberts. High temperature annealing of ion irradiated tungsten [J]. Acta Materialia, 2015, 90: 380–393.
[77] K.M. Bowkett, B. Ralph. The annealing of radiation damage in tungsten investigated by field-ion microscopy [J]. Proceedings of the Royal Society of London. Series A, 1969, 312: 51–63.
[78] L.K. Keys, J. Moteff. Neutron irradiation and defect recovery of tungsten [J]. Journal of Nuclear Materials, 1970, 34: 260–280.
[79] L. Keys, J. Smith, J. Moteff. Stage III recovery in neutron irradiated tungsten [J]. Scripta Metallurgica, 1967, 1: 71–72.
[80] L. Keys, J. Smith, J. Moteff. High-temperature recovery of tungsten after neutron irradiation [J]. Physics Review, 1968, 176(3): 851–856.
[81] L. Keys. Comparison of the recovery of damage in W and Mo after neutron irradiation [J]. Journal of Applied Physics, 1969, 40(9): 3866.
[82] F. Liu, Y. Xu, H. Zhou, X. Li, Y. Song, C. Zhang, Q. Li, C. He, G. Luo. Defect production and deuterium retention in quasi-homogeneously damaged tungsten [J]. Nuclear Instruments and Methods in Physics Research Section B: Beam Interactions with Materials and Atoms, 2015, 351: 23–26.
[83] H. Watanabe, N. Futagami, S. Naitou, N. Yoshida. Microstructure and thermal desorption of deuterium in heavy-ion-irradiated pure tungsten [J]. Journal of Nuclear Materials, 2014, 455: 51–55.
[84] M. Shimada, Y. Hatano, Y. Oya, T. Oda, M. Hara, G. Cao, M. Kobayashi, M. Sokolov, H. Watanabe, B. Tyburska-Püschel, Y. Ueda, P. Calderoni, K. Okuno. Overview of the US–Japan collaborative investigation on hydrogen isotope retention in neutron-irradiated and ion-damaged tungsten [J]. Fusion Engineering and Design, 2012, 87: 1166–1170.
[85] B. Tyburska, V.K. Alimov, O.V. Ogorodnikova, K. Schmid, K. Ertl. Deuterium retention in self-damaged tungsten [J]. Journal of Nuclear Materials, 2009, 395(1–3): 150–155.
[86] O.V. Ogorodnikova, B. Tyburska, V.K. Alimov, K. Ertl. The influence of radiation damage on the plasma-induced deuterium retention in self-implanted tungsten [J]. Journal of Nuclear Materials, 2011, 415(1): S661–S666.
[87] M. Shimada, Y. Hatano, Y. Oya, T. Oda, M. Hara, G. Cao, M. Kobayashi, M. Sokolov, H. Watanabe, B. Tyburska-Püschel, Y. Ueda, P. Calderoni, K. Okuno. Overview of the US-Japan collaborative investigation on hydrogen isotope retention in neutron-irradiated and ion-damaged tungsten [J]. Fusion Engineering and Design, 2012, 87(7–8): 1166–1170.
[88] Y. Hatano, M. Shimada, T. Otsuka, Y. Oya, V.K. Alimov, M. Hara, J. Shi, M. Kobayashi, T. Oda, G. Cao, K. Okuno, T. Tanaka, K. Sugiyama, J. Roth, B. Tyburska-Püschel, J. Dorner, N. Yoshida, N. Futagami, H. Watanabe, M. Hatakeyama, H. Kurishita, M. Sokolov, Y. Katoh. Deuterium trapping at defects created with neutron and ion irradiations in tungsten [J]. Nuclear Fusion, 2013, 53: 073006.
[89] Y. Oya, M. Shimada, M. Kobayashi, T. Oda, M. Hara, H. Watanabe, Y. Hatano, P. Calderoni, K.Okuno. Comparison of deuterium retention for ion-irradiated and neutron-irradiated tungsten [J]. Physica Scripta, 2011, T145: 014050–014050

Chapter 6
Transient thermal shock behavior and damage of tungsten-matrix materials for fusion reactors

6.1 Introduction

As the PFMs at the divertor intersect with the magnetic field lines, a large number of high-energy particles directly hit the surface of the material along those lines, resulting in enormous energy being deposited on the target plate, which is subjected to a high thermal load. During typical ITER steady-state operation, the maximum surface heat flux is expected to reach about 5–10 MW/m^2 and the steady-state thermal load have a long duration and occur over many cycles. In addition, there exists a type of heat load with duration of only several seconds, known as slow transient thermal load with a size of up to 20 MW/m^2. Although the duration of the slow transient thermal load is shorter than that of the steady-state heat load, its thermal load is much larger than former, hence much greater damage to PFMs.

In addition, the PFMs are subject to high current transient thermal loads of different types due to the frequent occurrence of some uncontrollable electromagnetic disturbances during plasma operation and the inherent instability of the plasma, which mutually reinforce each other under the action of positive feedback. The transient thermal loads fall into three categories, namely, plasma disruptions, edge-localized modes (ELMs), and vertical displacement events (VDEs), which can cause material damage, such as surface plastic deformation, cracking, melting of the tungsten materials. All types of transient thermal loads generate a huge thermal load. Plasma disruption can cause the surface thermal load of the inner and outer target plates of the divertor to reach ~7–40 MJ/m^2 and ~7–40 MJ/m^2 in a short period (0.1–3 ms), respectively. The occurrence of VDEs leads to significant plasma interaction with PFMs, causing the surface heat load on the outer wall blanket module to increase to about ~20–30 MJ/m^2 and ~7–40 MJ/m^2 within 0.1–0.3 μs. ELMs cause a loss of plasma energy at the edges, creating substantial instantaneous thermal loads on the divertor targets. ELMs, under control, would impose a thermal load of 0.5 MJ/m^2 and 0.3 MJ/m^2 on the surface of the inner and outer target plates of the divertor, respectively. If the ELMs are out of control, the surface thermal load of the inner and outer targets of the divertor will reach 10 MJ/m^2 and 6 MJ/m^2, respectively, within 0.25–0.5 μs [1, 2]. Typical thermal load conditions in ITER are listed in Table 6.1 [3].

Prolonged exposure to steady-state or transient thermal loads reduces the strength and toughness of tungsten and decreases the thermal shock resistance. Due to the pulse operation mode, the thermal load will change periodically, which is likely to cause thermal fatigue at the joining interface. Comparatively, the damage of transient thermal shock to materials is particularly serious [4]. During the service life of ITER, the materi-

 | https://doi.org/10.1515/9783111556987-006

als are subjected to approximately 10^6 ELM shocks [5, 6], with burst frequencies as high as 1 Hz. Under the action of transient thermal shock, the surface of the material will heat up and cool down rapidly in a short time, which will form a high-temperature gradient and produce high thermal stress, accompanied by the emergence of the surface cracking, melting, sputtering, protruding structures, recrystallization and grain growth and other phenomena of the material. Therefore, studying and understanding the thermal and irradiation behaviors of W under transient thermal shock is beneficial for the future design and manufacture of divertor components.

6.2 Transient thermal load experiments on tungsten and tungsten alloys

6.2.1 Transient thermal load experimental simulation device

Currently, the transient thermal load test of materials is primarily carried out by using experimental devices to simulate the conditions. The results are then subject to theoretical analysis and evaluation. At present, due to the limited experimental resources of Tokamak devices and the narrow range of plasma parameters, the simulation of material performance under ELM-like transient thermal shocks mainly utilizes energy sources such as electron beams, lasers, and plasma, which have abundant experimental resources [7–9]. Table 6.2 shows the equipment and parameters currently in service that can meet the requirements for thermal shock experiments of ITER components [10].

Table 6.1: Typical thermal load conditions in ITER [3].

Event	Divertor	The first wall
Steady-state operation		
Maximum surface heat flux (MW/m²)	5–10 (W/CFC)	0.5 (Be)
Time (s)	≤450	≤450
Number of cycles	3,000	30,000
Slow transient state		
Maximum surface heat flux (MW/m²)	20 (CFC)	——
Time (s)	10	——
Frequency (%)	10	——
Disruption		
Maximum surface heat flux (MW/m²)	10–30	1
Time (ms)	0.1–0.3	0.1–0.3
Frequency (%)	10	10

Table 6.1 (continued)

Event	Divertor	The first wall
Vertical displacement events		
Maximum surface heat flux (MW/m^2)	60	–
Time (ms)	100–300	–
Frequency (%)	1	
Edge localized modes		
Maximum surface heat flux (MJ/m^2)	1	–
Time (ms)	0.1–0.5	0.1–0.5
Frequency (Hz)	1	–

Table 6.2: Instrument parameters for different heat sources [10].

Heat source	Facility	Power, (kW)	Max, power density (GW/m^2)	Pulse duration (s)	Max, loaded area (m^2)	Impact energy (keV)	Institute country
Electron beam	JUDITH	60	15	0.001-CW	0.01	150 (e)	FZ Juelich, DE
Electron beam	JEBIS	400	2	0.0001-CW	0.18	100 (e)	JAERI, Japan
Electron beam	FE200	200	60	0.001-6000	1	200 (e)	CEA Le Cresout, France
Electron beam	EB1200	1200	10.6	0.0005-CW	0.27	40 (e)	Sandia, USA
Electron beam	EBTS	30	9.5	0.002-CW	0.01	30 (e)	Sandia, USA
Electron beam	TSEFFY	60	0.2	0.01-CW	0.25	30 (e)	Efremov Inst. Russia
Electron beam	ACT	50	1	CW	0.04	30 (e)	NIFS, Japan
Electron beam	Electron beam	60	3	0.3-CW	2.5×10^{-5}	60 (e)	SWIP, China
Electron beam	EBTH	3	0.35	0.1-CW	5×10^{-5}	20 (e)	Kyushu Univ., Japan
Particle beam	PBEF	1,500	60/10	0.01-1000	0.1	50 (H, He)	JAERI, Japan
Particle beam	JET NBTB	7,500	0.25	0.005-20	0.08	130 (D)	JET, UK

Table 6.2 (continued)

Heat source	Facility	Power, (kW)	Max, power density (GW/m^2)	Pulse duration (s)	Max, loaded area (m^2)	Impact energy (keV)	Institute country
IR heater	CEF 1-2 EDA-BETA	100	8×10^{-4}	~300	<0.11 × 0.08	Infrared	Brasimone, Italy
Plasma gun	MK200UG	10^6	300	5×10^{-5}	$<3.9 \times 10^{-5}$	1–3 (ion)	TRINITI, Russia
Plasma gun	QSPA-T	3.5×10^{-5}	125	$2–5 \times 10^{-4}$	$<2.8 \times 10^{-5}$	0.1 (ion)	TRINITI, Russia

These devices are found to vary greatly in parameters and ion types, with each having different specifications and focusing on different aspects of simulation experiments. The JUDITH-1 and JUDITH-2 electron beam devices [11] of the Jülich Research Center in Germany are capable of simultaneously simulating the influence of power flux and ELMs on materials under steady-state plasma operation. Additionally, several electron beam devices have been constructed both domestically and internationally to conduct simulations of ITER's steady-state thermal load fatigue and transient thermal shock experiments, such as TSEFFY [12] of Efremov in Russia, FE200 [13] of CEA in France, and EMS-60 of the Southwest Institute of Physics of China National Nuclear Corporation. The GLADIS ion beam device at the Max Planck Institute for Plasma Physics [14, 15] in Germany can simulate the irradiation damage of a divertor under plasma; The beam parameters of the Russian steady-state plasma-like plasma acceleration device QSPA-T [16] are close to the ITER transient thermal load, allowing it to simulate the ablation mechanism of materials under the transient thermal loads. Additionally, the Russian pulse plasma gun device MK200UG [17] can measure the melting and boiling energy thresholds of W. The linear plasma device NAGDIS-II in Nagoya, Japan, conducts irradiation damage experiments on W by low-energy. Furthermore, the combination of laser devices [20] can be used to investigate the synergistic mechanism of helium plasma and transient thermal loads, as shown in Figure 6.1. In addition, the existing Tokamak facilities, such as JET and EAST, are undertaking some of the research tasks, and the under-construction ITER device will also provide scientific and technical support for fusion material research and development [21].

6.2.2 Transient thermal load damage behavior of commercial tungsten

Under the impact of transient heat loads (THLs), the tungsten surface is susceptible to a range of issues, including recrystallization, grain growth, cracking, melting, and erosion. Understanding the response mechanism of tungsten in THL experiments can

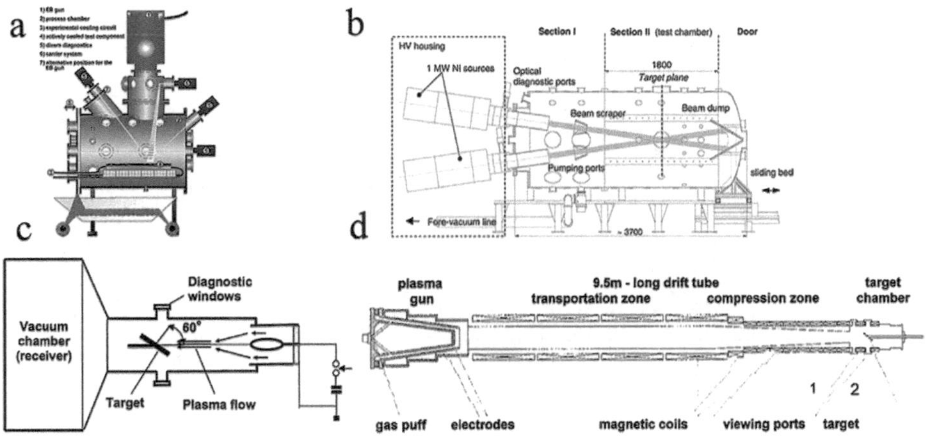

Figure 6.1: Schematic diagram of various thermal load simulation devices: (a) JUDITH-2 [11]; (b) GLADIS [14]; (c) QSPA [16]; and (d) MK200UG [17].

promote the control of plasma confinement to mitigate or even eliminate the THL in Tokamak. The THL bombardment of ELMs can lead to cracks and even failure of the tungsten surface, during which the service temperature, thermal load parameters, material types and preparation processes of the materials all play a role. As shown in Figure 6.2, the divertor covers a wide range of service temperatures from the heat sink material interface to the plasma surface, and the combined effect of high heat load and THL may cause recrystallization of the divertor component material [22], leading to a decrease in material strength, which is also one of the main causes of macroscopic cracks. Figure 6.3 shows the typical microstructure damage caused by the THL of ELMs to tungsten-based materials/components. It can be observed that cracks, whether in equiaxed or elongated grains, always originate at grain boundaries and propagate along them [23], with crack depths ranging from tens to hundreds of micrometers.

A single or small amount of thermal shock can cause surface cracks, but not necessarily thermal fatigue. The threshold for heat flux-induced cracking in tungsten (99.97 wt%, forged state) at room temperature following 100 thermal shock pulses (1 ms each) ranges from 0.19–0.38 MJ/m^2 (F_{HF} = 6–12 MW/m^2 s$^{0.5}$) to 0.5–0.6 MJ/m^2 (F_{HF} = 16–19 MW/m^2 s$^{0.5}$), contingent upon the microstructure and composition of material. Initial crack widths (<1 μm to >10 μm) typically expand with increasing thermal shock incidents but quickly reach saturation. The crack distance/density remains constant after the crack network is formed. When thermal shock cycles surpass 10^3, thermal fatigue becomes pronounced. Under these conditions, for matrix temperatures of 200 and 700 °C, the threshold for macroscopic surface damage diminishes significantly to 0.13 MJ/m^2 (F_{HF} = 6 MW/m^2 s$^{0.5}$) after 10^6 cycles [22].

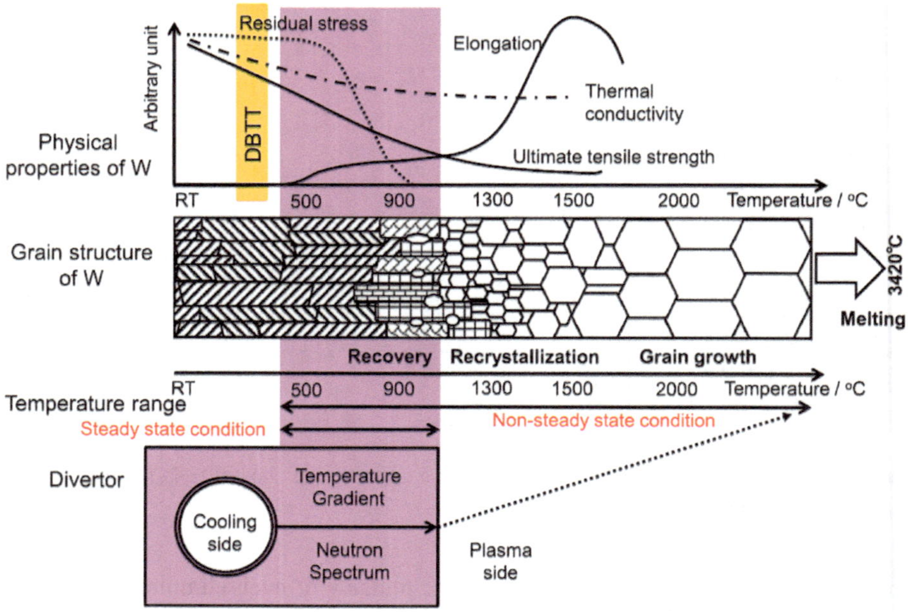

Figure 6.2: Schematic diagram of the service temperature conditions of the divertor from the heat sink material interface to the plasma surface [22].

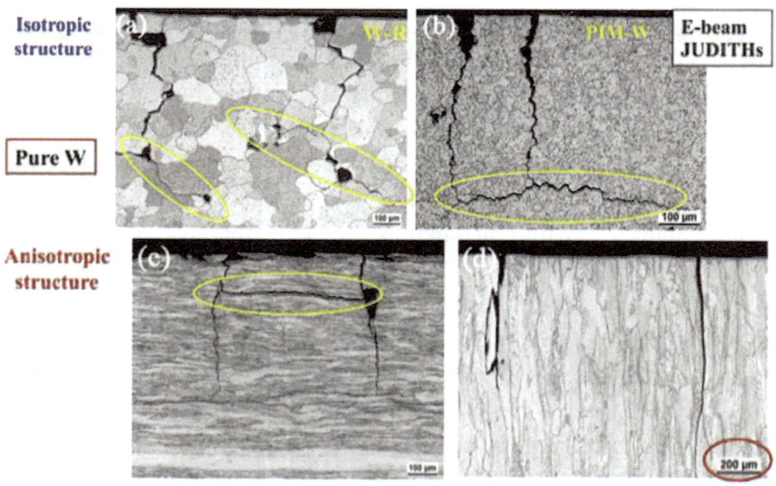

Figure 6.3: Cracks in tungsten materials under electron irradiation under simulated heat load conditions of ELMs: (a) recrystallized tungsten; (b) injection-molded tungsten; (c) rolled tungsten plate; and (d) rolled tungsten rod [23].

Figure 6.4 shows the crack formation process and mechanism during thermal shock. During the heating process, the surface temperature of tungsten rises rapidly, surpassing its DBTT (about 400 °C). At this point, the material transitions into a more ductile state. In addition, the local transient heating generates significant thermal stress. If this stress exceeds the yield strength of material, plastic deformation ensues; upon the cessation of thermal shock, the material rapidly cools, transitioning back to a brittle state below the DBTT. The repetitive nature of this deformation cycle can directly cause surface cracking. The surface of the material may become roughened with a thermal load similar to the first ELM, and the stress does not consistently result in crack release through plastic deformation. Such observations have been made in various tungsten materials subjected to thermal shock, particularly when the substrate temperature is sufficiently high to permit plastic deformation (T_{base} >100–300 °C) [24, 25].

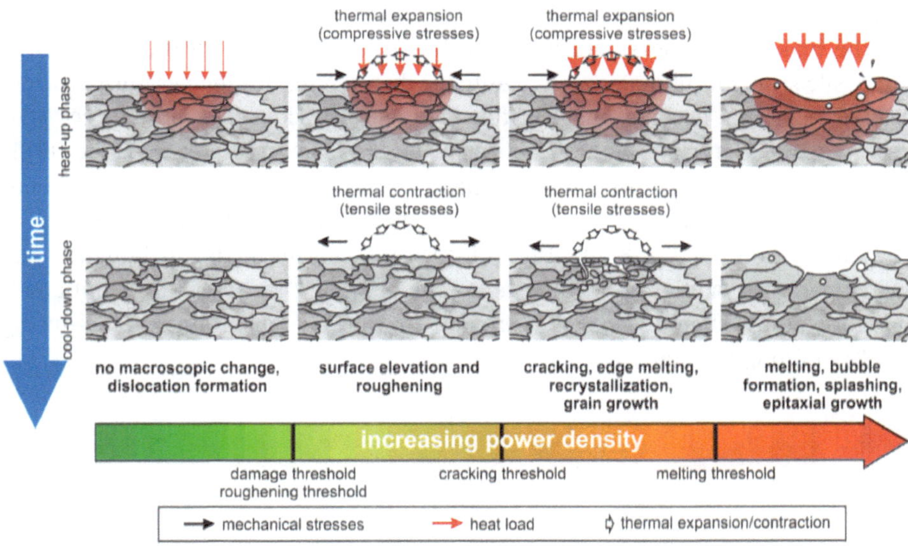

Figure 6.4: Schematic diagram of the effect of different impact strengths on the tungsten surface under thermal shock [22].

6.2.2.1 Electron beam thermal shock

Under the transient electron beam thermal shock, materials inevitably undergo processes such as cracking, melting, formation of protruding structures, recrystallization and grain growth, with crack generation being an unavoidable outcome. To simulate transient thermal shock, electron beams and lasers have been employed as simulation methods for more than 20 years [22, 26]. The diffusion depth of electrons generated by the existing electron beam devices in tungsten is limited to a few μm, which does not significantly affect the thermal load response of the material. Hirai et al. [27] studied the surface morphology of tungsten materials subjected to a single 5 ms thermal shock at different energy densities and matrix temperatures, and found that there were two main types of tungsten cracks after electron beam irradiation: main cracks and secondary cracks. As the number of irradiation pulses increases, scattered main cracks emerge on the surface. Subsequently, these scattered cracks begin to interconnect, forming a dense network of cracks. With further increases in the number of irradiation pulses, smaller cracks start to form between the main cracks.

The main crack occurs due to the brittleness of tungsten at room temperature. These cracks are typically deep and the fracture begins to occur along the grain boundaries, releasing part of the stress. Up to a certain number of irradiation pulses, the surface grain boundaries are almost completely cracked; the stress is largely relieved, resulting in minimal further changes in the density of main cracks. Secondary cracks on the other hand, are caused by thermal stress and are relatively shallow. Under the action of THLs, the surface of the material experiences rapid heating and cooling within a short period, forming a high-temperature gradient, which results in high thermal stress. Cracks naturally form when the stress exceeds the compressive and tensile strengths of material. In addition, when the service temperature is below DBTT, the material becomes more brittle, and dense cracks appear in the recrystallized area. These cracks tend to disappear when the temperature exceeds the DBTT. The surface morphology in Figure 6.5 reflects the crack situation; however, it is not cracking itself but rather material erosion due to melting that affects the lifespan of the material. Linke et al. [28, 29] also confirmed this conclusion, highlighting that the low-temperature brittleness of tungsten is the dominant factor in crack initiation. Therefore, efforts should focus on improving the working temperature and reducing the DBTT of tungsten materials.

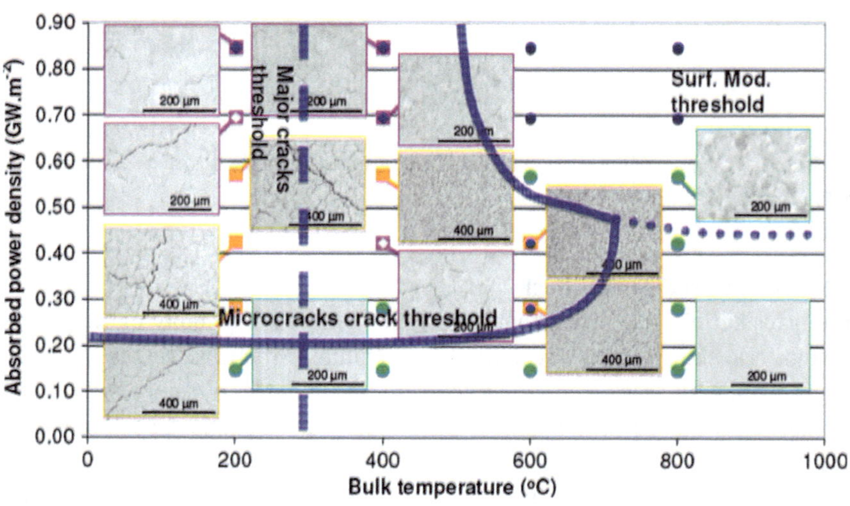

Figure 6.5: Microstructure of tungsten surface at different matrix temperatures and energy densities [27].

Wang et al. [30] studied the effects of different power and number of impacts on the surface morphology of the recrystallized tungsten under electron beam thermal shock. The thermal load power increased from 24 to 48 MW/m^2, and the change of impact fequency ranged from 100 to 1,000. The material surface became rough and the stepped protruding structure formed, which was related to the plastic deformation caused by thermal fatigue. As the number of impact cycles increased, the density of the intragranular shear bands also increased, with varying directions observed. Additionally, the grain boundary protrusion phenomenon became more pronounced. The shear bands are parallel to the grain orientation, leading to the formation of protruding structures, as illustrated in Figure 6.6. With increasing thermal shock power, the extent of surface damage intensified, primarily due to slip plastic deformation. Under transient conditions, the rapid temperature fluctuations on the material surface induce both compressive and tensile stresses. The varying orientations of different grains result in distinct slip planes, which subsequently introduce shear stress. When the shear stress reaches a critical threshold, repeated slip occurs, and the plastic deformation of the crystal slip surface ultimately leads to the formation of surface protrusions and collapsed structures.

Loewenhoff et al. [31] utilized the JUDITH 2 device to perform electron beam thermal shock experiments on tungsten at higher temperatures (1,200 °C, 1,500 °C) and higher energy densities (0.2, 0.6 GW/m^2). They introduced the average roughness (R_a) to access the thermal shock resistance of the material under different conditions. For the number of impacts from 18,000 to 100,000 with an interval of 0.48 ms, the variation trend of thermal shock damage degree of materials is almost of the similarity to the above results, with more severe damage observed at high temperatures. In the case of 100,000 impacts, experimental temperature of 1,500 °C and energy density of 0.6 GW/m^2, the surface roughness of tungsten reaches its maximum value of 119 μm, as shown in Figure 6.7. The higher the temperature, the lower the yield strength of tungsten, and the more severe the plastic deformation under the same conditions.

6.2.2.2 Laser thermal shock

The laser could be used to help with high-energy instantaneous local high-speed heating, suitable for simulating the impact resistance of materials under transient thermal shock conditions. Dai et al. [32] conducted related experiments using laser beam thermal shock. As shown in Figure 6.8, the energy distribution of the laser beam follows a Gaussian distribution, and the smaller the beam spot, the more concentrated the energy and the greater the energy density. Figure 6.9 depicts the surface topography of the tungsten specimen after the laser impact experiment, where the impact area is visibly melted and exhibits large cracks throughout. The impact process subjects the material's radial stress and circumferential stress to a compressive stress state. During cooling, these stresses shift to a tensile stress state, which attains the material's maximum tensile strength and subsequently triggers the formation of cracks.

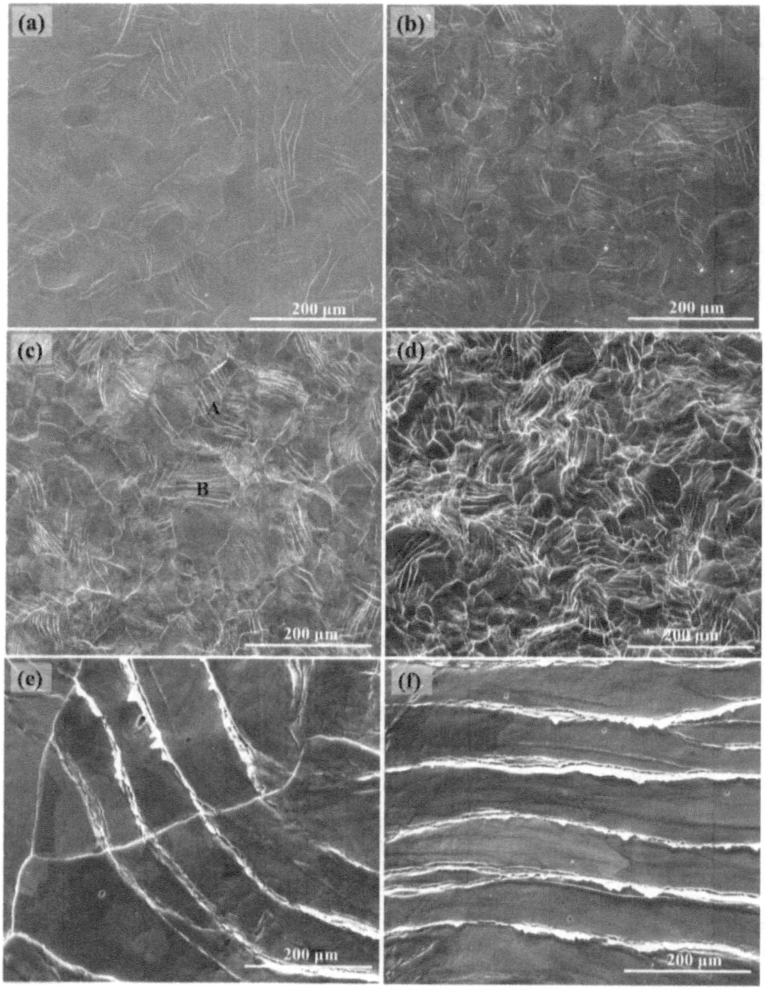

Figure 6.6: SEM diagram of surface morphology of recrystallized tungsten with an absorption power of 48 MW m^{-2} with different impact times: (a) 100 times; (b) 300 times; (c) 500 times; (d) 1,000 times; (e, f) are the high-magnification plots of the grains in (c) [30].

Farid et al. [33] studied the surface morphology and sputtering behavior of tungsten under the action of high-energy laser pulses to simulate the conditions experienced by the first wall tungsten material during ELMs. It was found that some areas of the tungsten surface melted and subsequently recrystallized, and cracks with widths ranging from 150 to 500 nm appeared on the tungsten surface under 50 laser pulses with a beam energy density of 0.46 MJ/m^2, as shown in Figure 6.10. The formation of these cracks is attributed to the combined effects of the thermal fatigue stress and the high DBTT of tungsten.

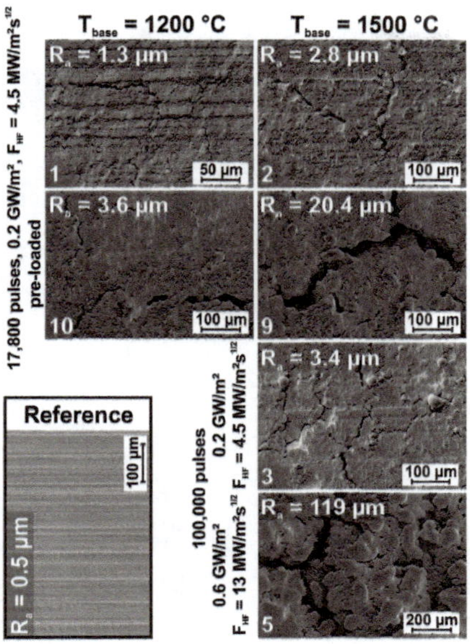

Figure 6.7: SEM images of the surface morphology of the material under different experimental conditions (the lower left is the reference image without thermal shock, and R_a is flat average roughness) [31].

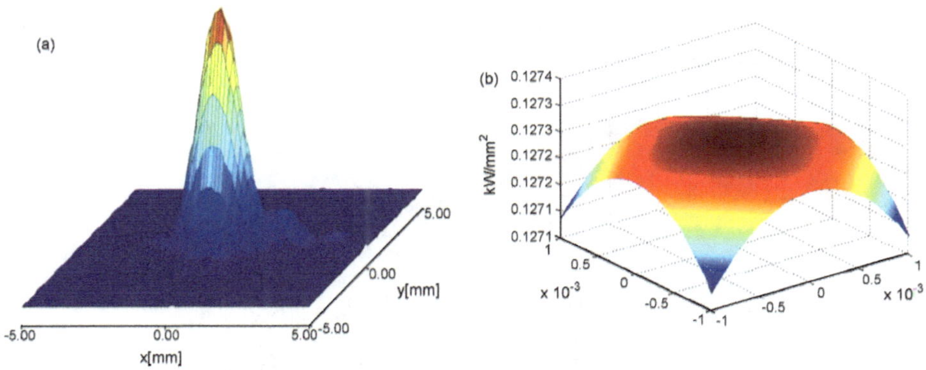

Figure 6.8: The results of laser impact simulation: (a) spatial morphology of the laser beam and (b) spatial morphology of laser energy density [32].

Wirtz et al. [34] studied the thermal load damage behavior of ultra-high-purity single-wrought tungsten (surface area of 12 mm × 12 mm, the thickness of 5 mm) under irradiation of an electron beam and a pulsed laser beam. The pulse duration was 1 ms, and 100 pulses were applied at power densities of 0.19, 0.38, 0.76, 1.14, and 1.51. Figure 6.11 shows the surface morphology after the thermal load experiment, indicating that the damage thresholds for tungsten material are within the range of 0.19–0.38 GW/m^2.

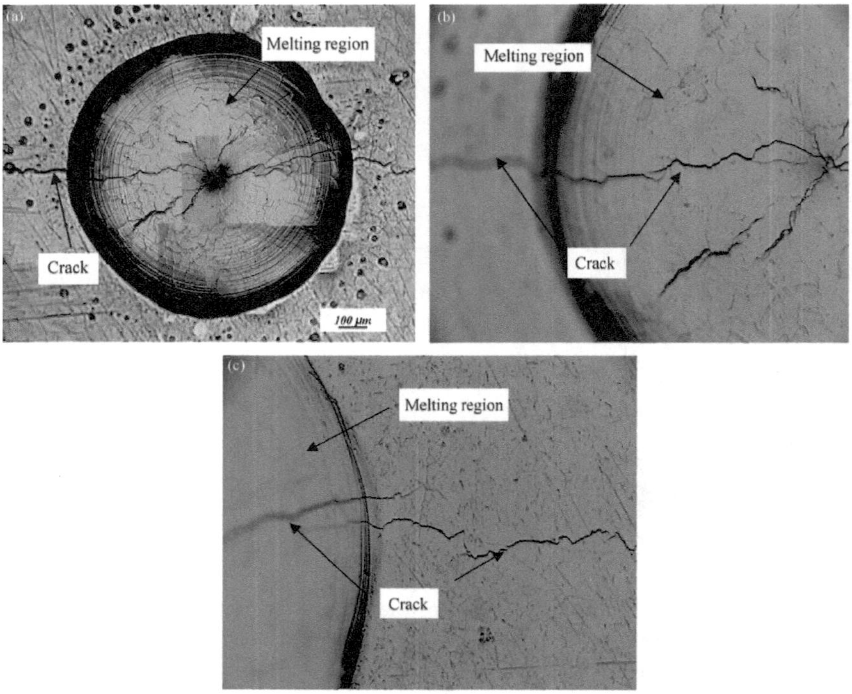

Figure 6.9: SEM diagram of tungsten specimen after laser impact test: (a) cracks in the center of the melting zone; (b) cracks at the edge of the melt zone; (c) cracks outside the melting zone, the laser energy $P_0 = 2$ kW, the laser beam radius $r_0 = 0.539$ mm, and the duration $t_0 = 1$ s [32].

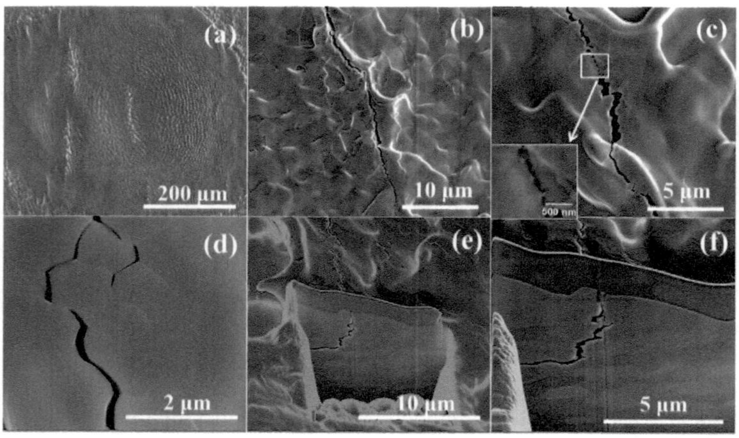

Figure 6.10: (a) Surface topography of tungsten under 50 laser pulses of 0.46 MJ/m^2; (b) cracks generated near the melting region with crack widths of 150–500 nm; (c, d) cracks generated in the recrystallization region; and (e, f) FIB cross sections of cracks in the recrystallization region [33].

However, there is a difference in the crack distribution between the two methods, which may be attributed to the size of the loading area and the energy factor carried. The loading area of the focused electron beam is 4 mm × 4 mm, while that of the pulsed laser beam is a 2 mm diameter circle. Furthermore, only 27% of the laser energy from the pulsed laser beam can be absorbed by the sample, due to losses in laser fiber transmission and reflection from the sample surface.

In addition to analyzing the crack parameters, Wirtz et al. also examined the crack extension mode. The cross-sectional morphology of the sample after laser beam and electron beam loading at the same power density is shown in Figure 6.12. Under both loading modes, the crack initially propagates perpendicular to the surface of the material sample and then terminates at a certain depth. Subsequently, the cracks propagate parallel to the surface within the material. The crack formation is due to the elongated grain structure parallel to the surface, which is produced during material processing. Compared to the electron beam, the laser beam has a smaller loading area, resulting in finer cracks within the material. This, in turn, leads to changes in the distribution of thermal stresses, including the temperature gradient distribution and the location of maximum stress.

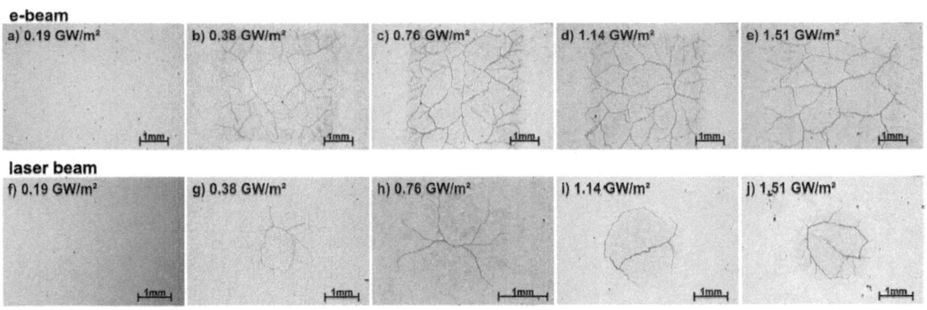

Figure 6.11: SEM images of electron and laser beams simulating the loading surface of ELMs after thermal shock [34].

6.2.2.3 Plasma thermal shock

Budaev et al. [35] utilized a tritium plasma beam to conduct a simulated heat load test on rolled commercial tungsten (sample surface area of 100 mm × 10 mm, thickness of 2 mm). They applied a heat flux of 2 MJ m^{-2} for 0.5 ms, corresponding to a heat flux factor of 90 MJ/($m^2 \cdot s^{0.5}$), which exceeded the material's melting threshold. Figure 6.13 presents the SEM image of the tungsten cross section after two pulses, revealing a notable columnar crystal structure on the surface. Moreover, the resolidified layer on the surface has a thickness of approximately 50 μm. This layer gives rise to submicron-scale particles as the consequence of brittle damage from fracture.

Unlike the case when the number of applied pulses is 2 when the number of applied pulses exceeds 50, the corresponding image is shown in Figure 6.14. Arrow 1 in-

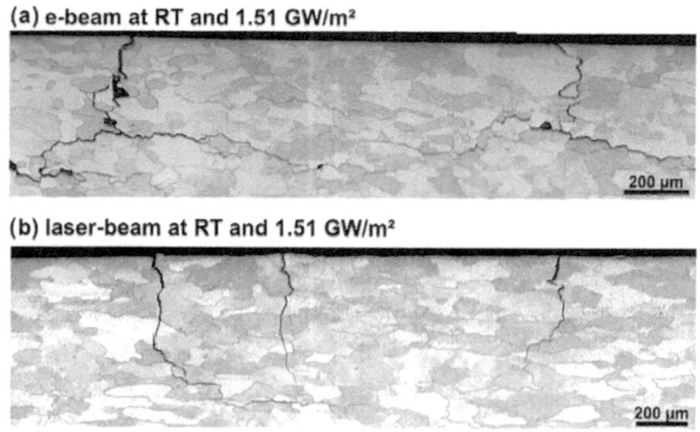

Figure 6.12: Cross-sectional morphology of the sample after loading with an electron beam and a laser beam at 1.51 GW/m^2 [34].

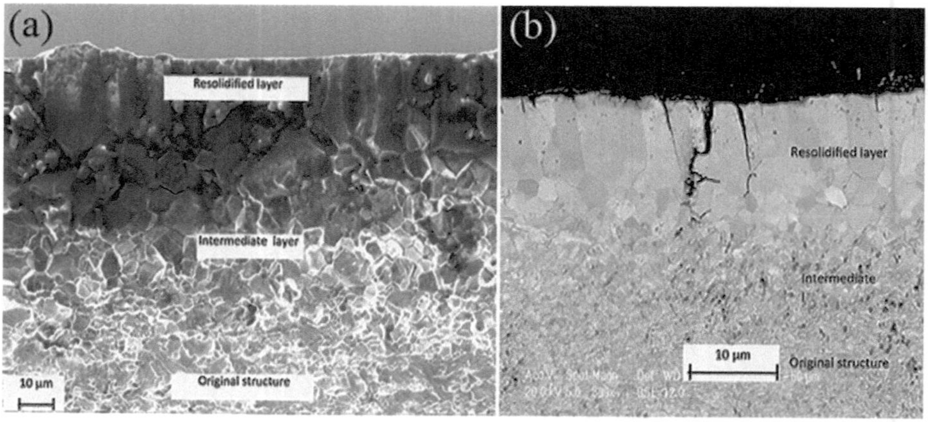

Figure 6.13: Metallographic cross section of W surface after energy density with 2 MJ/m^2, $F_{HF} \approx 90$ MJ/m^2 s$^{0.5}$ tritium plasma impacts on a QSPA-T device, with the number of impacts $N = 2$ and the duration $\Delta t = 0.5$ ms [35].

dicates, the crack initiates in the resolidification layer on the surface and extends along the interface between the surface and the middle layer, as shown in arrow 2. Additionally, there is a large-size crack inside the substrate (about 300 μm from the surface). The rapid solidification of the material between pulses leads to longitudinal cracking of the reconsolidated layer under thermal load pulses, which result in further damage of the sample after a certain number of pulses, and in turn produces visible transverse cracks between the reconsolidated layer and the intermediate layer. Furthermore, the movement of dislocations under stress results in their linear arrangement, leading to obvious transverse cracking within the matrix.

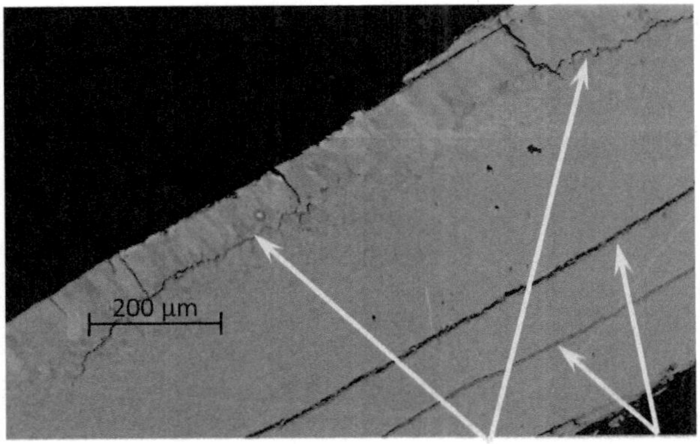

Figure 6.14: Metallographic cross section of tungsten surface on a QSPA-T device with an energy density of 2 MJ/m^2, $F_{HF} \approx 90$ MJ/m^2 s$^{0.5}$ tritium plasma, the number of impacts $N > 50$, and the duration $\Delta t = 0.5$ ms [35].

6.2.2.4 Plasma/laser synergistic thermal shock

Compared with the simple electron beam and laser thermal shock, the combination of background plasma and laser simulates the THL, closely mimicking the actual service environment of the fusion device. Since the plasma gun does not provide steady-state background plasma, the linear plasma device can be coupled with the laser to simulate the transient thermal load of ELMs, for example, using PISCES, and PSI-2. Alternatively, the linear plasma device itself can be upgraded to obtain a pulsed plasma source, as seen in advanced systems like Magnum-PSI, or Pilot-PSI [36, 37]. An alternative approach is to study the synergistic effects between the linear plasma device and the laser by sequentially loading plasma and transient heat flux (implemented on different devices if necessary) [38–40].

Huber et al. [41] utilized the PSI-2 device (Figure 6.15) to study the surface morphology changes of commercial tungsten, with a purity of 99.5% (sample surface area of 12 mm × 12 mm and a thickness of 5 mm) under the simultaneous and alternating application of laser beams and deuterium plasma. Due to the forging and rolling process, the grains are preferentially oriented, with a grain length of approximately 110 µm and a width of 40 µm, perpendicular to the loading surface. According to Zhang et al. [42], cracking behavior tends to occur perpendicular to grain orientation under a certain heat load. Figure 6.16 shows the surface morphology of a commercial tungsten sample after applying 1,000 pulses with pulse duration of 1 ms, a power density of 0.3 GW/m^2 laser beam, along with a deuterium ion stream of 6.0×10^{21} m^{-2} s^{-1} for 2,000 s. The left figure shows that there are no cracks on the surface after the laser beam and deuterium plasma are applied sequentially. The middle and right figures reflect the changes in surface morphology when deuterium plasma and laser beam are applied together and when deuterium plasma is pre-applied before the laser beam, respectively, both scenarios promote crack formation and surface roughening due to the diffusion of the pre-applied deuterium plasma into the tungsten lattice. Lattice defects such as vacancies, dislocations, and grain boundaries become pinned, leading to lattice distortion and the generation of internal stresses. These additional stresses reduce the critical stress for crack formation and propagation.

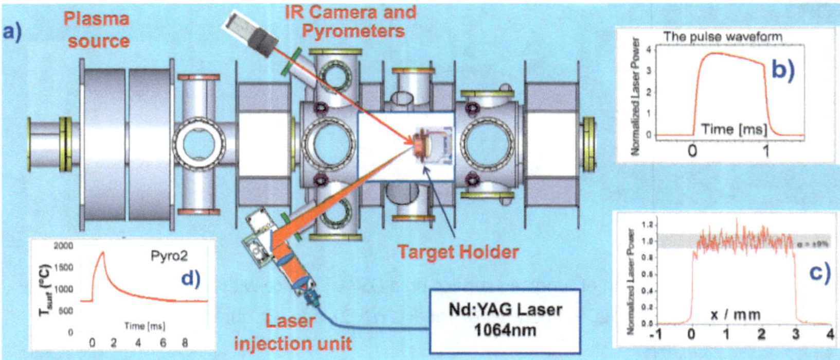

Figure 6.15: Schematic diagram of a device for simulating ELMs by connecting the PSI-2 laser irradiation of the FZJ to a plasma source [41].

6.2.2.5 Plasma/electron beam synergistic thermal shock

Tamura et al. [43] studied the damage behavior of high-purity tungsten under the interaction of hydrogen plasma beam and electron beam. The damage behavior such as cracking and roughening was more significant than that of the sample under electron beam load, as shown in Figure 6.17. After 100 times pulse irradiation, no obvious damage was observed on the surface of the two samples. However, after 250 times irradiation, a large crack with a length of 4 mm appeared on the surface of the sample, while only a fine crack with a length of 50 μm appeared on the surface of the sample irradiated by the electron beam. In addition to performing relevant physical experiments, Pestchanyi et al. [44] used the PEGASUS-3D thermo-mechanical module to simulate the cracking behavior of tungsten surface due to thermal stress in the resolidified surface layer after irradiation. The morphology of the experimental and simulated crack propagation profiles is qualitatively consistent, both characterized by the obvious primary crack network with the interior filled by a finer secondary crack network, as shown in Figures 6.18 and 6.19.

Pestchanyi et al. also analyzed the crack distribution of the material in the vertical cross section. The cross-sectional morphology of the tungsten tile after the experiment is shown in Figure 6.20(a), where the characteristic depth of the main crack is about 500 μm, and the average distance between the cracks is 1–2 mm. As shown in Figure 6.20(b), the average depth of secondary cracks is about 50 μm, and the average distance between cracks is 100–200 μm. Figure 6.20(c) shows the simulation results, which fully reproduce the crack distribution characteristics of the cross-section.

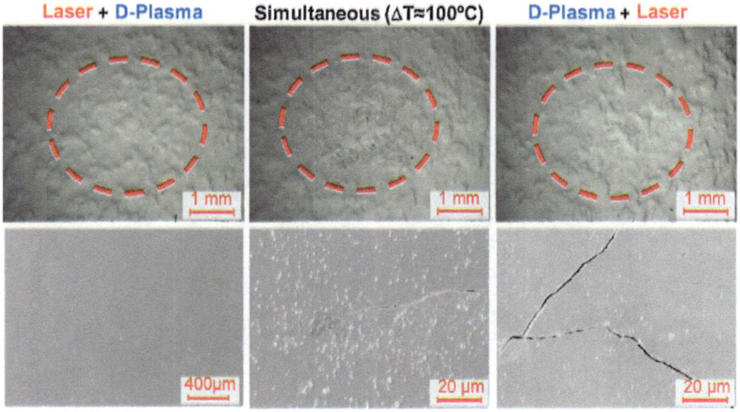

Figure 6.16: Surface morphology of tungsten irradiated by laser and plasma in different irradiation sequences and at the same time [41].

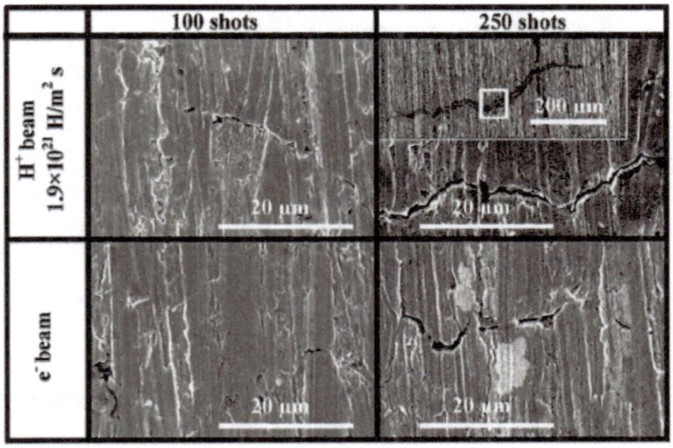

Figure 6.17: Surface morphology of tungsten under cyclic heat load of hydrogen beam and electron beam [43].

6.2.2.6 Laser/high magnetic field synergistic thermal shock

Considering that PFMs need to work in a high magnetic field environment in an actual fusion reactor, Qu et al. [45] incorporated high magnetic field interference into the laser impact experiment. The experimental conditions included an energy density of 3.4 GW/m^2, several 1,200 shocks of 3 ms each, a frequency of 10 Hz, and the magnetic field strength parameters of 0, 4, and 8 T, respectively. In addition to observing the melting zone and cracks on the surface of the specimens after laser impact, the depth and shape of the melting zone varied under different magnetic field strengths. As the magnetic field strength increased, the edge depth and melting area decreased, as shown in Figures 6.21 and 6.22. The presence of a strong magnetic field inhibits the

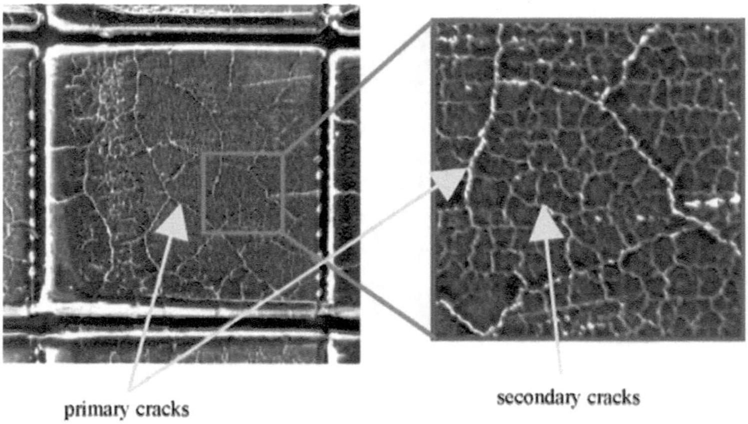

Figure 6.18: Surface morphology of tungsten tile after heat load of 0.5 ms, 100 plasma impacts, and power density of 0.9 MJ/m^2 [44].

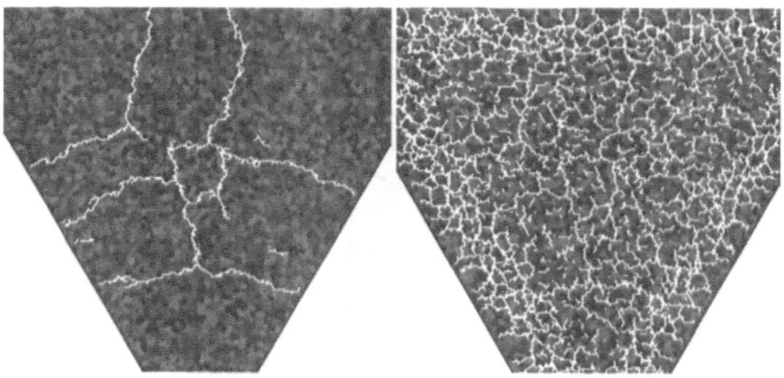

Figure 6.19: Simulated crack propagation morphology of a pre-irradiated tungsten sample after a heat load of 0.5 ms [44].

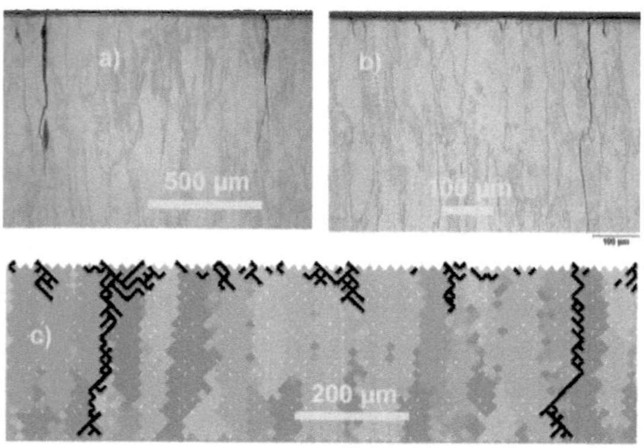

Figure 6.20: Cross-sectional morphology of the pre-irradiated tungsten sample after 0.5 ms and 100 times of plasma shock heat load (power density of 0.9 MJ/m^2) [44].

heat transfer in the horizontal direction of the material, altering the temperature field, which in turn affects the stress field and thermal diffusion of the material.

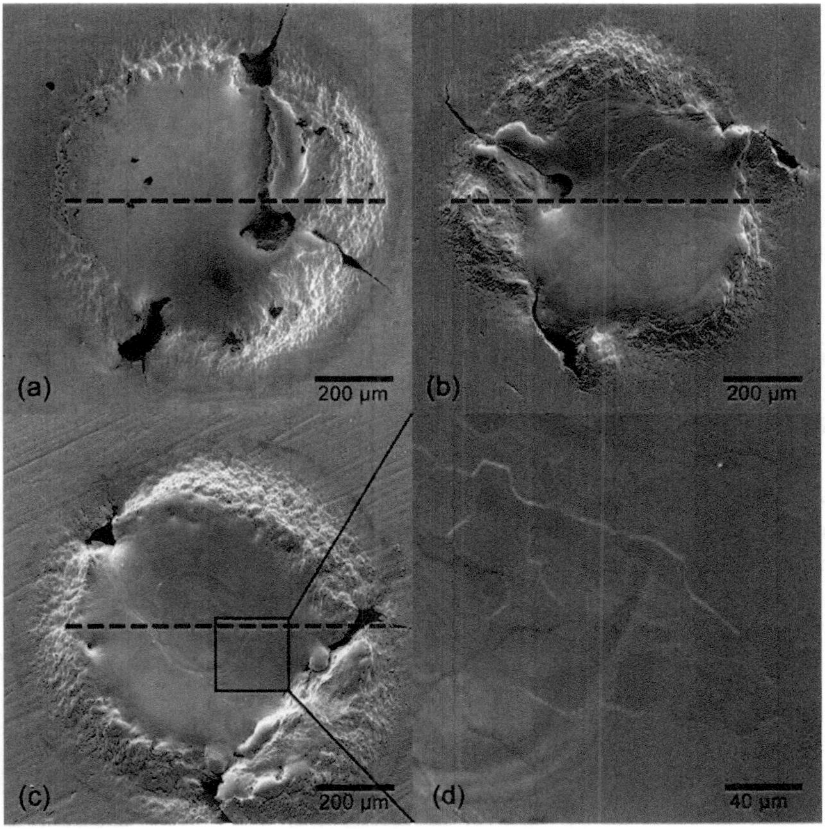

Figure 6.21: Surface diagram of the melting zone of the material under different magnetic field strengths: (a) 0 T; (b) 4 T; (c) 8 T; and (d) coarse grains [45].

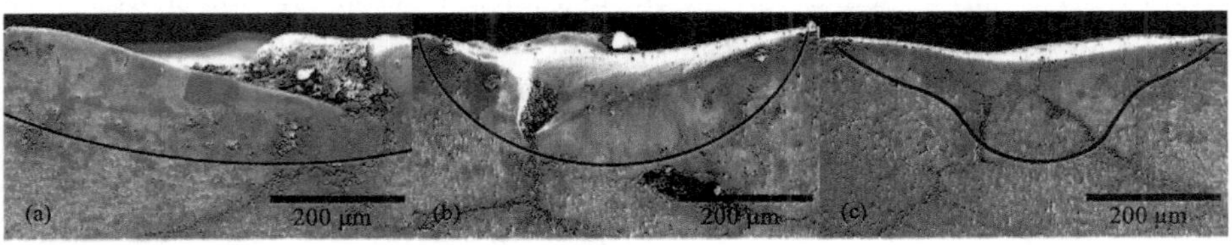

Figure 6.22: Cross-sectional view of the melting zone of the material under different magnetic field strengths: (a) 0 T; (b) 4 T; and (c) 8 T [45].

6.2.3 Transient thermal load damage behavior of ultrafine-grained tungsten

The development of ultrafine-grained tungsten is proposed to improve the irradiation resistance of tungsten. Ackland [46] proposed the self-healing mechanism of nanomaterials under irradiation damage, providing valuable insights and a reference for the study of tungsten irradiation damage. Further research is needed to investigate the thermal load behavior of ultrafine-grained tungsten.

Zhou et al. [47] prepared ultrafine-grained tungsten samples with average grain sizes of 0.2, 1, and 10 μm using resistance sintering under ultra-high pressure. They studied the ability of tungsten samples with different grain sizes to resist thermal loads. Thermal load conditions: sample size of 15 mm × 5 mm, single pulse application with a pulse duration of 5 ms, and a spot size of 4 mm × 4 mm. As the tungsten grain size decreased, the microhardness and bending strength increased significantly, while the thermal conductivity and density showed a decreasing trend. Compared with commercial tungsten, ultrafine-grained tungsten exhibits a more severe cracking phenomenon at the same power density [48, 49]. As shown in Figure 6.23, W02 and W10 exhibit more significant cracking at the same power density compared with W100, and droplet splashing is evident on the surface. The reason is that there are large number of gases and impurities remaining in the ultrafine-grained tungsten. Under heat load, the expansion of these residual gases causes very high stress inside the material. Additionally, impurities tend to enrich at the grain boundaries after sintering, lowering the melting point of the grain boundaries. This reduces the material's ability to resist thermal loads, leading to local overheating. Consequently, surface melting and droplet splashing occur even when the surface heat flux is below the melting threshold.

Zhang et al. [50] studied the surface cracking phenomenon of two kinds of grain size sintered pure tungsten (PW_1 and PW_2) under high heat loads. PW_1 has a larger grain size with a bimodal distribution of 1.85 μm and 0.47 μm, while PW_2 has an average grain size of 0.33 μm. As shown in Figure 6.24, only the surface of the PW_2 sample formed microcracks. The grain separation at the main crack of the PW_2 sample increased the crack width, whereas PW_1 exhibited a narrower crack width compared with PW_2. The surface damage of the small-grained material PW_2 was more severe after the heat load test, which is consistent with the results of Figure 6.25.

In addition to the experimental investigation, Du et al. [51] used the transient thermal simulation module of ANSYS software to analyze tungsten with different grain sizes. The fatigue life of a material, defined as the number of cycles during which a material can withstand the corresponding THL without surface cracking, is shown in Table 6.3. These results are purely computational, based on the assumptions of relevant properties, and show that the fatigue life of the material decreases with the increase of grain size, that is, the ability of the material to resist the THL decreases, which is the opposite of the experimental phenomenon. However, the corresponding fatigue life shows an upward trend with the increase of the preheating tem-

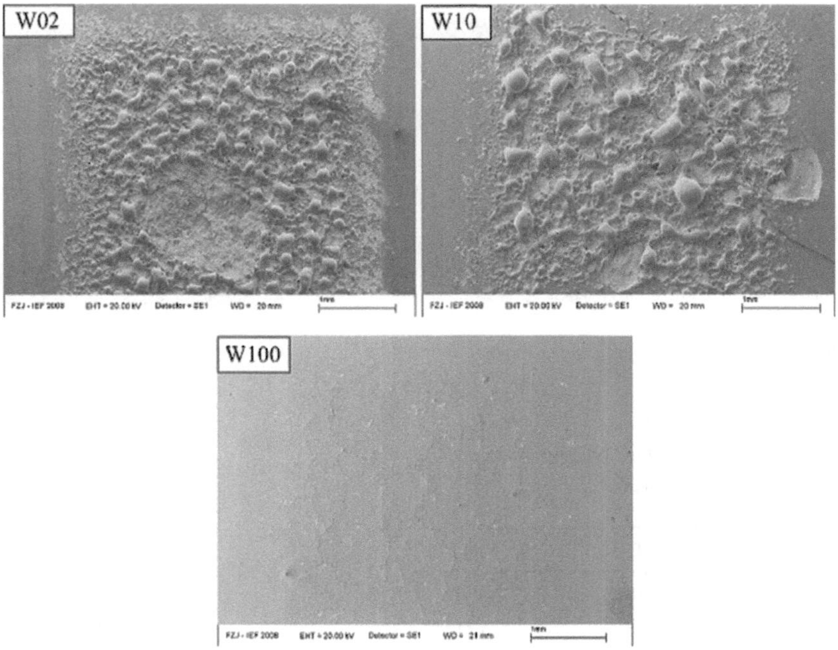

Figure 6.23: SEM image of tungsten with different grain sizes after being subjected to a heat load of 0.55 GW/m^2 [47].

perature, that is, as the temperature rises, the capability of the material to resist the THL improves and is consistent with the results of the study of Wirtz et al. [52].

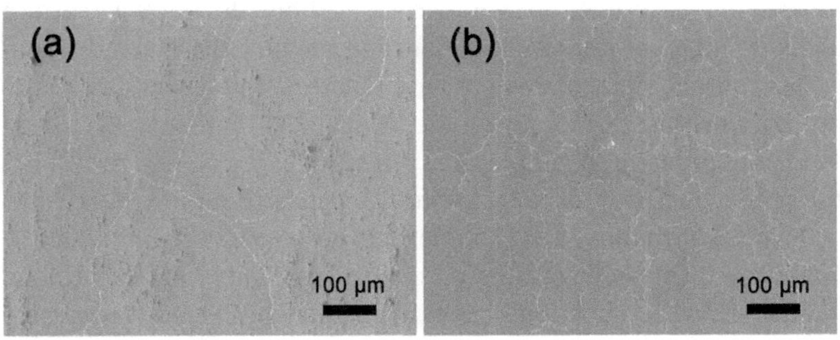

Figure 6.24: Low-magnification surface morphology of one time 0.22 GW/m^2 electron beam thermal shock at room temperature for 5 ms: (a) PW_1 and (b) PW_2 [50].

It can be concluded that ultrafine-grained pure tungsten exhibits poor resistance to thermal loads.

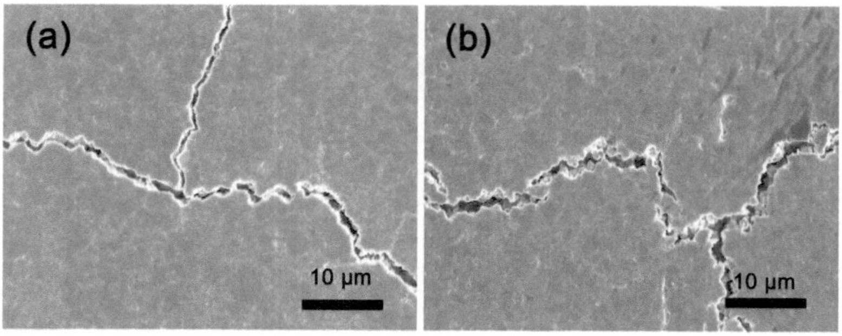

Figure 6.25: High-magnification surface morphology of one time 0.22 GW/m^2 electron beam thermal shock at room temperature for 5 ms: (a) PW_1 and (b) PW_2 [50].

Table 6.3: Prediction of fatigue life of different types of tungsten after applying a heat load of 1.14 GW/m^2 [51].

	RT	400 °C
Rolled tungsten	928	1141
50 µm tungsten	736	913
100 µm tungsten	182	260

6.2.4 Transient thermal load damage behavior of tungsten alloy

Tungsten added with certain alloying elements would obtain a strengthening effect, enhancing the material's density, strength, toughness, and thermal stability, as well as improving its irradiation performance. High thermal load hits the tungsten surface layer, generating a strong temperature field that causes severe thermal stress and rapid cracking. This cracking situation becomes even more severe if tungsten is subjected to a heat load when the temperature is below its DBTT. Tungsten has a high DBTT, so it's especially prone to strong thermal stresses during plasma operation, which may cause severe cracking. One notably effective approach for improvement is alloying, which involves forming a solid solution between tungsten and atoms of other elements that have high melting points, good toughness, and low reactivity. The following sections are about the impact of several representative alloying elements on the thermal load performance of tungsten materials.

6.2.4.1 W-K alloys

Potassium-doped tungsten exhibits excellent creep and relaxation resistance properties and enhances the strength and high-temperature toughness of tungsten. These improvements are attributed to the formation of potassium bubbles within tungsten-doped materials. Potassium bubbles may react with tungsten lattice defects, and pinning dislocations hinder the movement of grain and subgrain boundaries [53]. Linke et al. [54] performed 100 pulses of ELM simulation experiments on W-K alloy with 15–40 ppm potassium content, using ultra-high-purity tungsten (W-UHP) and forged tungsten for comparison. As shown in Figure 6.26, both the W-K alloy and pure tungsten exhibited the same roughening threshold of 0.2 GW/m^2 (green bar). Pintsuk et al. [55] performed a high heat flux testing on a fully recrystallized W-K alloy (with a potassium content of less than 22 ppm) annealing at 1,800 °C for 2 h. Under a power density of 0.32 GW/m^2 applied over 100 pulses, no damage was observed. However, after 1,000 pulses, the material began to crack, and by 10,000 pulses, the crack density had stabilized, as illustrated in Figure 6.27. Zhang et al. [56] prepared W-K alloys (with a potassium content of 70–75 ppm) using forging and rolling methods. When a single pulse was applied to the W-K sample at room temperature, the cracking threshold of the W-K alloy was found to be between 0.44 and 0.66 GW/m^2.

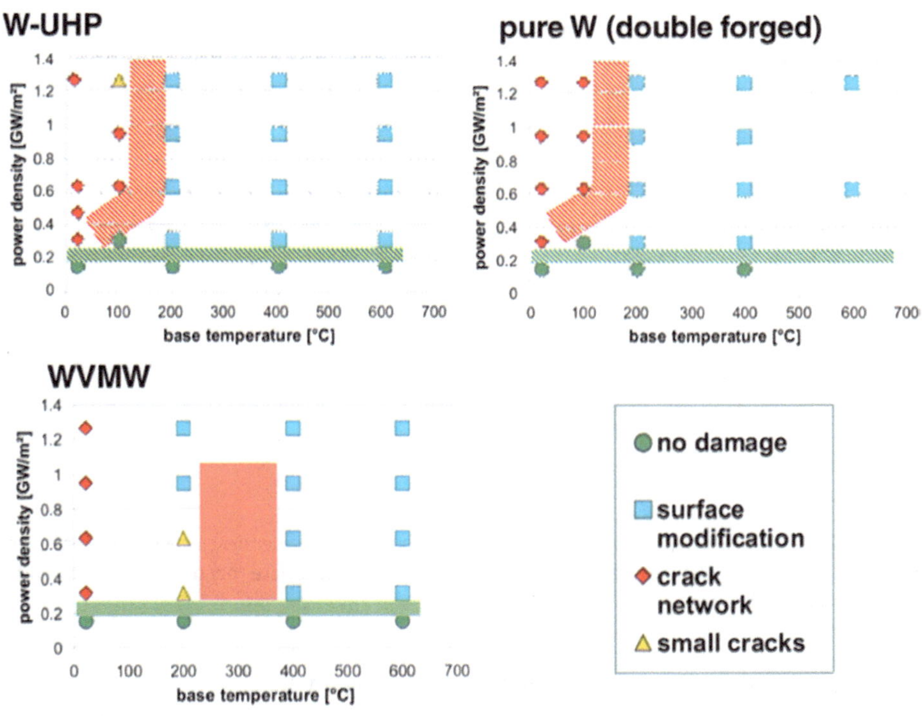

Figure 6.26: Surface damage of W-K alloy after 100 electron beam thermal shocks at different matrix temperatures on JUDITH1 equipment [54].

Huang et al. [57] studied the thermal load properties of W-K alloys with different compositions prepared by conventional sintering and Spark Plasma Sintering (SPS) method. The W-K alloy with a potassium content of 37–82 ppm was sintered by conventional sintering. In a single pulse experiment with a power density of 0.37 GW/m^2

and pulse duration of 5 ms, the sample dimensions were 15 mm × 2 mm and the spot size was 4 mm × 4 mm. As the potassium content increased, the depth and width of the crack decreased, as shown in Figure 6.28. Although potassium doping does not significantly increase the cracking threshold of tungsten, the damage after thermal shock becomes smaller as the amount of potassium doping increases. This indicates that the potassium bubble hinders the damage of tungsten under thermal load and cannot completely prevent the cracking of the material under thermal load.

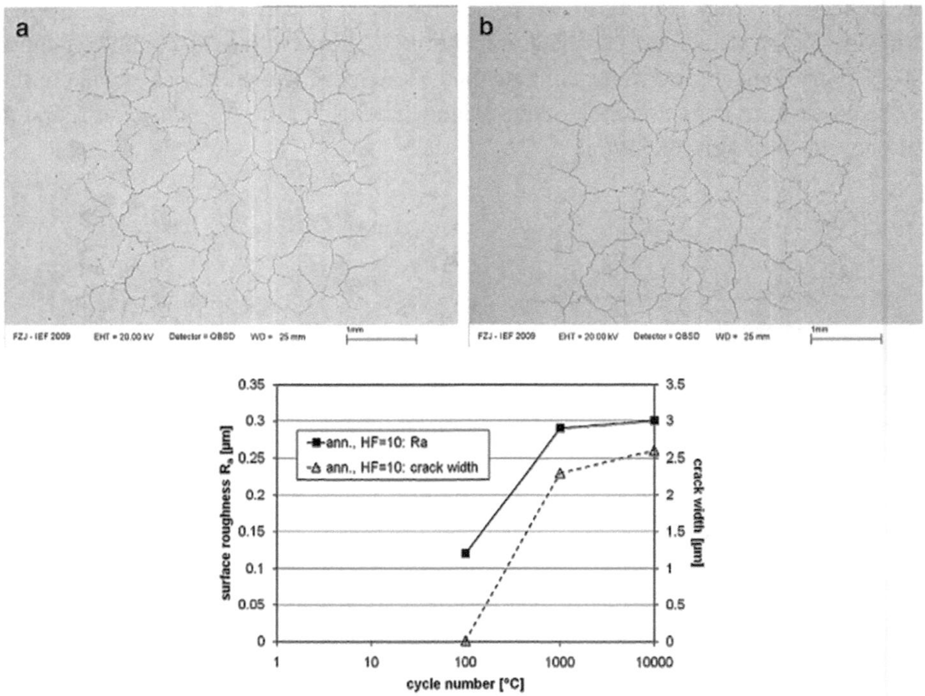

Figure 6.27: SEM images of electron beam after thermal shock with a power density of 0.32 GW/m^2, pulse duration of 1 ms: (a) 1,000 times; (b) 10,000 times; and (c) variation of surface roughness and crack width with the number of pulses [55].

To compare with the traditional sintering, W-K alloy a potassium content of 60 ppm was prepared by SPS sintering and the thermal load test was conducted under the same experimental parameters. Figure 6.29 shows the fracture morphology and surface morphology of the SPS sintered W-K alloy after a thermal shock of 0.37 GW/m^2. The area marked by the black line in Figure 6.29a shows an obvious transgranular fracture. In Figure 6.29b, the distribution of potassium bubbles inside the grain can be observed, which differs from traditional sintering. Because the traditional sintering holds for a long time, which makes the potassium bubbles migrate to the grain boundaries, making it difficult to observe the presence of potassium bubbles in the

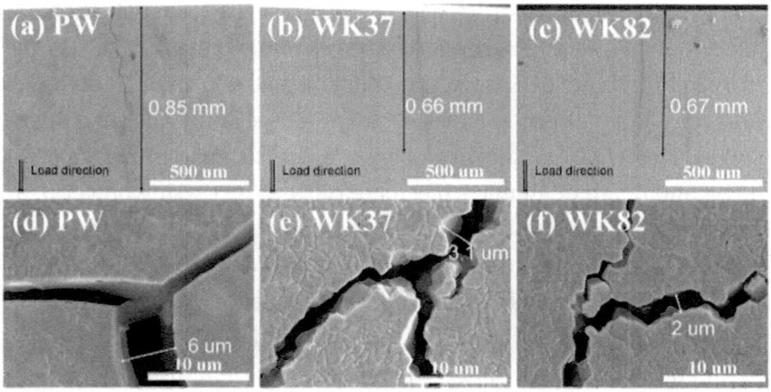

Figure 6.28: Cross-sectional and cracked surface morphology of a traditional W-K alloy after a single pulse impact of 0.37 GW/m^2 with a pulse duration of 5 ms [57].

grain [58]. Figure 6.29c shows the surface morphology after the thermal load test, indicating that there is no obvious change on the surface of the sample and the increased cracking threshold of the material surface. Compared with the W-K alloy prepared by traditional sintering, SPS sintering has improved resistance to thermal load, which has valuable insights for future efforts aimed at enhancing the thermal load performance of W-K alloy.

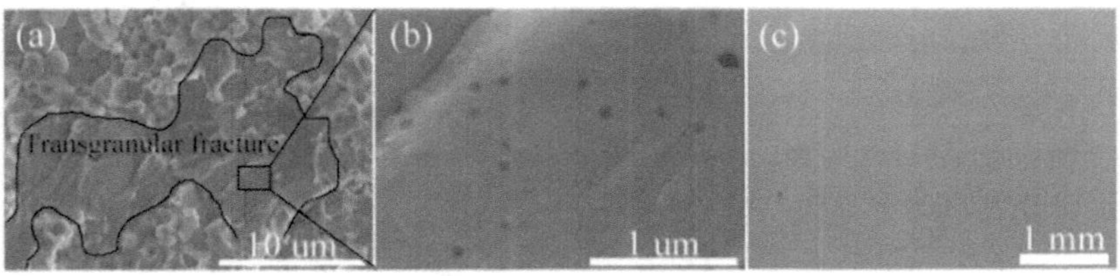

Figure 6.29: Fracture morphology of SPS sintered W-K alloy and surface morphology after 0.37 GW/m^2 thermal shock [57].

6.2.4.2 W-Ta alloy

Tantalum (Ta) element exhibits good plasticity, low activation, and high irradiation resistance, and is not easy to generate a brittle phase during irradiation. The addition of a small amount of Ta can significantly enhance the toughness of tungsten materials, making W-Ta alloys highly promising for use in the divertor of future fusion devices.

Linke et al. [54] performed 100 pulses of ELMs simulation experiments on W-Ta alloys with different Ta contents and the results are shown in Figure 6.30. WTa1 and WTa5 represent the mass fractions of Ta of 1% and 5%, respectively. The results show that WTa1 and W-K alloys exhibit the same roughness threshold of 0.2 GW/m^2. However, as the Ta content increases, the ability of WTa5 samples to resist transient thermal shock is significantly enhanced, with the roughness threshold exceeding 0.4 GW/m^2 and the cracking threshold exceeding 0.6 GW/m^2. Wirtz et al. [59] investigated the crack parameters and types after thermal shock at room temperature for ultra-high-purity W (W-UHP), WTa1, and WTa5, as shown in Figure 6.31. As shown in Figure 6.31a and b, the crack spacing and depth of all materials increase with increasing power density. The crack parameters of W-UHP and WTa1 are essentially identical, but the crack spacing and depth of WTa5 are significantly smaller in each power density. Figure 6.31c shows that the cracks of both W-UHP and WTa5 propagate perpendicularly to the thermal loading surface after the thermal shock test of 1.27 GW/m^2, and then stop at a certain depth and subsequently propagate parallel to the thermal loading surface. The crack propagation depends on the grain size. W-UHP, with its larger grain size, primarily exhibits grain boundary cracks that propagate along the grain boundaries, while WTa5, with its smaller grain size, predominantly features transgranular cracks, as shown in Figure 6.32.

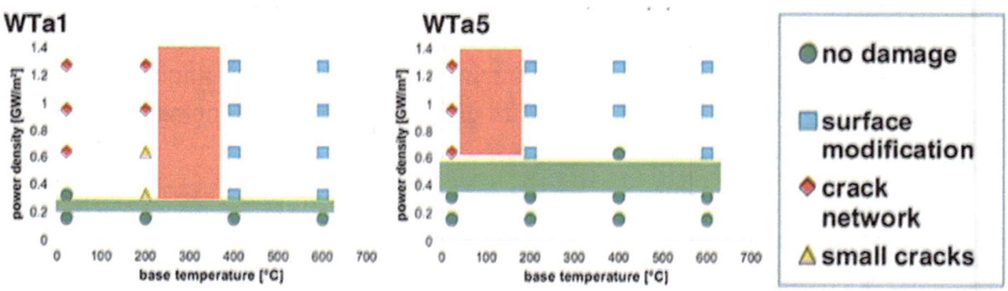

Figure 6.30: Surface damage of W-Ta alloy after 100 electron beam thermal shocks at different matrix temperatures on JUDITH1 equipment [54].

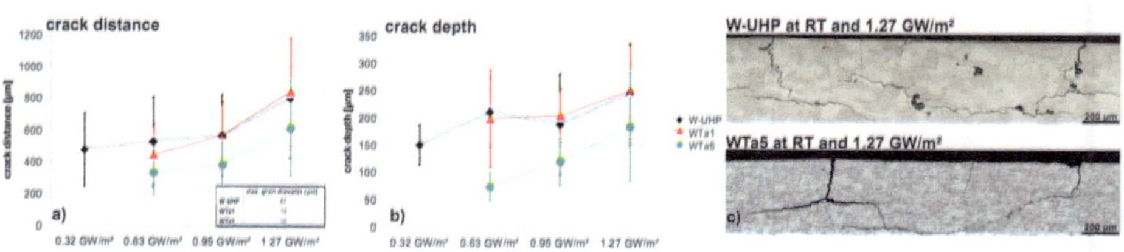

Figure 6.31: Crack parameters and types of ultra-high-purity tungsten (W-UHP), WTa1, and WTa5 after thermal shock at room temperature: (a) crack spacing and (b) crack depth change with power density; and (c) cross section after thermal shock at 1.27 GW/m^2 for 100 times and pulse duration of 1 ms [59].

Wirtz et al. believed that the thermal and mechanical properties of the material affect the thermal shock behavior of tungsten alloy. On the one hand, alloying improves the compressive and tensile strength of the material, thereby increasing the damage threshold during the thermal shock. As the power density increases, the surface tem-

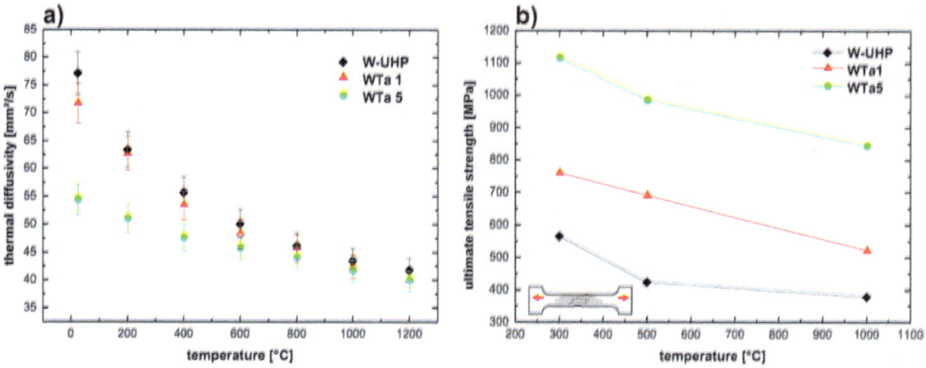

Figure 6.32: Variation curves of the thermal diffusion coefficient and tensile strength with temperature for W-UHP, WTa1, and WTa5 [59].

perature and strain rate of the sample also rise during each thermal shock process, and the increased fracture toughness of the material can offset some of the thermal stresses. On the other hand, the reduced thermal diffusivity of the material leads to an extremely high temperature gradient at the surface, preventing thermal stresses from penetrating to deeper regions during the thermal shock process, thus reducing the depth of crack propagation.

6.2.4.3 W-V alloys

Tungsten doped with vanadium can greatly improve the material properties, such as the microstructure of the tungsten matrix, increase the density and enhance mechanical properties. With the increase of vanadium content, the fracture toughness of W-V alloy gradually increases. Tungsten materials exhibit significant degradation of their mechanical properties at high temperatures, especially above recrystallization temperatures. The addition of vanadium is beneficial for improving the thermal stability of tungsten materials [59].

Arshad et al. [60] studied the thermal load damage behavior of tungsten-vanadium alloys, with vanadium doping amounts of 1%, 5%, and 10% (mass fraction). The sample dimensions of 12 mm × 2mm, and a single pulse with a pulse duration of 5 ms, and spot sizes of 4 mm × 4 mm. The parameters related to surface cracking of the W-V alloys after transient thermal loading are shown in Table 6.4. As given in Table 6.4, W-5V and W-10V exhibit higher cracking thresholds compared to W-1V. Both W-5V and W-10V produce similar crack widths, but W-5V has a lower crack density, suggesting that W-5V is more resistant to cracking than W-10V. Because the thermal conductivity of V is lower than that of tungsten, the thermal conductivity of W-V alloys decreases as the vanadium content increases. This reduction in thermal conductivity directly affects the surface temperature distribution, leading to the highest surface temperature in W-10V, which results in more severe damage.

Table 6.4: Cracking of W-V alloy surfaces after high transient heat loads [60].

Samples	Power density (MW/m^2)	Avg. crack width (μm)	Avg. distance (μm)	Crack number density ($cracks/mm^2$)
W-1V	155	0.6	90.9	127
	207	1.0	89.3	131
	311	2.7	101.5	110
W-5V	159	No cracks	No cracks	No cracks
	212	1.3	571.4	18
	317	2.9	589.2	8
W-10V	162	No cracks	No cracks	No cracks
	216	1.0	112.7	59
	324	2.5	197.1	23

6.2.5 Transient thermal load damage behavior of dispersion-strengthened tungsten composites

The dispersion strengthening of tungsten refers to the addition of second-phase particles to tungsten to form a dispersion phase in the material. These particles play a pinning role in grain boundaries and dislocations, hindering their movement, thereby inhibiting grain growth. This process increases the deformation resistance of tungsten matrix, thus enhancing the material's strength.

6.2.5.1 Oxide dispersion-strengthened tungsten-based materials

ODS-W alloys are prepared by using a small amount of nano-oxides as the strengthening phase, which are uniformly dispersed throughout the tungsten matrix. The goal is to utilize these ultrafine oxides to hinder the movement of grain boundaries and dislocations, thereby improving the mechanical properties and irradiation resistance of tungsten-based materials at high temperatures.

WL10 (W-1 wt% La_2O_3) is a hot topic of research as a tungsten alloy that can be produced commercially. Zhang et al. [61] studied the thermal load damage behavior of the rolled WL10 alloy under an electron beam. When a single pulse was applied, circumferential cracks appeared on the surface of the samples with power densities ranging from 0.22 to 0.88 GW/m^2, and these cracks completely disappeared at power densities of 1.1 GW/m^2, as shown in Figure 6.33. However, cracks were observed in the cross section and propagated along the direction of electron beam loading, indicating that the cracking threshold of the material is below 0.22 GW/m^2.

Lian et al. [62] prepared W-0.7% Y_2O_3 (volume fraction) powder by chemical method, and then fabricated W-Y_2O_3 block with a diameter of 6.5 cm by rapid forging, with a threshold of 0.22–0.33 GW/m^2 for 100 cycles thermal fatigue cracking at room temperature. As shown in Figure 6.34a, no crack formation was observed on the sample surface below the damage threshold. A crack network appeared on the surface when a heat flux of 0.33 GW/m^2 was applied at room temperature (Figure 6.34b), and the crack width increased with higher heat flux intensity (Figure 6.34c). When the matrix temperature exceeded the cracking temperature, crack formation was inhibited, and only severe plastic deformation occurred on the surface (Figure 6.34d–f). Tested for the mechanical properties of the material, W-Y_2O_3 is brittle at room temperature with low tensile strength, leading to a low cracking threshold. W-Y_2O_3 deforms plastically at 100 °C, indicating that the cracking threshold temperature is below 100 °C. When the temperature exceeds the cracking threshold temperature, the thermal stress can be released by plastic deformation.

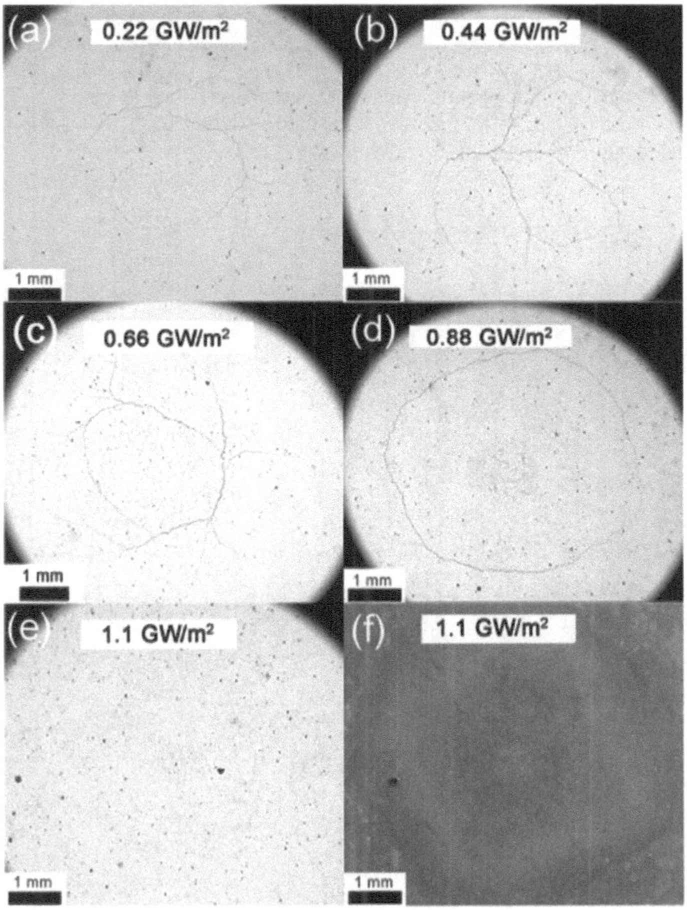

Figure 6.33: Surface morphology of WL10 alloy in the rolled state under a single pulse electron beam thermal shock [61].

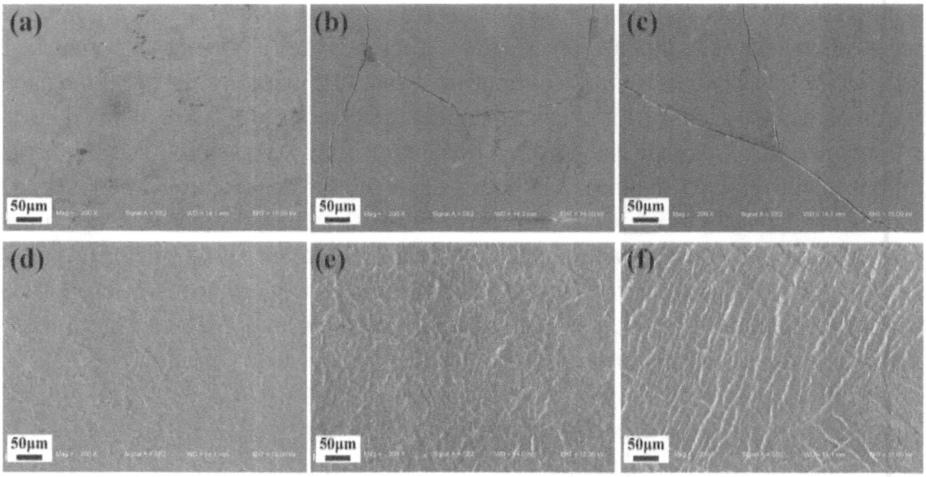

Figure 6.34: Surface morphology of W-Y_2O_3 forging surface under 100 electron beam thermal shocks simulated under ELMs thermal load conditions: (a) RT, 0.22 GW/m^2; (b) RT, 0.33 GW/m^2; (c) RT, 0.44 GW/m^2; (d) 100 °C, 0.44 GW/m^2; (e) 100 °C, 1 GW/m^2; and (f) 200 °C, 1 GW/m^2 [62].

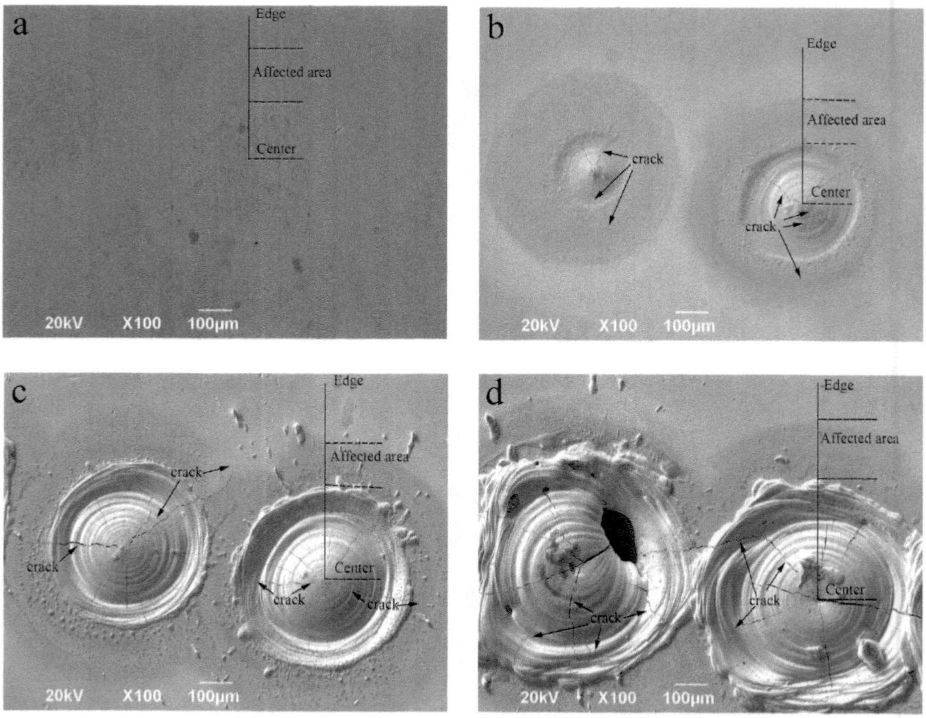

Figure 6.35: Surface SEM of W-0.5 wt% Y_2O_3 under laser thermal shock with different heat flow factors (a) F1 = 0.45 MW/m^2 s$^{1/2}$, (b) F2 = 0.59 MW/m^2 s$^{1/2}$, (c) F3 = 0.75 MW/m^2 s$^{1/2}$, (d) F4 = 0.90 MW/m^2 s$^{1/2}$ [63].

Zhao et al. [63] investigated the behavior of W-Y_2O_3 composites prepared by the wet chemical method and SPS technique under laser thermal shock, as shown in Figure 6.35. Under the impact of a high-energy pulsed laser, thermal cracks are generated on the surface of W-Y_2O_3 and propagate along the grain boundaries. Heat-affected zone, tungsten undergoes recrystallization, forming columnar grains. Pintsuk et al. [64] studied the effect of different Y_2O_3 doping on the thermal shock resistance of tungsten, and the results showed that the Y_2O_3 doping with a volume fraction of 1% had the best thermal shock resistance, as shown in Figure 6.36. There are no cracks appearing on the surface of W-1 vol% Y_2O_3 after thermal shock with a power density of 1.13 GW/m^2 electron beam at room temperature, only surface roughening (Figure 6.36b). In contrast, both W-5 vol% Y_2O_3 and pure tungsten show significant crack formation on their surfaces. This result is contrary to the results of mechanical properties, where Charpy impact experiments show that Y_2O_3 doping increases the DBTT of W. The exact reason for this is unclear, but Pintsuk et al. believed that it may be due to the microstructure of the material, including factors such as grain size and particle distribution.

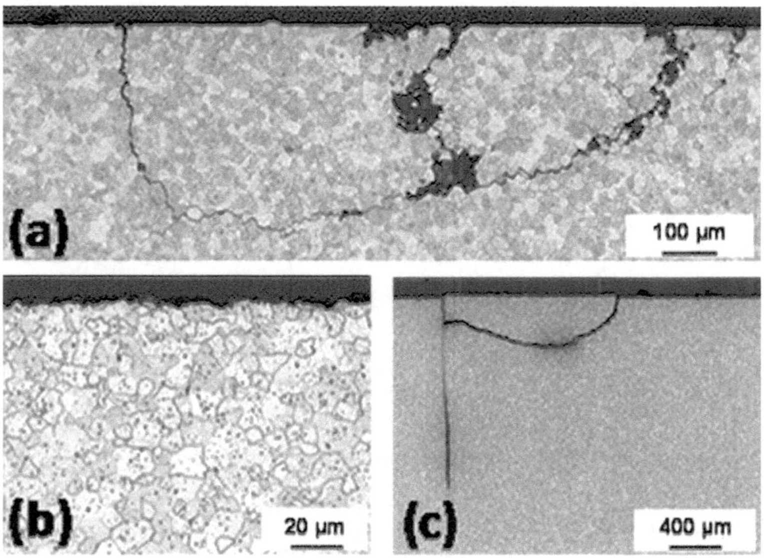

Figure 6.36: Crack formation and surface roughening after 1,000 electron beam thermal shocks at room temperature on JUDITH I: (a) Pure W, 0.38 GW/m^2; (b) W-1 vol% Y_2O_3, 1.13 GW/m^2; and (c) W-5vol% Y_2O_3, 1.13 GW/m^2 [64].

6.2.5.2 Carbide dispersion-strengthened tungsten-based materials

Compared with oxides, carbides have a higher melting point, greater thermal stability, and are less likely to aggregate and grow during high-temperature sintering, allowing them to maintain their nanostructures.

Kurishita et al. [65] prepared W-1.1TiC with DBTT below room temperature at 235 K. No cracks were observed on the surface after electron beam bombardment, indicating that the formation of tungsten surface cracks is due to the combined effect of fatigue thermal stress under thermal fatigue load and high DBTT of pure tungsten (Figure 6.37). Fan et al. [66] performed transient thermal load tests on W-1 wt% TiC and W-3 wt% ZrC prepared by micro-nano composite technology. The results showed that both materials could withstand a high thermal flux of 200 MW/m^2 (5 ms), demonstrating more than double the thermal load resistance compared to traditionally sintered pure tungsten. As shown in Figure 6.38a and b, no cracks are developed at power densities of 100–200 MW/m^2, while small cracks begin to appear as the power density is increased to 300 MW/m^2 (Figure 6.38c). As the power density increases to 400 MW/m^2 (Figure 6.38d), the crack expands and widens, and when the power density reaches 600 MW/m^2, local melting occurs (Figure 6.38e and f).

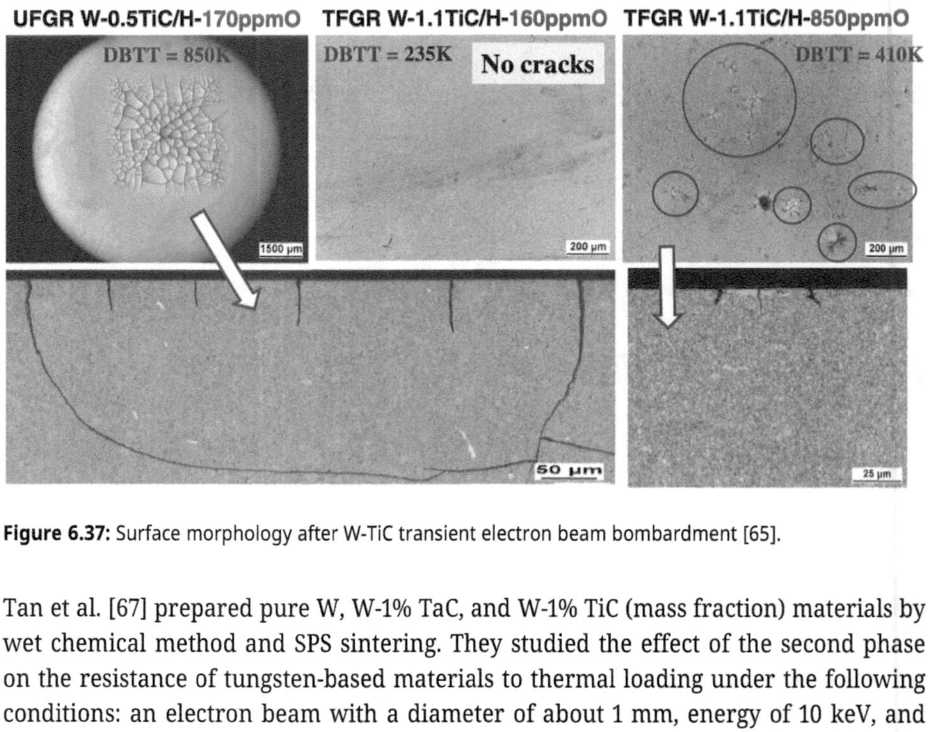

Figure 6.37: Surface morphology after W-TiC transient electron beam bombardment [65].

Tan et al. [67] prepared pure W, W-1% TaC, and W-1% TiC (mass fraction) materials by wet chemical method and SPS sintering. They studied the effect of the second phase on the resistance of tungsten-based materials to thermal loading under the following conditions: an electron beam with a diameter of about 1 mm, energy of 10 keV, and power of 8 kW was used to heat the samples at room temperature for 2 s. The total pulse duration was 9.5 s, and 100 pulses were applied. Figure 6.39 shows the surface morphology of the material after thermal load loading. Severe cracking and plastic deformation were found on the surface of the thermal load area of pure W. In contrast, few cracks and slight plastic deformation were observed on the surface of W-1% TaC, while only plastic deformation was found on the surface of W-1% TiC. Both TiC and TaC can effectively improve the material's resistance to thermal loads, with W-1%

TiC showing superior performance. Since the second-phase nanoparticles are dispersed in grain boundaries, TaC and TiC can enhance the strength of grain boundaries and prevent crack formation. Under the same doped mass fraction, the volume fraction of TiC is higher than that of TaC, indicating that the size of TiC particles is smaller, which can more effectively enhance the grain boundaries and improve the resistance to crack formation. Moreover, the plastic deformation of the second-phase particles inside the grain hinders the grain boundary failure after thermal loading. Specifically, the thermal

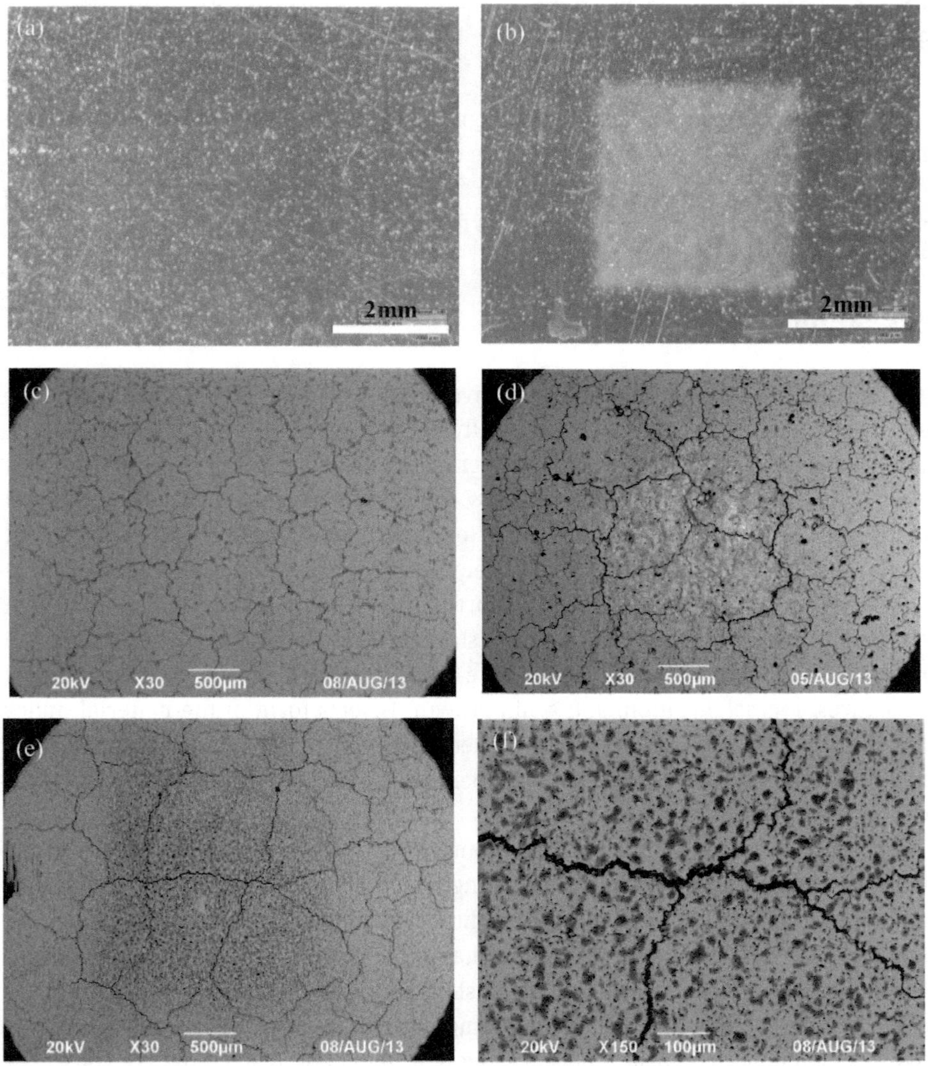

Figure 6.38: Surface morphology of W-3wt% ZrC under different heat flux intensities: (a) 100 MW/m^2; (b) 200 MW/m^2; (c) 300 MW/m^2; (d) 400 MW/m^2; (e) 600 MW/m^2; and (f) high magnification images of (d) [66].

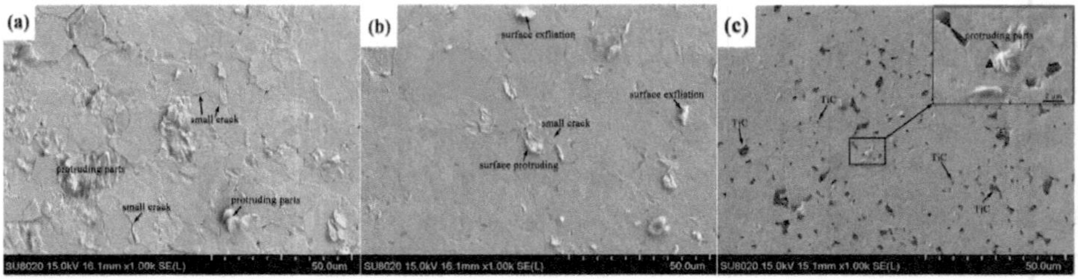

Figure 6.39: SEM image of the sample surface after thermal load loading: (a) pure W; (b) W-1 wt% TaC; and (c) W-1 wt% TiC [67].

stress generated by the high-temperature gradient can be released by plastic deformation. Therefore, the doping of second-phase particles TiC and TaC has a beneficial effect on improving the thermal load performance of tungsten-based materials. These materials can be considered as candidate systems for PFM in fusion reactors.

6.3 Summary

The maximum heat flux of PFMs at the divertor site is about 5–10 MW/m^2 during the steady-state operation of ITER. Additionally, the duration of the slow transient thermal load is relatively short, where the thermal load can reach up to 20 MW/m^2. This type of thermal load is classified as a slow transient thermal load. Transient thermal loads are mainly plasma rupture, edge localized mode and vertical displacement mode. Under the influence of these transient thermal shocks, the material surface rapidly heats up and cools down over a short period, leading to the formation of high temperature gradients and high thermal stresses. These conditions are accompanied by various phenomena, including surface cracking, melting, sputtering, protruding structures, recrystallization, and grain growth. Defects form in the material under thermal load, significantly reducing its thermal conductivity. This reduction in thermal conductivity causes the surface temperature to rise sharply, and the cyclic process generates new defects.

Long-term steady-state and/or transient thermal loads weaken tungsten's resistance to thermal shock and reduce its strength and toughness. Due to the pulsed operation mode, the thermal load varies periodically, leading to thermal fatigue at the connection interface. The study of the thermal and irradiation behaviors of tungsten-based materials under transient thermal shock, such as surface plastic deformation, cracking, and melting, is crucial. Simulating thermal shock scenarios using methods like lasers and electron beams helps to theoretically obtain results that are consistent with experimental simulations. These findings are beneficial for the design and manufacturing of future fusion reactors, particularly for divertor components.

References

[1] A.R. Raffray, R. Nygren, D.G. Whyte, S. Abdel-Khalik, R. Doerner, F. Escourbiac, T. Evans, R.J. Goldston, D.T. Hoelzer, S. Konishi, P. Lorenzetto, M. Merola, R. Neu, P. Norajitra, R.A. Pitts, M. Rieth, M. Roedig, T. Rognlien, S. Suzuki, M.S. Tillack, C. Wong. High heat flux components-Readiness to proceed from near term fusion systems to power plants [J]. Fusion Engineering and Design, 2010, 85(1): 93–108.

[2] O.A. Waseem, H.J. Ryu. Tungsten-based composites for nuclear fusion applications [M]. Nuclear Material Performance, 2016.

[3] J. Linke. Plasma facing materials and components for future fusion devices-development, characterization and performance under fusion specific loading conditions [J]. Physica Scripta, 2006, T123: 45–53.

[4] C. Lei, L. Xiang, L. Youyun. Damage behavior of high-purity tungsten under transient thermal shock [J]. Nuclear Fusion and Plasma Physics, 2014, 34(1): 53–56.

[5] B. Riccardi, R. Giniatulin, N. Klimov, V. Koidan, A. Loarte. Preliminary results of the experimental study of PFCs exposure to ELMs-like transient loads followed by high heat flux thermal fatigue [J]. Fusion Engineering and Design, 2011, 86(9–11): 1665–1668.

[6] X. Liu, Y. Lian, L. Chen, Z. Chen, J. Chen, X. Duan, J. Fan, J. Song. Experimental and numerical simulations of ELM-like transient damage behaviors to different grade tungsten and tungsten alloys [J]. Journal of Nuclear Materials, 2015, 463: 166–169.

[7] A. Huber, A. Burdakov, M. Zlobinski, M. Wirtz, J.W. Coenen, J. Linke, P. Mertens, V. Philipps, G. Pintsuk, B. Schweer, G. Sergienko, A. Shoshin, U. Samm, B. Unterberg. Investigation of the impact on tungsten of transient heat loads induced by laser irradiation, electron beams and plasma guns [J]. Transactions of Fusion Science and Technology, 2013, 63: 197–200.

[8] T. Loewenhoff, J. Linke, G. Pintsuk, C. Thomser. Tungsten and CFC degradation under combined high cycle transient and steady state heat loads [J]. Fusion Engineering and Design, 2012, 87(7–8): 1201–1205.

[9] B. Riccardi, P. Gavila, R. Gliniatulin, V. Kuznetsov, R. Rulev, N. Klimov, D. Kovalenko, V. Barsuk, V. Koidan, S. Korshunov. Effect of stationary high heat flux and transient ELMs-like heat loads on the divertor PFCs [J]. Fusion Engineering and Design, 2013, 88: 1673–1676.

[10] T. Hirai, K. Ezato, P. Majerus. ITER relevant high heat flux testing on plasma facing surfaces [J]. Materials Transactions, 2005, 46(3): 412–424.

[11] P. Majerus, R. Duwe, T. Hirai, W. Kühnlein, J. Linke, M. Rödig. The new electron beam test facility JUDITH II for high heat flux experiments on plasma facing components [J]. Fusion Engineering and Design, 2005, 75–79(1): 365–369.

[12] A. Gervash, R. Giniyatulin, V. Komarov, I. Mazul, N. Litunovsky, A. Ganenko, A. Vainerman, V. Fedotov, D. Davydov, R. Zalavutdinov. Comparative thermal cyclic testing and strength investigation of different Be/Cu joints [J]. Fusion Engineering and Design, 1998, 39–40(1): 543–549.

[13] I. Bobin-Vastra, F. Escourbiac, M. Merola, P. Lorenzetto. Activity of the European high heat flux test facility: FE200 [J]. Fusion Engineering and Design, 2005, 75–79(1): 357–363.

[14] H. Greuner, H. Bolt, B. Böswirth, T. Franke, P. McNeely, S. Obermayer, N. Rust, R. Süß. Design, performance and construction of a 2 MW ion beam test facility for plasma facing components [J]. Fusion Engineering and Design, 2005, 75–79: 345–350.

[15] H. Greuner, B. Boeswirth, J. Boscary, P. McNeely. High heat flux facility GLADIS: Operational characteristics and results of W7-X pre-series target tests [J]. Journal of Nuclear Materials, 2007, 367: 1444–1448.

[16] A. Zhitlukhin, N. Klimov, I. Landman, J. Linke, A. Loarte, M. Merola, V. Podkovyrov, G. Federici, B. Bazylev, S. Pestchanyi, V. Safronov, T. Hirai, V. Maynashev, V. Levashov, A. Muzichenko. Effects of ELMs on ITER divertor armour materials [J]. Journal of Nuclear Materials, 2007, 363: 301–307.

[17] N.I. Arkhipov, V.P. Bakhtin, S.M. Kurkin, V.M. Safronov, D.A. Toporkov, S.G. Vasenin, A.M. Zhitlukhin, H. Würz. Material erosion and erosion products in disruption simulation experiments at the MK-200 UG facility [J]. Fusion Engineering and Design, 2000, 49–50: 151–156.
[18] M.Y. Ye, H. Kanehara, S. Fukuta, N. Ohno, S. Takamura. Blister formation on tungsten surface under low energy and high flux hydrogen plasma irradiation in NAGDIS-I [J]. Journal of Nuclear Materials, 2003, 313: 72–76.
[19] D. Nishijima, M.Y. Ye, N. Ohno, S. Takamura. Formation mechanism of bubbles and holes on tungsten surface with low-energy and high-flux helium plasma irradiation in NAGDIS-II [J]. Journal of Nuclear Materials, 2004, 329–333: 1029–1033.
[20] S. Kajita, S. Takamura, N. Ohno, D. Nishijima, H. Iwakiri, N. Yoshida. Sub-ms laser pulse irradiation on tungsten target damaged by exposure to helium plasma [J]. Nuclear Fusion, 2007, 47(9): 1358.
[21] H. Shukai. Study on the properties of plasma-oriented tungsten-based materials under transient heat load[D]. Beijing: Peking University, 2012.
[22] Y. Ueda, K. Schmid, M. Balden, J.W. Coenen, T. Loewenhoff, A. Ito, A. Hasegawa, C. Hardie, M. Porton, M. Gilbert. Baseline high heat flux and plasma facing materials for fusion [J]. Nuclear Fusion, 2017, 57: 092006.
[23] H. Kurishita. Recent progress in toughening nanostructured W alloys through domestic and international research collaboration. Charleston, SC, USA, October 17–21, 2011.
[24] G. Pintsuk, A. Prokhodtseva, I. Uytdenhouwen. Thermal shock characterization of tungsten deformed in two orthogonal directions [J]. Journal of Nuclear Materials, 2011, 417(1–3): 481–486.
[25] M. Wirtz, J. Linke, T. Loewenhoff, G. Pintsuk, I. Uytdenhouwen. Thermal shock tests to qualify different tungsten grades as plasma facing material [J]. Physica Scripta, 2016, T167: 014015.
[26] M. Akiba, M. Araki, S. Suzuki, S. Tanaka, M. Dairaku, K. Yokoyama, M. Seki. Performance of JAERI electron beam irradiation stand [J]. Plasma Devices and Operations, 1991, 1: 205–212.
[27] T. Hirai, G. Pintsuk, J. Linke, M. Batilliot. Cracking failure study of ITER-reference tungsten grade under single pulse thermal shock loads at elevated temperatures [J]. Journal of Nuclear Materials, 2009, 390–391: 751–754.
[28] J. Linke, R. Duwe, A. Gervash. Material damage to beryllium, carbon, and tungsten under severe thermal shocks [J]. Journal of Nuclear Materials, 1998, 258–263(1): 634–639.
[29] J. Linke, M. Akiba, R. Duwe. Material degradation and particle formation under transient thermal loads [J]. Journal of Nuclear Materials, 2001, 290–293(3): 1102–1106.
[30] L. Wang, B. Wang, S.D. Li, D. Ma, Y.H. Tang, H. Yan. Thermal fatigue mechanism of recrystallized tungsten under cyclic heat loads via electron beam facility [J]. International Journal of Refractory Metals and Hard Materials, 2016, 61: 61–66.
[31] T. Loewenhoff, J. Linke, G. Pintsuk, R.A. Pitts, B. Riccardi. ITER-W monoblocks under high pulse number transient heat loads at high temperature [J]. Journal of Nuclear Materials, 2015, 463: 202–205.
[32] Z.J. Dai, Y. Mutoh, S. Sujatanond. Numerical and experimental study on the thermal shock strength of tungsten by laser irradiation [J]. Materials Science and Engineering A, 2008, 472: 26–34.
[33] N. Farid, S.S. Harilal, O. El-Atwani, H. Ding, A. Hassanein. Experimental simulation of materials degradation of plasma-facing components using lasers [J]. Nuclear Fusion, 2014, 54: 012002.
[34] M. Wirtz, J. Linke, G. Pintsuk, L. Singheiser, M. Zlobinski. Comparison of thermal shock damages induced by different simulation methods on tungsten [J]. Journal of Nuclear Materials, 2013, 438: S833–S836.
[35] V.P. Budaev, Y.V. Martynenko, A.V. Karpov, N.E. Belova, A.M. Zhitlukhin, N.S. Klimov, V.L. Podkovyrov, V.A. Barsuk, A.B. Putrik, A.D. Yaroshevskaya, R.N. Giniyatulin, V.M. Safronov, L.N. Khimchenko. Tungsten recrystallization and cracking under ITER-relevant heat loads [J]. Journal of Nuclear Materials, 2015, 463: 237–240.

[36] J.H. Yu, M.J. Baldwin, R.P. Doerner, T. Dittmar, A. Hakola, T. Höschen, J. Likonen, D. Nishijima, H.H. Toudeshki. Transient heating effects on tungsten: Ablation of Be layers and enhanced fuzz growth [J]. Journal of Nuclear Materials, 2015, 463: 299–302.
[37] A. Huber, A. Arakcheev, G. Sergienko, I. Steudel, M. Wirtz, A.V. Burdakov, J.W. Coenen, A. Kreter, J. Linke, P. Mertens, V. Philipps, G. Pintsuk, M. Reinhart, U. Samm, A. Shoshin, B. Schweer, B. Unterberg, M. Zlobinski. Investigation of the impact of transient heat loads applied by laser irradiation on ITER-grade tungsten [J]. Physica Scripta, 2014, T159: 014005.
[38] M. Wirtz, J. Linke, G. Pintsuk, G. De Temmerman, G.M. Wright. Thermal shock behaviour of tungsten after high flux H-plasma loading [J]. Journal of Nuclear Materials, 2013, 443(1–3): 497–501.
[39] N. Lemahieu, H. Greuner, J. Linke, H. Maier, G. Pintsuk, G.V. Oost, M. Wirtz. Synergistic effects of ELMs and steady state H and H/He irradiation on tungsten [J]. Fusion Engineering and Design, 2015, 98–99: 2020–2024.
[40] T. Loewenhoff, S. Bardin, H. Greuner, J. Linke, H. Maier, T.W. Morgan, G. Pintsuk, R.A. Pitts, B. Riccardi, G. De Temmerman. Impact of combined transient plasma/heat loads on tungsten performance below and above recrystallization temperature [J]. Nuclear Fusion, 2015, 55: 123004.
[41] A. Huber, M. Wirtz, G. Sergienko, I. Steudel, A. Arakcheev, A. Burdakov, H.G. Esser, M. Freisinger, A. Kreter, J. Linke, C. Linsmeier, P. Mertens, S. Möller, V. Philipps, G. Pintsuk, M. Reinhart, B. Schweer, A. Shoshin, B. Unterberg. Combined impact of transient heat loads and steady-state plasma exposure on tungsten [J]. Fusion Engineering and Design, 2015, 98–99: 1328–1332.
[42] X. Zhang, Q. Yan. The thermal crack characteristics of rolled tungsten in different orientations [J]. Journal of Nuclear Materials, 2014, 444(1–3): 428–434.
[43] S. Tamura, K. Tokunaga, N. Yoshida, M. Taniguchi, K. Ezato, K. Sato, S. Suzuki, M. Akiba, Y. Tsunekawa, M. Okumiya. Damage process of high purity tungsten coatings by hydrogen beam heat loads [J]. Journal of Nuclear Materials, 2005, 337–339: 1043–1047.
[44] S.E. Pestchanyi, J. Linke. Simulation of cracks in tungsten under ITER specific transient heat loads [J]. Fusion Engineering and Design, 2007, 82(15): 1657–1663.
[45] S. Qu, S. Gao, Y. Yuan, C. Li, Y. Lian, X. Liu, W. Liu. Effect of high magnetic field on the melting behavior of W-0.5wt% La_2O_3 under high heat flux [J]. Journal of Nuclear Materials, 2015, 463: 189–192.
[46] G. Ackland. Controlling radiation damage [J]. Science, 2010, 327: 1587.
[47] Z. Zhou, G. Pintsuk, J. Linke, T. Hirai, M. Rödig, Y. Ma, C. Ge. Transient high heat load tests on pure ultra-fine grained tungsten fabricated by resistance sintering under ultra-high pressure [J]. Fusion Engineering and Design, 2010, 85: 115–121.
[48] I. Uytdenhouwen, M. Decréton, T. Hirai, J. Linke, G. Pintsuk, G. Van Oost. Influence of recrystallization on thermal shock resistance of various tungsten grades [J]. Journal of Nuclear Materials, 2007, 363–365: 1099–1103.
[49] G. Pintsuk, W. Kühnlein, J. Linke, M. Rödig. Investigation of tungsten and beryllium behavior under short transient events [J]. Fusion Engineering and Design, 2007, 82(15–24): 1720–1729.
[50] X. Zhang, Q. Yan, S. Lang, M. Xia, C. Ge. Thermal shock performance of sintered pure tungsten with various grain sizes under transient high heat flux test [J]. Journal of Fusion Energy, 2016, 35: 666–672.
[51] J. Du, Y. Yuan, M. Wirtz, J. Linke, W. Liu, H. Greuner. FEM study of recrystallized tungsten under ELM-like heat loads [J]. Journal of Nuclear Materials, 2015, 463: 219–222.
[52] M. Wirtz, G. Cempura, J. Linke, G. Pintsuk, I. Uytdenhouwen. Thermal shock response of deformed and recrystallised tungsten [J]. Fusion Engineering and Design, 2013, 88(9–10): 1768–1772.
[53] P. Schade. Potassium bubble growth in doped tungsten [J]. International Journal of Refractory Metals & Hard Materials, 1998, 16(1): 77–87.

[54] J. Linke, T. Loewenhoff, V. Massaut, G. Pintsuk, G. Ritz, M. Rödig, A. Schmidt, C. Thomser, I. Uytdenhouwen, V. Vasechko, M. Wirtz. Performance of different tungsten grades under transient thermal loads [J]. Nuclear Fusion, 2011, 51(7): 600–606.
[55] G. Pintsuk, I. Uytdenhouwen. Thermo-mechanical and thermal shock characterization of potassium doped tungsten [J]. International Journal of Refractory Metals and Hard Materials, 2010, 28(6): 661–668.
[56] X. Zhang, Q. Yan, S. Lang, M. Xia, C. Ge. Basic thermal-mechanical properties and thermal shock, fatigue resistance of swaged+rolled potassium doped tungsten [J]. Journal of Nuclear Materials, 2014, 452(1/3): 257–264.
[57] B. Huang, Y. Xiao, B. He, J. Yang, J. Liao, Y. Yang, N. Liu, Y. Lian, X. Liu, J. Tang. Effect of potassium doping on the thermal shock behavior of tungsten [J]. International Journal of Refractory Metals and Hard Materials, 2015, 51: 19–24.
[58] B.P. Bewlay, N. Lewis, K.A. Lou. Observations on the evolution of potassium bubbles in tungsten ingots during sintering [J]. Metallurgical and Materials Transactions A, 1992, 23(1): 121–133.
[59] M. Wirtz, J. Linke, G. Pintsuk, L. Singheiser, I. Uytdenhouwen. Comparison of the thermal shock performance of different tungsten grades and the influence of microstructure on the damage behavior [J]. Physica Scripta, 2011, T145: 014058.
[60] K. Arshad, M.Y. Zhao, Y. Yuan, Y. Zhang, Z.J. Zhou, G.H. Lu. Thermal stability evaluation of microstructures and mechanical properties of tungsten vanadium alloys [J]. Modern Physics Letters B, 2014, 28(26): 1450207.
[61] X. Zhang, Q. Yan. Morphology evolution of La_2O_3 and crack characteristic in W-La_2O_3 alloy under transient heat loading [J]. Journal of Nuclear Materials, 2014, 451(1–3): 283–291.
[62] Y. Lian, X. Liu, F. Feng, J. Song, B. Yan, Y. Wang, J. Wang, J. Chen. Mechanical properties and thermal shock performance of W-Y_2O_3 composite prepared by high-energy-rate forging [J]. Physica Scripta, 2017, T170: 014044.
[63] M. Zhao, L. Luo, J. Lin, X. Zan, X. Zhu, G. Luo, Y. Wu. Thermal shock behavior of W-0.5wt%Y_2O_3 alloy prepared via a novel chemical method [J]. Journal of Nuclear Materials, 2016, 479: 616–622.
[64] G. Pintsuk, D. Blagoeva, J. Opschoor. Thermal shock behavior of tungsten based alloys manufactured via powder injection molding [J]. Journal of Nuclear Materials, 2013, 442: S282–S286.
[65] H. Kurishita, H. Arakawa, S. Matsuo, T. Sakamoto, S. Kobayashi, K. Nakai, G. Pintsuk, J. Linke, S. Tsurekawa, V. Yardley, K. Tokunaga, T. Takida, M. Katoh, A. Ikegaya, Y. Ueda, M. Kawai, N. Yoshida. Development of nanostructured tungsten based materials resistant to recrystallization and/or radiation induced embrittlement [J]. Materials Transactions, 2013, 54: 456–465.
[66] J. Fan, Y. Han, P. Li, Z. Sun, Q. Zhou. Micro/nano composited tungsten material and its high thermal loading behavior[J]. Journal of Nuclear Materials, 2014, 455(1–3): 717–723.
[67] X. Tan, P. Li, L. Luo, Q. Xu, K. Tokunaga, X. Zan, Y. Wu. Effect of second-phase particles on the properties of W-based materials under high-heat loading [J]. Nuclear Materials and Energy, 2016, 9: 399–404.

Chapter 7
Manufacture and evaluation of tungsten-copper modules for fusion reactor divertor

7.1 Introduction

The first wall of a magnetic confinement fusion reactor is composed of many blanket modules connected together. As the component of the test blanket module that directly faces the plasma, it must withstand high thermal loads. The outermost layer of the first wall of ITER is the armor material, and beryllium (Be) is selected as the PFM in the first wall area of ITER because it is a low-*Z* material, with no chemical sputtering and strong oxygen absorption capability. Adjacent to the armor is the heat sink, which dissipates heat. The material is made of alloys such as CuCrZr. The next layer connected is the structural material such as austenitic stainless steel [1].

With the ongoing development of magnetic confinement fusion reactor experimental devices, divertors have evolved from water-cooled to helium-cooled systems, leading to ongoing changes and advancements in divertor materials [2–4]. In ITER, under the extreme plasma operating conditions in the deuterium and tritium stages, the surface heat flux of the first wall reaches 0.1–1.0 MW/m^2, while the heat flux of the divertor is 5–20 MW/m^2. In the case of Experimental Advanced Superconducting Tokamak (EAST), the vertical target plate of the divertor is designed to withstand 1,000 times of irradiation with a heat load of 20 MW/m^2 [5]. Therefore, the other side of the PFM must be forced to cool, which, together with the heat sink materials, form plasma-facing components (PFCs). The connections between the PFM and heat sink materials, as well as between the heat sink and structural materials, have become critical manufacturing steps in the development of fusion devices. Tungsten, due to its high melting point, high thermal conductivity, and low-tritium retention, has been selected as the PFM for the divertor in ITER [6] and as the alternative material for the divertor and the first wall in the DEMO and future fusion reactors. Copper alloys have been used as candidate materials for the divertor heat sink in modern Tokamak devices and ITER due to their high thermal conductivity and strength, excellent thermal stability, and resistance to neutron irradiation activation,

7.2 Divertor for fusion reactor

7.2.1 Design of the divertor

The fusion reactor is a huge structure composed of numerous functional components, among which the divertor is one of the key PFCs. The ITER divertor primarily consists

 | https://doi.org/10.1515/9783111556987-007

of inner and outer target (OT) plates (mono-block structure), a dome (flat structure), reflector(flat structure) support structures, and 54 cassette bodies (CB), as illustrated in Figure 7.1 [7]. The steady-state heat flux in the dome region of the ITER divertor is expected to be ~5 MW/m^2, while at the inner and OT surfaces, it is expected to be ~10 MW/m^2. Each CB of the ITER divertor consists of an inner target (IT) plate component and an OT plate component, and each vertical target plate component comprises two similar but independent PFCs, with each made up of many long plasma-facing units (PFU). The OT plate component of the ITER divertor consists of 22 PFUs and 2 support structures, and the IT plate part consists of 16 PFUs and 2 support structures. At the end of 2012, the tasks of ITER were allocated as follows: the European Union was responsible for the construction of the IT plate and the CB [8], Japan was responsible for the OT plate, and Russia was responsible for the dome, the particle reflector [9, 10] as well as the construction of the ITER Divertor Test Facility (IDTF).

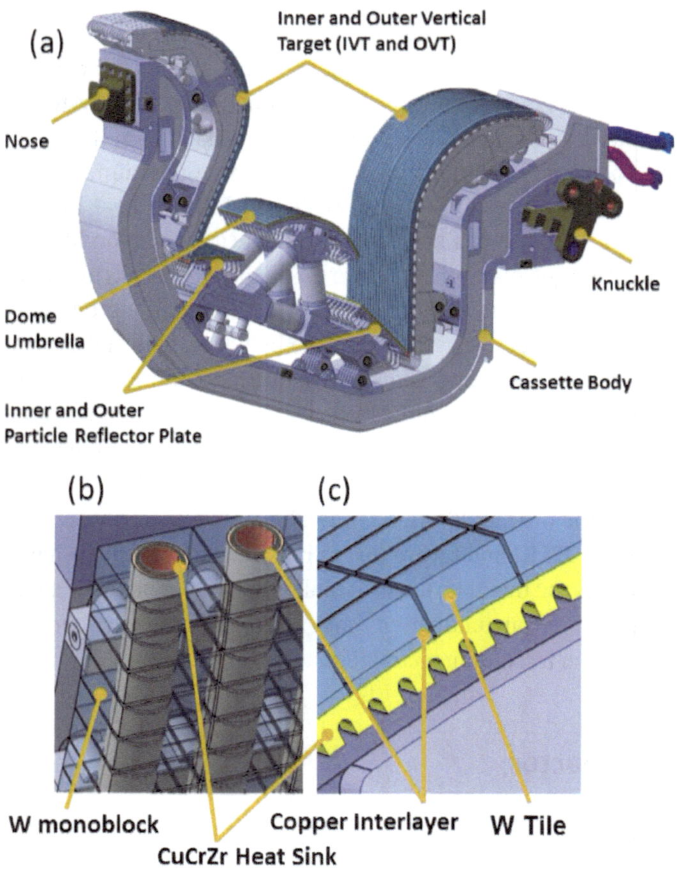

Figure 7.1: Schematic diagram of the structure of the ITER divertor [7].

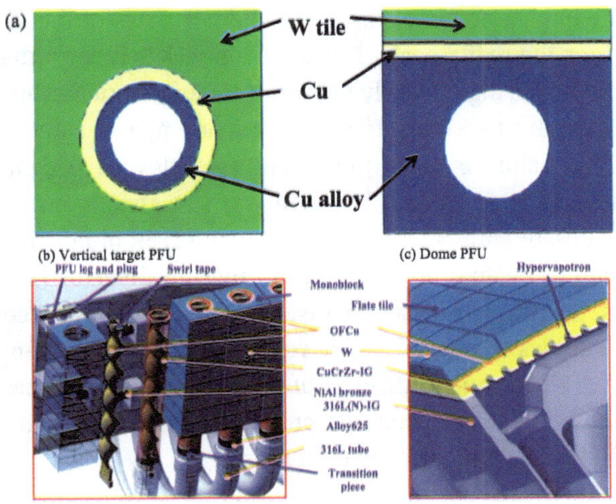

Figure 7.2: (a) Schematic diagram of mono-block and flat-type PFCs; (b) mono-block; and (c) flat-type [11].

The structure of PFCs, as shown in Figure 7.2, mainly consists of two types: flat-type and mono-block [11]. The flat-type PFC connects the PFM directly to the CuCrZr alloy heat sink using a pure copper interlayer. This approach offers advantages such as structural simplicity, ease of manufacturing, and reduced costs. The mono-block PFC can form a cooling water pipeline in the PFM. It connects the PFM and the heat sink using methods like brazing, hot isostatic pressing (HIP), and hot radial pressing (HRP) [12, 13]. The basic principle of HRP is to add high-pressure gas to the CuCrZr cooling pipeline and, at the same time, heat the PFM and CuCrZr pipeline. This induces slight plastic deformation in the copper pipeline to achieve the diffusion welding between the CuCrZr pipeline and the PFM. Compared with the flat-type, the mono-block PFC, with its unique symmetrical structure and circular connection interface, can reduce the concentration and eliminate the ambiguity of thermal stress. And the W/Cu interface is closer to the cooling water, resulting in lower temperatures and reduced thermal stress, thus providing better resistance to thermal loads [7].

7.2.2 Selection of materials for divertor components

(1) Plasma-facing materials

The divertor components mainly include PFM, heat sink, and support structures, with variations in material compositions and structural forms across different types. Tungsten, owing to its superior properties, is widely acknowledged as the most promising material for PFM. The initial design of the ITER was constrained by the manufacturing technologies available at the time, leading to a two-phase procurement process for the divertor components. In the first phase, carbon fiber composite (CFC) material was chosen for the PFM near the impact points of the inner and OT plates of the divertor, while tungsten was used for the baffle parts. This selection was driven by the high incidence of plasma disruptions during the early phases of ITER operations, during which CFC demonstrated significant resistance to transient load impacts without melting. The second phase involved a shift to an all-tungsten divertor design. As component manufacturing techniques advanced, the full W-PFM was successfully tested in the ASDEX Upgrade test [14]. Subsequently, in November 2011, the ITER organizing committee initiated the certification evaluation for the full-tungsten divertor [15], which, after 2 years of rigorous testing and assessments, was officially approved in November 2013 [16]. Table 7.1 shows the performance requirements (purity, density, hardness, grain size, etc.) of ITER for rolled tungsten plate. Table 7.2 summarizes the properties of PFC tungsten materials for divertors purchased from different manufacturers, all of which meet ITER's stringent performance criteria.

Table 7.1: ITER tungsten plate material inspection specifications [7].

Examination	Applicable standards	Acceptance criteria
Chemical composition	To be agreed with supplier	W ≥ 99.94% C ≤ 0.01wt% O ≤ 0.01wt% N ≤ 0.01wt% Fe ≤ 0.01wt% Ni ≤ 0.01wt% Si ≤ 0.01wt%
Density	ASTM B311	≥19.0 g/cm^3
Hardness (Vickers HV30)	ASTM E92 or EN ISO 6507-1	≥410
Grain size	ASTM E112	Grain size number 3 or finer
Microstructure		Elongated grain

(2) Heat sink materials

Heat sink materials are required to possess several key properties, including high thermal conductivity, high strength at elevated temperatures, good processability, excellent water, and air tightness, as well as strong resistance to corrosion at high temperatures. Among these, thermal conductivity stands out as the most critical criterion for selecting a heat sink material. Commonly used materials include CuCrZr and oxide dispersion strengthened copper (ODS-Cu), with CuCrZr emerging as the preferred choice for ITER heat sink applications. The CuCrZr alloy features several advantages such as age-hardening capability, superior heat resistance, excellent electrical and thermal transport properties, and cost-effective fabrication, making it widely applicable in fields like welding, nonferrous metallurgy, and electrical engineering. Additionally, CuCrZr alloy is also used in the ITER's first wall, divertor, and heating systems.

Table 7.2: W products purchased for divertor [11].

Supplier	Form	Chemical composition	density	Hardness HV30	Gain size number[a]	Procured amount[b]
Plansee	Bar (cross section of 36 × 36 mm²)	As required	19.24–19.25	435–455	10–11.6	1,536 kg (gross)
Polema	Plate (12 mm thick) Plate (24 mm thick)	As required	19.1–19.15 19.0–19.14	437–446 419–450	8.0–8.5 8.0–8.5	712 kg (gross) 3,400 kg (gross)
ALMT	Plate (12 mm thick)	As required	19.14–19.20	443–470	9.5–12.5	376 kg (net)
AT&M	Plate (12 mm thick)	As required	19.2–19.28	435–468	4.5–7.0	90 kg (net)
XRTM	Plate (12 mm thick)	As required	19.13–19.28	415–495	6.5–7.0	181 kg (net)
TLWM	Plate (12 mm thick)	As required	19.2	420–450	7.0–8.0	18 kg (net)

[a]Grain size conforms to ASTM E112. The face tested is perpendicular to the direction of deformation.
[b]"Gross" is the amount of tungsten material purchased and "net" is the weight of the final shape.

ITER's requirements for the chemical composition of CuCrZr alloy are more stringent to reduce fluctuations in material properties and ensure that the materials meet the design requirements during the component preparation process. For ITER applications, the design code for heat sink CuCrZr material is CuCrZr-IG (ITER grade) [17]. During the ITER engineering design phase, the thermal/plastic processing techniques for CuCrZr-IG differ accordingly due to varying applications and the distinct fabrication processes of different components. Numerous studies have been conducted on the properties of CuCrZr alloys. Table 7.3 presents the tensile properties of CuCrZr-IG after heat treatment, which can basically meet the ITER requirements compared with the ITER requirements for the strength of the parts (yield strength > 175 MPa and tensile strength > 280 MPa).

Compared with water-cooled divertors, helium-cooled divertors operate at higher temperatures and lower pressures. As the operating temperature in the fusion device increases, the cooling system develops from the water-cooled type (Figure 7.3) [18] to the helium-cooled type (Figure 7.4) [19], with the first wall of the divertor subjected to a heat flux of 10–15 MW/m^2, which accounts for about 15% of the thermal power of the reactor [20]. In the future, the increasing working temperatures of DEMO devices will position the helium-cooled divertor design as the mainstream approach. Correspondingly, the heat sink materials are advancing from copper and its alloys, as used in devices such as EAST and ITER to low-activation steel materials in future fusion demonstration power plants (DEMO). At present, commonly used materials such as CuCrZr and ODS-Cu exhibit obvious thermal creep when subjected to prolonged operation at 300–400 °C, and their mechanical properties deteriorate under high-temperature conditions. Therefore, the development of novel copper alloys with high strength, high

Table 7.3: Tensile properties of CuCrZr-I G after heat treatment [11].

Manufacturing process	Tensile properties ($YS_{0.2\%}$ [MPa], UTS [MPa], elongation [%]) before manufacturing heat cycle		Tensile properties ($YS_{0.2\%}$ [MPa], UTS [MPa], elongation [%]) after manufacturing heat cycle	
	Room temperature	250 °C	Room temperature	250 °C
HIP (~100 MPa, 550 °C for 6 h)	153, 316, 13	265, 286, 7	197, 345, 23	251, 303, 17
Hot radial pressing (60 MPa, 580 °C for 2 h)	411, 492, 22	332, 368, 15	230, 340, 28	205, 266, 25
Brazing (980 °C for 60 min → gas quench → aging at 480 °C for 3 h)	259, 420, 27	228, 318, 25	223, 379, 28	186, 281, 26
Brazing (970 °C for 45 min → gas quench → aging at 480 °C for 2.5 h)	421, 468, 18	383, 390, 19	216, 341, 22	199, 275, 21

thermal conductivity, and resistance to thermal creep and radiation at elevated temperatures is critical to meeting the urgent requirements for key heat sink materials in future fusion reactor construction.

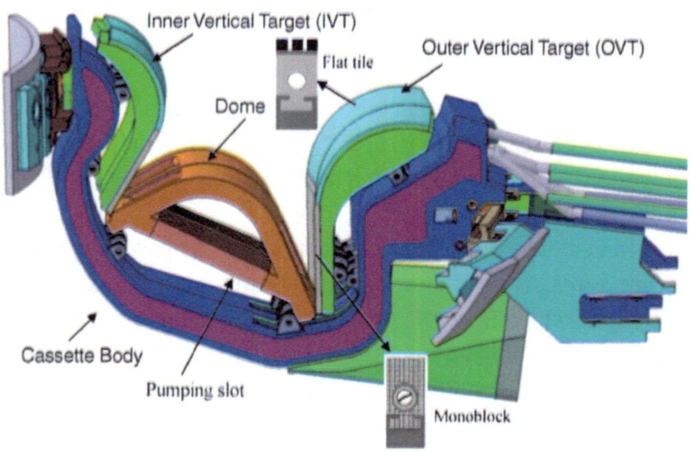

Figure 7.3: Schematic diagram of the water-cooled divertor [18].

(3) Structural materials

Structural materials, while not directly plasma-facing, must maintain good compatibility with PFM, tritium breeders, neutron multipliers, and coolants. In future fusion reactors, the production of high-energy neutrons in the WeV range will result in sig-

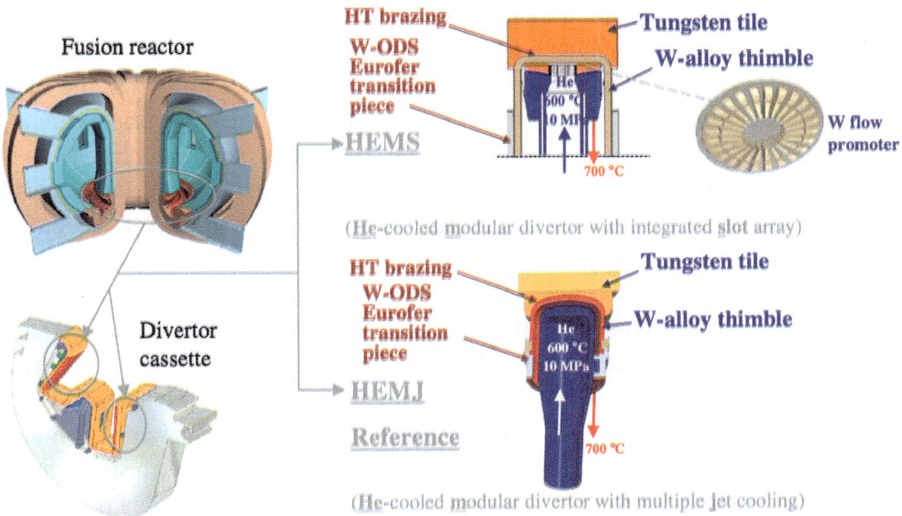

Figure 7.4: Schematic diagram of the helium-cooled divertor [19].

nificant irradiation damage to structural materials. The structural materials are expected to evolve from 316SS (austenitic stainless steel) used in ITER test reactors to low-activation materials for DEMO and commercial reactors such as low-activation steel, ODS steel (oxide dispersion strengthen steel), vanadium (V) alloys, and SiC/SiC composites. V-alloy and SiC/SiC composites exhibit excellent high-temperature performance, though their reliability requires further investigation. Therefore, low-activation steels have been selected as the primary material for DEMO demonstration reactors. Table 7.4 lists the composition of several typical low-activation steels.

Table 7.4: Composition of several typical low-activation steels [21].

Trademark	Main element (mass%)						
	Fe	Cr	C	W	Ta	Mn	V
CLF-1	Bal.	8.5	0.11	1.5	0.1	0.5	0.3
EUROFER97	Bal.	8.0–9.0	0.1–0.12	1.0–1.2	0.06–0.10	0.4–0.6	0.2–0.3
F82H	Bal.	8.0	0.10	2.0	0.04	0.50	0.20
CLAM	Bal.	9.0	0.10	1.5	0.15	0.45	0.20

7.2.3 Connection between the tungsten divertor and heat sink

PFCs play a critical role, primarily responsible for removing the thermal flux deposited on the surface of PFM through the heat sink, thereby ensuring the reliability of the components (materials) and the safety of the device. To design and manufacture a fully functional divertor component, it is essential that the key material units are interconnected effectively. During both the connection and service processes, the components will experience high thermal stress. Therefore, selecting appropriate connection technologies and processes to reduce thermal stress at the connection interface is crucial for enhancing the reliability and stability of the divertor. Material property differences between the connected components pose challenges to the connection interface, especially when joining dissimilar materials with vastly different properties. In such cases, the selection of connection technology and process parameters becomes critically important. The connection between the heat sink and the support structure has become a crucial technology in the fabrication of fusion device components. Methods for joining PFM and heat sink include bolt connection (such as graphite and chromium-zirconium-copper alloy) and welding (e.g., tungsten and chromium-zirconium-copper alloy). Among these, welding can achieve atomic-level contact, which significantly reduces the thermal resistance between the PFM and the heat sink, thereby improving thermal conductivity.

Table 7.5: Physical properties of tungsten, copper, and CLAM steel at room temperature [21, 22].

Material	Density (g/cm^3)	Melting point (K)	Coefficient of thermal expansion (10^{-6}/K)	Thermal conductivity (W/m•K)	Modulus of elasticity (GPa)
Tungsten	19.3	3,680	4.5	145	410
Copper	8.9	1,357	17	400	85
CLAM steel	7.8	1,723	11.1	24.5	218

The ITER fusion reactor employs a water-cooled divertor, with tungsten used as the armor material for the upper part and baffle of the inner and outer vertical targets, while copper alloys serve as the heat sink material. The properties of tungsten and copper vary greatly, with a fourfold difference in thermal expansion coefficient and a threefold difference in melting point, as shown in Table 7.5. These substantial differences make their joining relatively challenging. There are many effective methods for joining W/Cu, such as electron beam welding (EBW), brazing, casting, HIP, and vacuum plasma spray (VPS), but combinations of several techniques are often employed to optimize the connection.

(1) Electron beam welding

Accelerated and focused electron beam bombardment is applied to the welding surface, which can be in either a vacuum or a non-vacuum environment. This process melts the welded workpiece, forming a molten pool to achieve metallurgical bonding. EBW offers several advantages, including no need for welding rods, resistance to oxidation, good process repeatability, narrow weld seams, a small heat-affected zone, and small thermal deformation. These benefits make EBW widely applicable across various fields. Roedig et al. [23] used oxygen-free copper (OFC) as an intermediate layer to join W and CuCrZr using EBW. Their research demonstrated that the joint could withstand 1,000 cycles of high heat flux (HHF) testing at 13.7 MW/m^2. Smid et al. [24] utilized EBW to bond W-1%La_2O_3 to CuCrZr, and the samples remained intact after 1,000 thermal load testing at 16 MW/m^2. Among these, WL10 (W-1wt%La_2O_3) has received attention as a commercial tungsten material.

(2) Brazing connection

Brazing is a welding method where both the filler metal and the weldment are heated simultaneously to the melting temperature of the filler metal, which is below the melting point of the weldment. The liquid filler metal fills the gaps between the solid workpieces, resulting in metal bonding. Brazing minimally affects base materials' properties, with low residual stress and smooth joints, making it suitable for precision welding of dissimilar metals with significant property differences. During the brazing process, Ni-Cu-Mn was used as the solder [25], with OFC serving as the intermediate layer to successfully join W and CuCrZr alloy. Cu-Mn was also employed as the filler metal to bond W with the heat sink material CuCrZr alloy. Thermal cycling fatigue tests showed that the sample could withstand 1,000 times of 16 MW/m^2 heat load [26]. However, the choice of filler metal is critical, as it directly affects the quality and reliability of the brazed joint. It was found that the mechanical properties of joints containing Cu-Mn filler metal degrade significantly after neutron irradiation [27]. The development of new filler metals suitable for the operating environment of the divertor is particularly urgent.

(3) Casting connection

The liquid metal is cast into a mold cavity that matches the shape of the part, and after cooling and solidification, the corresponding part or blank is obtained. The casting method produces tungsten/OFC interfaces with high-bonding strength, making it one of the most commonly used methods at present. Under vacuum condition, the sample is heated to a temperature above the melting point of copper, allowing a layer of OFC to be cast onto the surface of W by using the wettability of copper. When the temperature exceeds 1,350 °C, the wetting angle of copper on W approaches zero, indicating complete wetting. However, tungsten recrystallization occurs when the temperature exceeds 1,200 °C. To maintain the mechanical properties of tungsten, the temperature during the bonding process should not exceed the recrystallization temperature of tungsten. Copper casting can be used to initially bond W and Cu due to the significant difference in the melting points of tungsten and copper. Subsequently, the cast copper can be connected to the heat sink material CuCrZr alloy via EBW or brazing [28].

(4) HIP connection

Under an inert gas atmosphere, the combination of pressure, temperature, and time applies equal inward extrusion force in all dimensions of the processed workpiece, which can effectively eliminate the internal porosity in materials such as metals, ceramics, and composite materials to improve the compactness, uniformity, and performance of the workpiece. The use of high pressure allows for a significant reduction in the sintering temperature, enabling the processed materials to maintain a fine microstructure and crystal structure. W and Cu were directly welded by HIP [29]. However, oxidation occurred at the interface, forming tungsten oxide, which led to crack initiation and reduced bond strength. It was found that adding a layer of OFC or another transition layer with a thickness greater than 0.3 mm can improve the bonding strength. The French Atomic Energy Commission (CEA) adopted a two-step HIP process, first using HIP with Ni as the transition layer to weld W-1%La_2O_3 to OFC and then using HIP to bond the copper to the heat sink CuCrZr alloy. This module was able to withstand 1,000 times of electron beam fatigue testing at 9.6 MW/m^2 [23].

(5) Vacuum plasma spraying
VPS technology mainly utilizes a plasma arc driven by direct current as a heat source to heat materials such as ceramics, alloys, and metals to a molten or semi-molten state. These materials are then sprayed at high speed onto the surface of a pretreated workpiece in a vacuum (less than 0.01 Pa) state to form a firmly attached surface layer. Plasma torch with their high temperature, fast speed, and atmospheric protection is optimal for preparing high-quality thick tungsten coating – a critical fabrication technology for fusion PFCs. VPS-W [30] was found to be able to withstand 1,000 times of >4.7 MW/m^2 fatigue experiments.

However, in the fabrication of plasma first-wall composites, different connection methods are typically employed based on the different thermal load areas of Tokamak devices. The VPS connection method is used for PFCs in low heat load areas due to the simplicity of the process and its low cost. In the area of medium and high heat load, several methods must be combined to enhance heat transfer and withstand higher thermal shocks. In comparison, these combined methods are more complex and costly [31].

In the future, the design of the DEMO device will adopt a helium-cooled divertor, and its design of connection is still in the early stages. Therefore, it is necessary to systematically study the structure, brazing filler metal, processes, etc. At present, two helium-cooled divertor structures have been proposed: the finger-type (Figure 7.5a) and the T-type (Figure 7.5b) [18, 32]. Among these, the helium-cooled modular divertor with multiple-jet cooling (HEMJ) proposed by KIT in Germany is widely recognized as the reference standard for future divertor design [33]. In the finger-type structure, a small 5-mm-thick tile-like tungsten block serves as a thermal-shielding sacrificial layer, which is welded into a tungsten alloy (WL10) sleeve to form a cooling finger unit. It is then connected to a low-activation steel support structure. Almost all future designs for high-temperature helium-cooled divertors will adopt the W-PFM scheme, while reduced activation ferritic/martensitic steels (RAFMs) are suitable for the supporting structural material. Both schemes involve the connection of tungsten and its alloys with low-activation steels and ODS steels. The significant thermal expansion coefficient difference between tungsten and steel can result in high thermal stress at the tungsten/steel joint interface during both the connection process and the component's service life. This stress may cause joint failure, and conventional fusion welding technologies struggle to produce high-quality tungsten/steel joints. At present, the connection of tungsten steel is mainly focused on diffusion welding and brazing, and the high-performance and reliable connection between tungsten and steel has become a key technical challenge in the development of divertors.

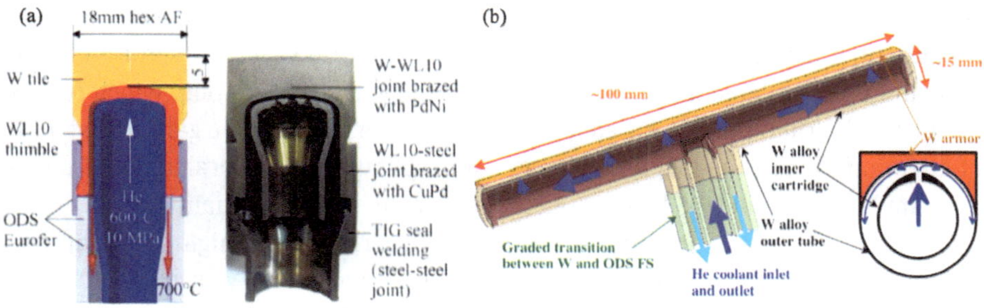

Figure 7.5: The He-cooled finger design of the KIT in Germany and the T-tube design of the UCSD in the United States [32, 33].

(6) Vacuum brazing

At present, the connection between tungsten and low-activation steel is primarily achieved through vacuum brazing technology. During brazing, the temperature is higher than the liquidus temperature of the filler metal but lower than the solidus temperature of the base metal. The molten filler metal wets, flows, fills, spreads on the surface of the base metal, and solidifies to form a dense joint. The brazing technology has the characteristics of simple process, low impact on the base metal, and low requirements for the size and shape of the weldment, which can be used in the connection of W/RAFM steel [34]. Chehtov et al. [34] successfully brazed the W-1%La_2O_3 alloy and EUROFER97 low-activation steel using a Ni-based filler metal. Similarly, Liu et al. [35] used Ni-based crystalline foil strip as filler metal for tungsten-steel brazing. de Prado et al. [36] pressed Cu and Ti powders into thin-film filler metals and realized brazing connection between W-based materials and EUROFER steels.

To improve the lifespan of brazed components, introducing optimized interlayers with thermal expansion coefficients matching both tungsten and low activation steel is essential. Most researchers mainly focus on designing composition and joint structure of filler metal. For example, Kalin et al. [37] used Ni-based amorphous foil as filler metal, with a 0.5-mm-thick 50Fe-50Ni alloy as an interlayer to braze polycrystalline tungsten and EP450 ferritic/martensitic steel, resulting in joints without pores or cracks. However, after 100 thermal cycles, cracks appeared, indicating poor resistance to thermal shock. In a later study, Kalin et al. [38] utilized Ti-based and Fe-based brazing filers, incorporating a 0.1-mm-thick tantalum sheet as an interlayer to braze tungsten and ODS-EUROFER steel. The experiment showed that the joints did not develop cracks after 30 thermal cycles, indicating that the structure had high thermal shock resistance.

For tungsten/steel welding, proper selection of amorphous high-temperature filler metals – mainly Fe-based, Ti-based, and Ni-based alloys – is critical. Rapidly solidified amorphous and microcrystalline foils are effectively methods for joining dissimilar materials. Compared with crystalline foil strips, amorphous alloy foil strips exhibit good uniformity in composition and structure, have a narrow liquid-solid phase range, and demonstrate excellent melt flow and spreading properties. They also possess good toughness, allowing them to bend up to 180° without fracturing, making them suitable for welding complex parts like divertors. However, Ni and Cu can induce helium embrittlement in brazed joints under neutron irradiation, which makes the joint brittle. Therefore, Fe-based low-activation filler metals have garnered significant attention. Oono et al. [39] used Fe-based amorphous strips as brazing filler metal to vacuum braze W and ODS high-Cr ferritic steel, achieving high-quality welds.

Currently, the brazing of tungsten and steel is performed at high temperatures, which leads to the grain growth of the base metal, particularly in low-activation steels, and causes significant tungsten diffusion into steel. As a result, future research should focus on investigating the recovery of base metal microstructures and developing strategies to prevent tungsten diffusion into steel during the brazing process.

(7) Vacuum diffusion welding

Vacuum diffusion welding is a process performed in a vacuum, where the weldments are closely pressed together. Under appropriate temperature, bonding pressure, and holding time, atomic diffusion occurs at the contact surfaces, resulting in a connection. Diffusion welding forms a welded joint without melting the metals, making it particularly suitable for materials that are difficult to weld using fusion welding or other methods. For dissimilar metals, an intermediate layer is usually added to prevent the formation of brittle intermediate metals at the joint or to reduce the mismatch in the thermal expansion coefficient between the two metal wires. In direct tungsten/steel welding, intermetallic compounds like FeW and Fe_2W form, which drastically deteriorate the joint's mechanical properties. Basuki et al. [40] performed direct diffusion welding of W with EUROFER97 low-activation steel at 1,050 °C, and there was significant thermal stress at the welded joint, which fractured directly after heat treatment.

To obtain high-quality tungsten/steel joints, intermediate layer materials are usually added to relieve residual thermal stress and inhibit the formation of brittle intermetallic compounds. The selection principle for interlayer material requires minimal difference with parent materials, no formation of brittle/eutectic phases, and sufficient deformability. Many materials can be selected as intermediate layers. Nb has a thermal expansion coefficient between that of tungsten and steel, while V and Ti have linear expansion coefficients between tungsten and steel. These materials also exhibit excellent high-temperature mechanical properties, and tungsten/steel can form continuous solid solutions with them. This helps to alleviate residual stress in the joint and prevent the formation of brittle phase. Ni has good plasticity and low yield strength, which helps to relieve stress. At high temperatures, Ni and Fe can form continuous solid solutions, preventing carbon diffusion from steel into tungsten [41]. Cu is a non-carbide forming element that does not form brittle phases with steel, which not only prevents the diffusion of carbon atoms in steel into V but also relieves the thermal stress caused by the difference in the coefficient of thermal expansion of the base metal. The tensile strength of welded joints is primarily affected by the difference in thermal expansion coefficient between tungsten and the intermediate layer as well as by the formation of brittle intermetallic compounds at the interface of the intermediate layer.

Most of the intermediate layers used in tungsten/steel diffusion welding contain high neutron-active elements such as Nb, Ni, and Cu. These elements tend to induce helium segregation under neutron irradiation, resulting in embrittlement of the welded joints. Diffusion welding usually requires the application of high pressure and demands high surface quality, making it is unsuitable for connecting the components with complex structures and shapes. These are the key factors restricting the application of diffusion welding technology on tungsten/low-activation steels.

Regardless of the welding method, good weldability alone is insufficient to meet the stringent service conditions of fusion reactors. Firstly, it is crucial to ensure that the connected parts maintain their overall mechanical properties, allowing them to withstand the impact of high thermal loads. Additionally, attention should be paid to changes in the microstructure properties of the welded parts (joints) after exposure to high-intensity irradiation. Although the inclusion of highly activated elements or components may enhance welding bonding, the use of such materials in the intermediate or transition layers should be approached with caution.

7.2.4 ITER full-tungsten divertor certification test

The certification test for the full-tungsten divertor consists of several steps: (1) technology development and verification. The feasibility of the scheme was verified by fabricating small-scale W/Cu modules (mock-up) and PFU, followed by thermal load irradiation tests to validate the concept and (2) full-scale demonstration. The design capabilities and feasibility of the scheme are validated through manufacturing of full-scale parts and performing HHF tests.

For ITER's small-scale tungsten-copper modules, the requirements are as follows: at least three groups of modules must be used for HHF testing, with each module containing no fewer than five tungsten-copper blocks. Given that the CuCrZr alloy (a high-strength and high-conductivity alloy) tends to soften at temperatures above 450 °C, and under a thermal load of 20 MW/m^2 combined with ITER's hydraulic conditions, premature material damage may occur if smooth pipelines are used for cooling. Consequently, the design of the mono-block-type tungsten copper module must incorporate spoilers. Similarly, in the full-scale tungsten-copper components, spoilers are also required in the vertical target plate area (near the strike points).

The HHF verification testing is preceded by nondestructive testing including (1) visual inspection; (2) dimensional control; (3) interfacial ultrasonic testing; (4) room-temperature helium detection; and (5) stress test. The specific HHF testing procedures for small-scale parts and vertical target plate areas: (1) thermal mapping under a thermal load of 5 MW/m^2; (2) 5,000 times of heat load testing at 10 MW/m^2; (3) 300 times of thermal cycling at 20 MW/m^2; and (4) thermal mapping again under a heat load of 5 MW/m^2. Heat load test steps for the bending section area (Baffle) of the full-scale components: (1) thermal mapping under a heat load of 5 MW/m^2; (2) 5,000 times of heat load testing at 5 MW/m^2; and (3) final thermal mapping under a heat load of 5 MW/m^2.

The qualification criteria for HHF testing are as follows: (1) no obvious melting; (2) no water leakage after irradiation; (3) no debonding at the welding interface; (4) no hot spot (hot spot: the temperature of the hottest tungsten block in the experiment is 30% higher than the temperature of other nondefective places) in the test; (5) the maximum surface temperature difference in thermal mapping before and after testing should not exceed 20%. The test is followed by the evaluation of room-temperature helium leak and hydrostatic testing [42].

7.3 High heat load testing of mono-block tungsten-copper module at home and abroad

To enhance the thermal load resistance of the core components in fusion reactors, the major participating countries in the ITER program are intensifying their research and development efforts, focusing on both material selection to manufacturing specifications for key components. Some progress has been made in the design and manufacture of tungsten connection modules, which has played a positive role in the advancement of the ITER program.

7.3.1 Japan W/Cu-mono-block test evaluation

The Japanese Domestic Agency (JADA) is working on the preparation of the ITER procurement package. Japan's W/Cu-Mono-block is mainly developed by Japan's Atomic Energy Agency (JAEA). Under the guidance of the full tungsten divertor certification project, the previous conceptual design of the CFC/W divertor was replaced, leading to the conceptual research and development of the full tungsten divertor [43].

Initially, a small-scale tungsten-copper module was developed, with its dimensions shown in Figure 7.6 [44]: the width of the W block is 27.8 mm, the height is

28 mm, and the axial length is 12 mm. The thickness between the plasma facing surface of the W-PFM and the upper surface of the CuCrZr cooling pipeline is 8 mm, while the interval between two tungsten blocks is 0.5 mm. The thickness of a pure Cu intermediate layer is 1 mm, which is used to mitigate thermal stress due to the difference in physical properties between the PFM and the heat sink. The inner and outer diameters of the CuCrZr alloy heat sink are 12 and 15 mm, respectively. The connection methods between W/Cu are HIP, casting, and unidirectional diffusion welding, respectively. The tungsten-copper block is brazed to the chromium-zirconium-copper using a silver-free filler metal Ni-Cu-Mn alloy (NiCuMan-37). The thickness of the solder is 50 μm and the brazing temperature is 980 °C. Subsequently, the CuCrZr alloy is degassed and aged at 480 °C.

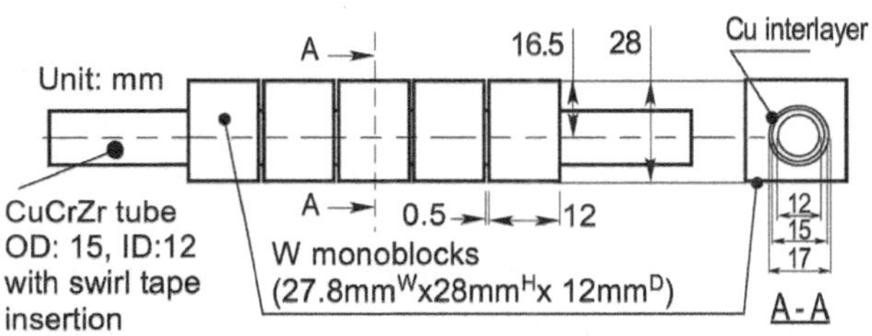

Figure 7.6: Schematic diagram of Japanese tungsten copper mono-block module [44].

With Russian IDTF equipment, a total of six sets of modules were used for high heat load testing. Test conditions: cooling water flow rate was 11 m/s, temperature at the entrance was 70 °C, and pressure was 3.9 MPa. The heat load conditions were 5,000 thermal cycling tests of 10 MW/m^2 and 300 thermal cycling tests of 20 MW/m^2 (IO certification), and an additional 700 heat load tests of 20 MW/m^2. The Japanese-made test modules passed the high heat load testing and met the high heat load requirements of the ITER organization, indicating that the module manufacturing and welding technology of JAEA in Japan are up to the mark for IO standards. As shown in Figure 7.7, after an additional 700 times of 20 MW/m^2 heat load, there is no macroscopic cracking (self-castellation) on the tungsten surface of the module, but obvious plastic deformation is found on the W surface.

Ultrasonic nondestructive testing of the W/Cu interface is required after high thermal load irradiation. The detection method involved probe scanning axially along the copper pipeline, followed by rotating the probe incrementally and continuing the axial scan until a full 360° rotation was completed. The data collected were used to generate ultrasonic C-scan images. Figure 7.8a shows the ultrasonic C-scan images of the Cu/CuCrZr and W/Cu interface of the module after a heat load of 5 MW/m^2. The scanning begins from 0°, with the vertical surface facing upward, and the θ angle increasing clock wise from 0° to 360°. Different colors represent varying echo values, with the yellow parts indicating locations of connection defects. Modules subjected to

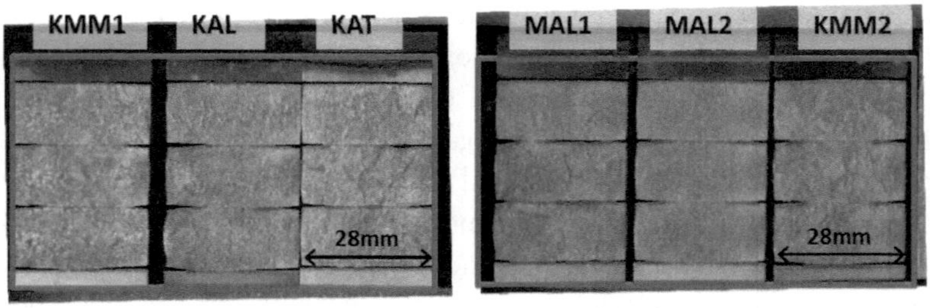

Figure 7.7: Surface morphology of the Japanese tungsten-copper module of JAEA after 5,000 times of 10 MW/m^2 and 300 times of 20 MW/m^2 heat loads [16].

heat load irradiation were assigned identification numbers. For example, after irradiation at a heat load of 5 MW/m^2, defects in module 1-5-4 were observed at angles between 0° and 180°, consistent with the overheated region (red area) observed on the right side of the specimen in the CCD image (Figure 7.8b). In several other modules, the locations of connection defects detected via ultrasound also matched the overheated regions shown in the CCD images.

In Japan, HHF tests were also carried out on small-scale modules (tungsten armor thickness of 7.7 mm) using JEBIS equipment. After 300 thermal cycles at heat loads exceeding 20 MW/m^2 (with peak values reaching approximately 23 MW/m^2), macroscopic cracking was observed on the surface of the tungsten. However, this cracking did not impair the module's heat transfer capability. The cracking may be caused by increased plastic deformation and the degradation of mechanical properties at elevated temperatures.

JAEA also manufactured a full-scale PFC, that is, CFC material is used in the strike point area, and W material (ITER early design) is used in other areas, and its structure is shown in Figure 7.9a. JAEA performed HHF testing on the tungsten section of the full-scale target plate using the IDTF. The tungsten-copper blocks in this area were of the same size as those used in the small-scale module, with a total of 5 × 4 tungsten-copper blocks, which were subjected to 5,000 times of 10 MW/m^2 and 1,000 times of 20 MW/m^2 heat load irradiation without any water leakage and continuous melting of the tungsten surface. Figure 7.9b shows the surface morphology of the tungsten irradiated by heat load, with grain growth on the surface, but no cracking. Compared with the small-scale module, the tungsten surface does not exhibit significant plastic deformation. The manufacturing and heat load testing of small-scale module and full-scale target plate show that the manufacturing technology of JAEA meets the requirements of ITER.

JADA fabricated 6 + 1 full-tungsten full-scale plasma-facing parts as shown in Figure 7.10. Six of the pieces were manufactured according to ITER quality standards, and the W mono-block was manufactured from two different manufacturers. The mono-block dimensions are 27.8 mm in width, 28 mm in height, and 12 mm in axial length. The thickness of tungsten armor and the pure Cu intermediate layer

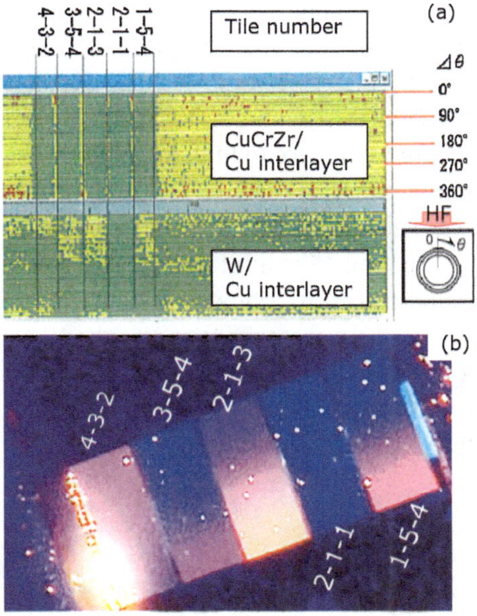

Figure 7.8: (a) Ultrasonic C-scan image of the connection interface of the 5 MW/m^2 module and (b) CCD image showing the overheating area in the heat load testing [44].

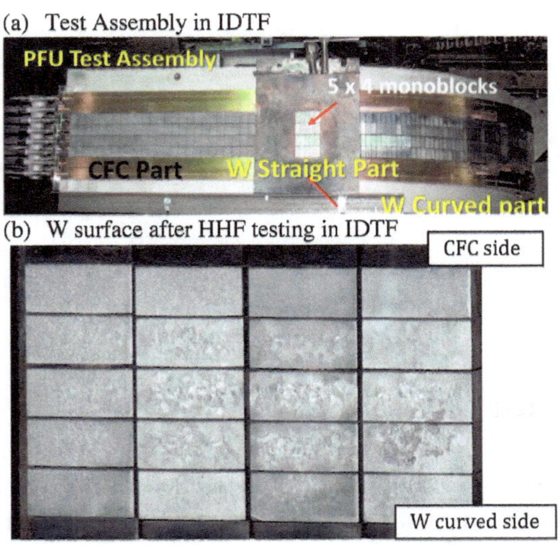

Figure 7.9: (a) The surface morphology of tungsten on a full-scale target plate mounted on an IDTF and (b) after 5,000 10 MW/m^2 and 1,000 20 MW/m^2 heat load tests [44].

is 7.7 mm and 1 mm, respectively. The connection method between W/Cu depends on the manufacturer of W; one utilizes casting, while the other employs unidirectional diffusion welding. The inner and outer diameters of CuCrZr cooling water pipelines are 12 and 15 mm, respectively [45]. Figure 7.11a shows ultrasonic C-scan images of the module's connection interface after exposure to high heat loads. The upper part of the image corresponds to the Cu/CuCrZr connection interface, while the lower part represents the W/Cu interface. The red regions indicate areas with connection defects, revealing that modules 1, 20, and 45 exhibit defects. These defect locations align with the infrared (IR) camera results, shown in Figure 7.11b, which highlight areas of overheating. These defects were only found at the W/Cu interface welded by diffusion welding, while no defects were found at the cast Cu/W interface. Therefore, the next step of work will focus on improving the quality of diffusion welding by optimizing the manufacturing process.

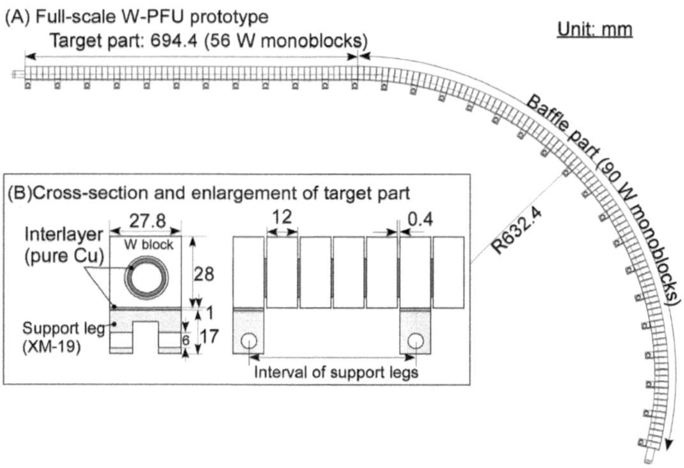

Figure 7.10: Schematic diagram of Japanese full-scale full-tungsten PFU [45].

7.3.2 Evaluation of W/Cu-mono-block test in Europe

The Efremov Research Institute in Russia undertook the purchase of ITER divertor dome parts and has successfully produced the dome QPs. These dome QPs withstood 1,000 times 3 MW/m^2 and 1,000 times 5 MW/m^2 thermal load shocks, passing certification of the IO. The European Union undertook the procurement contract for ITER's internal target plate components, and ANSALDO/ENEA in Italy and PLANSEE in Austria have successfully produced the QPs for internal target plates, and both have passed the verification and met the ITER acceptance standards. Europe also conducted large number of HHF tests on small-scale mono-block-type tungsten-copper modules and full-scale divertor target plate components. These tests were carried out using the Russian IDTF and the French FE200 facilities. The test conditions were as follows: the cooling water flow rate was 10.2 m/s, the temperature at the inlet was 100 °C, and the pressure was 3.3 MPa, all of which met the requirements of ITER hydraulic conditions. The heat load conditions consisted of 5,000 times at 10 MW/m^2 and 1,000 times at 20 MW/m^2, with the test time for both loading and halting the heat load set at 10 s.

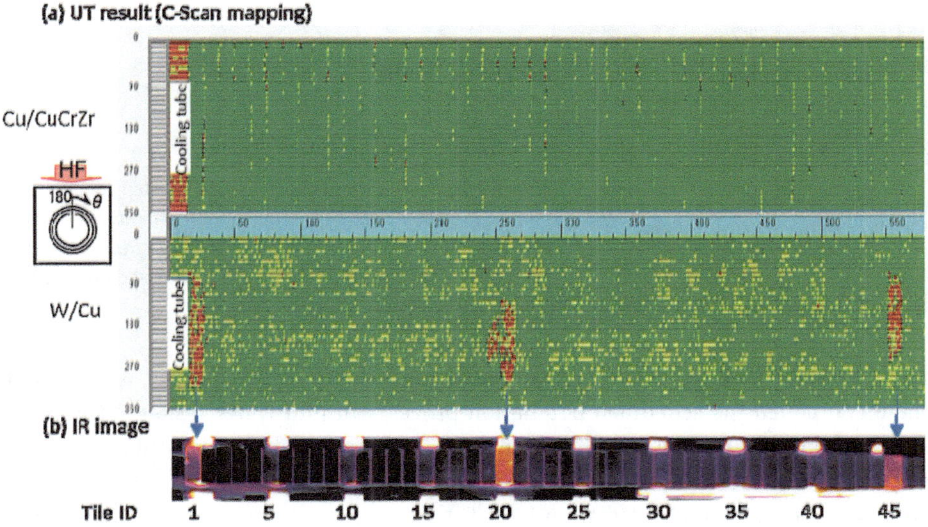

Figure 7.11: (a) Ultrasound C-scan image and (b) infrared camera image showing areas of overheating during the heat load testing [45].

The manufacture and high heat load testing of small-scale modules and full-scale tungsten-copper components were primarily conducted at F4E (Fusion for Energy) in Spain. A total of 26 small-scale tungsten-copper modules as well as 2 medium-sized components were produced by Plansee SE in Austria (10 pieces of size module and 1 piece of medium-sized target plate and HIP welding) and Ansaldo Nuclear in Italy (ANN, 16 pieces of size module, 1 piece of medium-sized target plate and hot Radial Press (HRP). Each small-scale tungsten-copper module consisted of seven tungsten-copper blocks, with a 0.8-mm-thick copper spoiler with a twist ratio of 2 inserted into the cooling pipeline. For all components produced by ANN, the inner and outer diameters of the cooling pipeline were 12 and 15 mm, respectively. The tungsten block dimensions were 28 mm wide and 12 mm long in the axial direction. The thickness of the armor (the minimum distance from the heated surface to the pure copper intermediate layer) was 6 mm. For PSE-produced components, there were three different geometries. For the first and second geometry, the sizes of the inner and outer diameters were 12 and 15 mm, respectively. The width of the tungsten block was 28 mm, the length was 12 mm in the axial direction, and the thickness of the armor was 7.5 and 5.5 mm, respectively. For the third geometry, the inner and outer diameters of the cooling pipeline were 10 and 12 mm, respectively, the width of the tungsten block was 23 mm, the length in the axial direction was 12 mm, and the thickness of the armor was 8 mm. The medium-sized target plate produced by PSE adopted the third geometry, incorporating three types of tungsten (rod, plate, and bar).

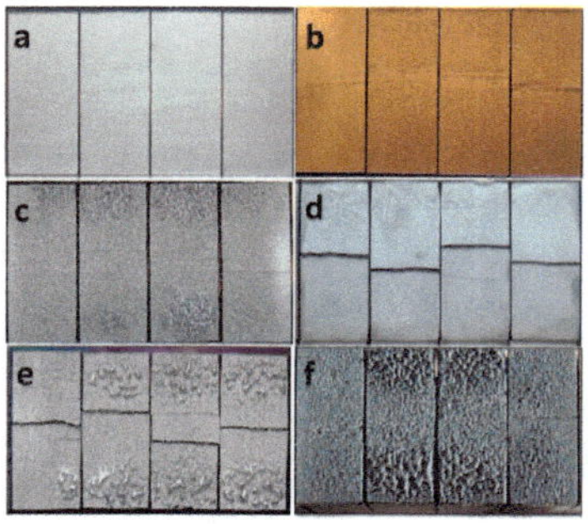

Figure 7.12: Surface morphology of PSE tungsten-copper module irradiated with high heat load: the armour thickness of a, b, d, and e is 5.5 mm, and the amour thickness of c and f is 7.5 mm. Among them, a, b, c, e, and f are tested on FE200 and d is IDTF [46].

PSE conducted HHF testing on six small modules using the FE200 facility, where all of them completed the scheduled irradiation process with no water leakage. The main forms of damage were local surface melting, roughness, plastic deformation at the edges, and macroscopic axial cracking, which developed after exposure to dozens of 20 MW/m^2 thermal load irradiation [46]. The surface morphology of the samples following high heat load irradiation is shown in Figure 7.12a–f, where (a) corresponds to 5,000 times at 10 MW/m^2, (b) represents 100 times at 20 MW/m^2, and (c) represents 300 times at 20 MW/m^2. Figure 7.12d–f shows the results after 1,000 times at 20 MW/m^2. The performance of the 10 small modules on the IDTF equipment met the minimum requirements given by the ITER organization (5,000 heat load irradiation of 10 MW/m^2 and 300 heat load irradiation of 20 MW/m^2). However, only seven modules successfully completed the full 1,000 cycles at 20 MW/m^2; the other three experienced water leakage during thermal cycling. Cracking occurred only in the first two structures and was unrelated to the thickness of the tungsten armor, consistent with the results obtained on the FE200. The damage on the IDTF device was less severe than that seen on the FE200, with no melting of tungsten, as shown in Figure 7.13. According to analysis, this may be due to the unique parameters of the FE200 device, that is, the diameter of narrow-focused electron beam spot (usually several millimeters) generates extremely high local thermal loads at high frequency (1–10 kHz) beam scanning. This high local heat load would result in microcracks after surface roughening and plastic deformation of the grains and eventually leading to local melting of the rough surface.

To better understand the specific damage to the module after the high heat load test, a damage analysis was performed on the PSE tungsten-copper module. The damage test of the module after high heat load on the IDTF equipment found that many impuri-

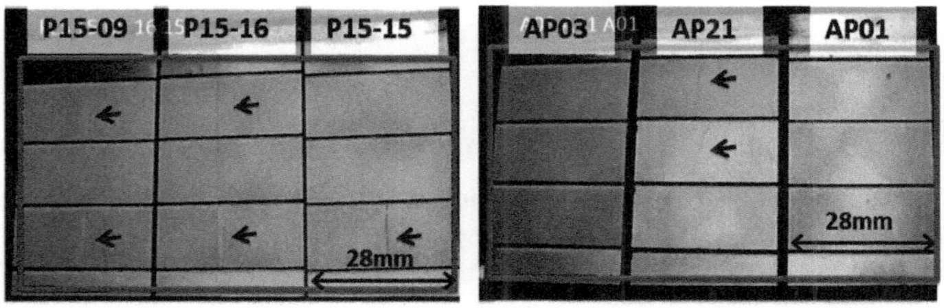

Figure 7.13: Surface morphology of a European small-scale mono-block type tungsten-copper module after 5,000 times of 10 MW/m^2 and 1,000 times of 20 MW/m^2 heat load tests on IDTF (armor thickness of 7.5 mm on the left and 6 mm on the right) [16].

ties were deposited on the internal surface of the heat sink, as shown in Figure 7.14a. It is speculated that the contamination was likely due to quality issues with the water used in the heat load test on the small-scale module, which increased the thermal resistance of the heat sink, led to greater thermal stress, and increased the plastic deformation of the CuCrZr water pipeline, as shown in Figure 7.14b. After the HHF test, cracks appeared on the inner surface of the cooling water pipeline in the PSE W/Cu mock-ups, and the length of the cracks was usually 20–100 µm, as shown in Figure 7.15. The impact of these cracks on the performance of the mock-ups is still unclear, and there is no significant change in the microstructure of the water pipelines at these locations. Interestingly, no scale formation was observed in the mock-ups tested on the FE200 unit. However, cracks with a length of 20–100 µm appeared on the inner wall of the water pipeline. These cracks were also present in unheated areas, which suggested that they might have been introduced during the manufacturing process.

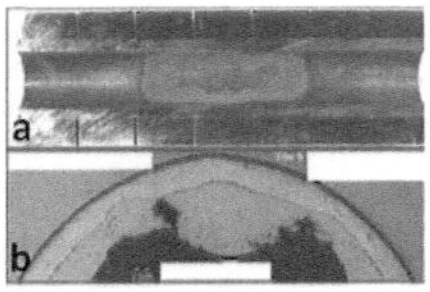

Figure 7.14: (a) Typical scale deposition and (b) deformation of the heat sink material [46].

To verify the critical thermal heat load resistance of the small tungsten-copper module, the critical heat flux (CHF) test was carried out on the module. During the test, the heat flux loaded on the sample was gradually increased until the sample failed. The CHF test was conducted on PSE three small modules (geometry 1 and armor thickness 7.5 mm) that were irradiated with the minimum heat load required by ITER, and the final critical heat load flux was about 28–30 MW/m^2. For the target plate tested on the FE200, there was no leakage or cracking in the minimum required test, but finally leakage occurred in the 616th heat load irradiation of 20 MW/m^2.

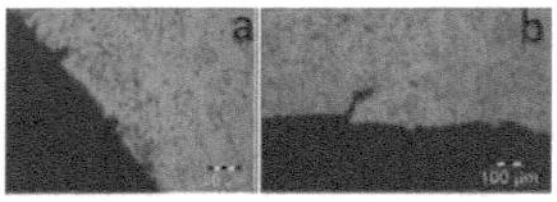

Figure 7.15: Cracks on the inner surface of the cooling water pipeline after the HHF test of the PSE module [46].

The ANN tungsten-copper modules had seven small modules and one vertical target prototypes completed 5,000 HHF tests at 10 MW/m^2 and 1,000 HHFs at 20 MW/m^2 without leakage. Five of the small modules and the vertical target plates were tested on the FE200 and two modules were tested on the IDTF equipment. With the similarity to the mock-ups produced by PSE, the surface of the ANN tungsten-copper modules appeared roughness and cracking; but in the end, there were two modules that did not appear surface cracking (Figure 7.16), which may result from the fact that tungsten material with better physical properties was used in these two modules. A destructive analysis of the ANN tungsten-copper modules showed that no cracks were found on the inner surface of the CuCrZr cooling water pipeline. Three of the modules were subjected to CHF tests after 300 heat load irradiation sessions at 20 MW/m^2, and the final heat load flux was about 30 MW/m^2.

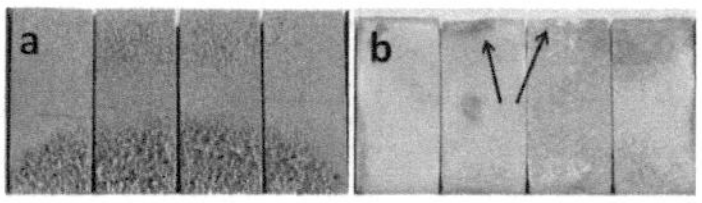

Figure 7.16: Surface morphology of the ANN tungsten-copper module (armor thickness of 6 mm) after 1,000 times of irradiation with a high heat load of 20 MW/m^2 wherein (a) tested on FE 200 and (b) IDTF [46].

7.3.3 China W/Cu-mono-block test evaluation

The EAST is the world's first fully superconducting Tokamak device developed by China. To withstand the impact of particles and energy under EAST long pulses and high-parameter plasma and to provide verification and accumulate experience for the design of ITER W/Cu divertor, the Institute of Plasma Physics of the Chinese Academy of Sciences started the EAST W/Cu divertor project in 2010. The plan was to realize the upper W/Cu divertor by the end of 2013 and the lower W/Cu divertor by the end of 2015, with the goal of gradually achieving the full-tungsten first wall structure [47]. Finally, in 2014, the upper divertor was converted into a full-tungsten divertor [48], as shown in Figure 7.17. Each EAST divertor module consists of a CB and three PFCs, namely an IT, an OT, and a Dome. The IT and OT were composed of four and five mono-block PFUs, respectively. The DOME consists of an upper plate and a lower plate, connected by EBW. Figure 7.18 shows a detailed diagram of the W/Cu PFCs of the EAST upper divertor [49].

Figure 7.17: Diagram of plasma materials and components of the EAST device after 2014 including full-tungsten upper divertor [49].

During the preparation and application process, high thermal stresses are naturally generated at the W/Cu interface, and these periodic alternating and superimposed thermal stresses can cause serious damage to W/Cu PFCs. Therefore, the preparation of W/Cu PFCs is highly challenging and represents one of the key technologies for the EAST W/Cu divertor engineering. The Institute of Plasma Physics of the Chinese Academy of Sciences developed three types of W/Cu PFCs, the first of which was a mono-block-type W/Cu PFCs with the similarity to the vertical target plate PFCs of ITER divertor, which will also be used in the area of the vertical target plate of the EAST divertor, and was designed to withstand a heat load of 7–10 MW/m^2. The second was flat-type W/Cu PFCs, which will be used in the EAST dome area and possibly in the first wall area other than the divertor, designed to withstand a heat load of 3–5 MW/m^2. The third was plasma-sprayed W/Cu PFCs, which will be used in the first wall area of the EAST in addition to the divertor and possibly in the divertor dome area, and also designed to withstand a heat load of 3–5 MW/m^2. The application distribution of the three types of W/Cu PFCs in the EAST device is shown in Figure 7.19 [50].

Hefei University of Technology provides tungsten-based powder materials with adjustable composition and scale, which are processed into larger tungsten blocks without cracking by Antai Technology Co., Ltd. The Institute of Plasma Physics of the Chinese Academy of Sciences and Antai Technology Co., Ltd. have developed the "HIP + HIP" welding process for tungsten and copper and for tungsten-copper blocks and copper alloys [51]. The process involves two stages of HIP: At 930 °C, 100 MPa for 2 h, a layer of OFC is applied as an intermediate layer on the inner wall of the hole of the W block. The W/Cu block and CuCrZr pipeline are then connected using a second HIP process at 600 °C and 100 MPa for 2 h. The photos of the W/Cu composite block and the SEM images of the W/Cu interface are shown in Figure 7.20 [13]. SEM shows that the W/Cu interface is well bonded, and there is no microcrack and

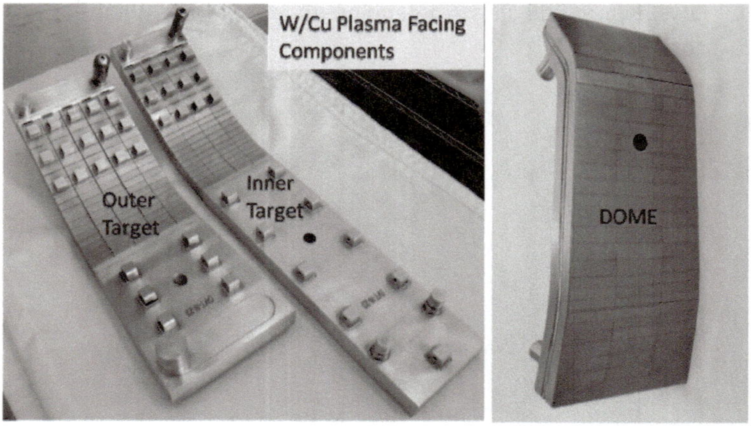

Figure 7.18: EAST W/Cu PFCs are composed of a single module, including nine mono-block type PFUs and three flat-type PFUs [49].

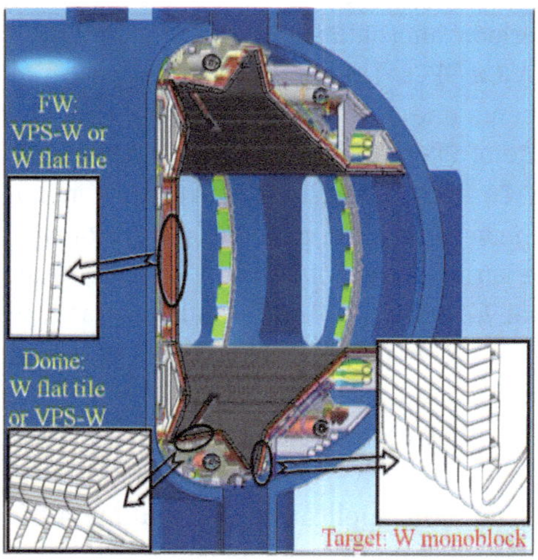

Figure 7.19: Distribution of three W/Cu PFCs in the EAST device [50].

interface separation. The interfacial bonding strength was measured by tensile tests upto ~150 MPa. In cooperation with Antai Technology Co., Ltd., the Institute of Plasma Physics has successfully developed a high-quality W/OFC block for mono-block-type W/Cu PFCs, which has reached the required level of ITER, and its physical properties are as follows: density ~19.2 g/cm^3, purity ~99.95%, hardness HV ~450, grain size ~50 μm (vertical rolling direction) × 200 μm (parallel rolling direction), tensile strength ~430 MPa (1,000 °C), and elongation (1,000 °C) >5%.

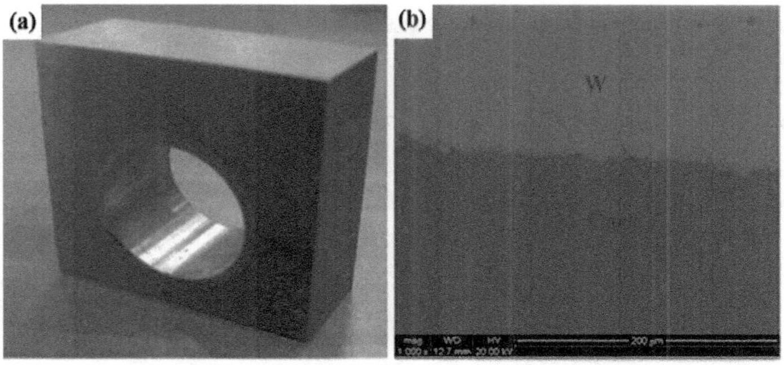

Figure 7.20: SEM photographs (b) of the interface between the W/Cu composite block (a) and the W/Cu made by HIP technology [13].

A W/Cu module containing five tungsten blocks was prepared using the “HIP + HRP” process [13], as shown in Figure 7.21(b). HRP is a method for preparing mono-block-type W/Cu PFCs developed by ENEA, Italy, which is a new technology combining diffusion welding and HIP. In cooperation with ENEA in Italy, the Institute of Plasma Physics has developed mono-block-type W/Cu PFCs for vertical target plate of EAST divertor using HRP technology. The W/Cu blocks were prepared by the HIP method and then connected with the OFC by HRP. The HRP parameters were as follows: temperature 600 °C, pressure 58 MPa, heat preservation, and pressure holding for 108 min. Ultrasonic nondestructive testing showed that the interface of Cu/CuCrZr was well bonded and no defects or cracks were found, as shown in Figure 7.21(a).

A high heat load testing was carried out at the Southwestern Institute of Physics (SWIP) of CNNC. Each tungsten block was 31 mm × 28 mm × 12 mm in size, with a 0.5 mm gap between the tungsten blocks. The equipment tested was the electron beam device EMS-60. The hydraulic parameters were as follows: the cooling water flow rate was 2 m/s, the inlet and outlet temperatures were 20 °C, and the pressure was 0.2 MPa. The heat load condition was 1,000, heat load irradiation of 8.4 MW/m^2, the acceleration voltage of the electron gun was 140 kV, and the electron beam current was 190 mA. The time for loading and unloading the heat load was 15 s each time. The temperature of the inlet and outlet of the cooling water was measured by two thermocouples, and the temperature of the module surface was measured by an IR camera and a high-temperature calorimeter. In addition, a CCD camera was used to monitor the test samples. The surface temperature of the module measured by the IR camera did not change during the evolution of the first and thousandth cycles, but the background temperature of the vacuum chamber increased, as shown in Figure 7.22. Figure 7.23 shows the photo of the module after the heat load shock, and there were no new defects on the module after ultrasonic nondestructive testing.

The Institute of Plasma Physics used the EBW-6MG electronic welding gun (the maximum power of electron gun is 6 kW, the acceleration voltage is 60 kV, and the vacuum degree of the vacuum chamber is 1×10^{-2} Pa) to irradiate medium-sized target

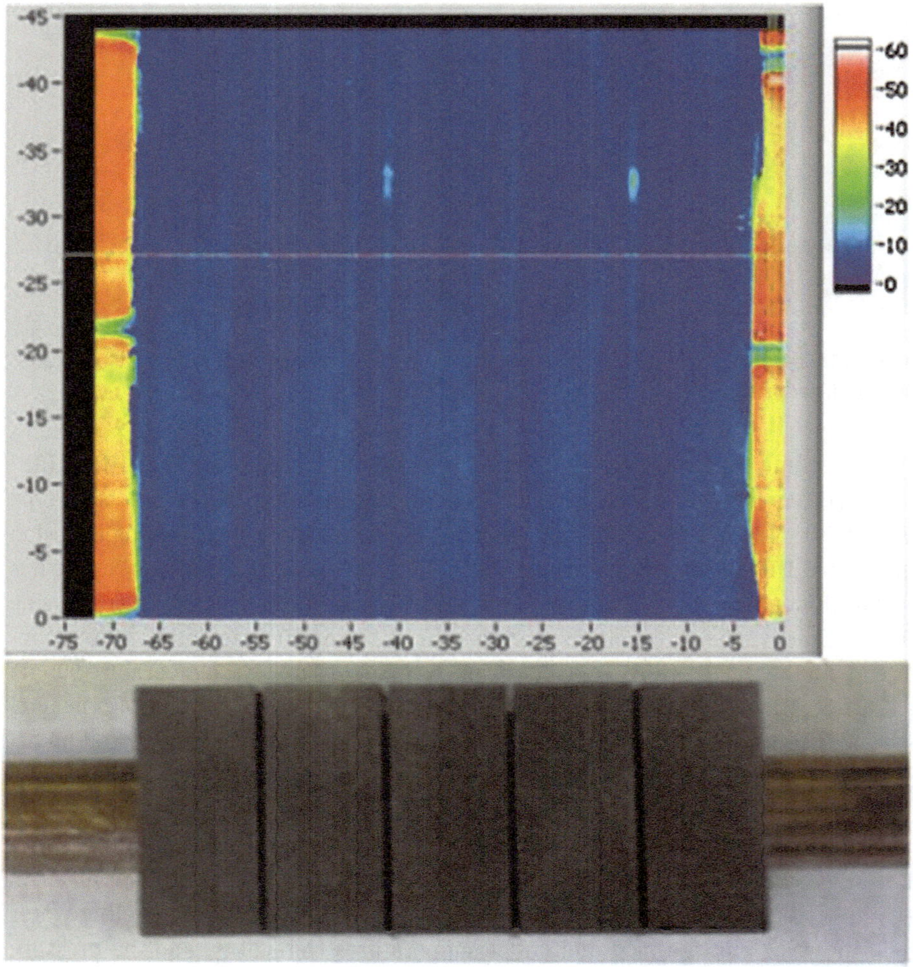

Figure 7.21: Ultrasonic nondestructive testing results of small-scale W/Cu module fabricated by HRP method and OFC/CuCrZr interface [13].

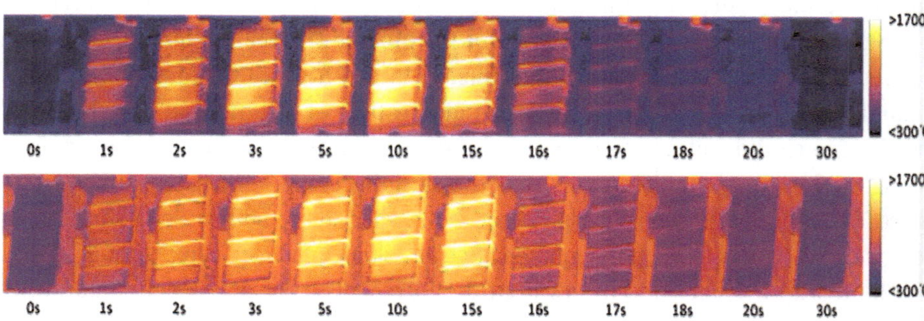

Figure 7.22: Temperature distribution under the first (a) and 1,000th (b) heat loads [13].

Figure 7.23: Surface morphology of the sample after the heat load test [13].

plate parts prepared by two steps HIP method with high heat load. The size of tungsten block was (28–31) mm × 28 mm × 12 mm, and the irradiated area was ~10 × 26 mm^2. To ensure a sufficiently long etching lifetime of the tungsten components, the distance from the top of the cooling water channel to the plasma-facing surface of tungsten block was 8 mm. The conditions of the test were the cooling water flow rate of 4 m/s and the inlet temperature of 25 °C. The electron beam irradiation power was 5.4 kW, the electron beam current was 90 mA, and the acceleration voltage was 60 kV. The absorption coefficient of tungsten surface to electron beam was generally 50%, that is, the heat load absorbed by tungsten surface was 10 MW/m^2, and the loading period was 15 s loading and 15 s unloading. The surface temperature of the module was observed in real time with an IR thermal imager. In the experiment, it was found that the highest surface temperature of the tungsten reached about 1,200 °C under the impact of the heat load, as shown in Figure 7.24. After 1,000 times of 10 MW/m^2 thermal irradiation, the maximum temperature of the sample surface did not change significantly, while the tungsten surface heated up rapidly during loading and cooled rapidly during unloading. Ultrasonic nondestructive testing was carried out on the parts before and after the experiment, and it was found that there were no new defects after irradiation, and the manufactured mono-block-type tungsten-copper parts could withstand the heat load of 10 MW/m^2.

In 2015, the Institute of Plasma Physics of the Chinese Academy of Sciences cooperated with the CEA-WEST team in France to provide the latter two pieces of mono-block-type tungsten-copper modules for high heat load irradiation tests. One of the modules was stopped due to water leakage caused by a failure of the cooling system during the experiment, and the other module was irradiated with 700 heat loads of 10 MW/m^2 on SWIP EMS-60, followed by 500 times of 10 MW/m^2 and 500 times of 20 MW/m^2 heat load irradiation on JUDITH-1, the specific irradiation conditions and irradiated surface are shown in Figure 7.25. After 500 and 300 times of 20 MW/m^2 heat load irradiation, a significant temperature gradient was observed on the tungsten side, likely due to recrystallization caused by high-surface temperatures. No macroscopic cracking of the tungsten was observed, and nondestructive testing revealed no

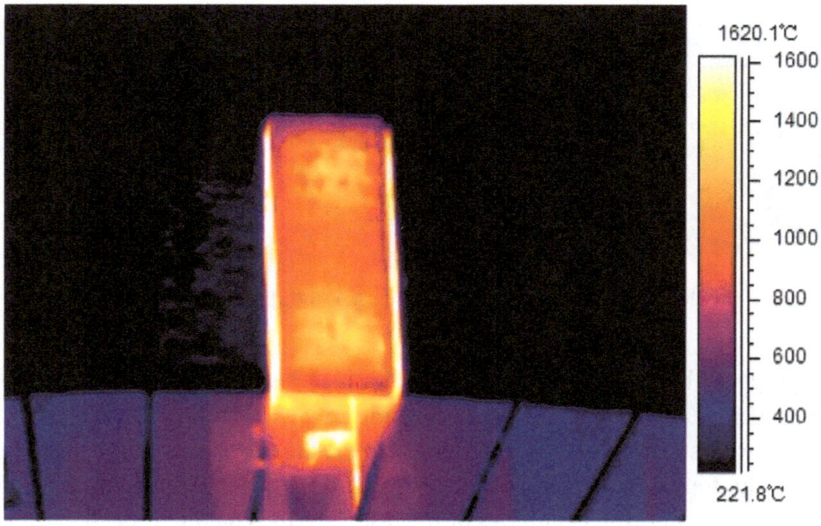

Figure 7.24: The temperature distribution of tungsten surface measured by an infrared camera [49].

significant defects. This shows that the mono-block-type tungsten-copper module can withstand 300 times of 20 MW/m^2 heat load irradiation.

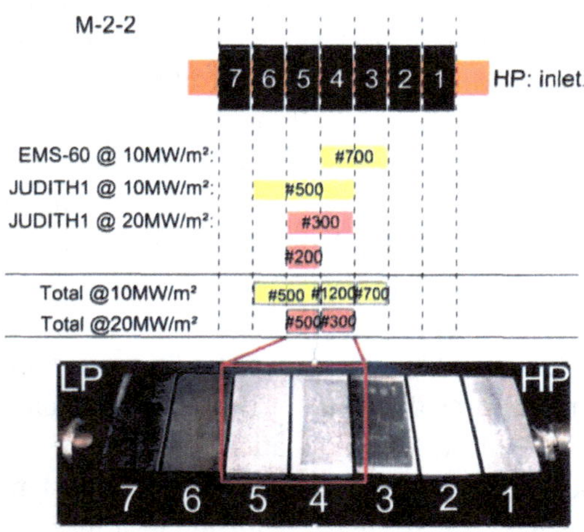

Figure 7.25: High heat load irradiation test for the CEA providing module [42].

In May 2015, the IO engineering department discussed with the Institute of Plasma Physics for several meetings, and it was finally decided that the Institute of Plasma Physics provide six pieces of W/Cu mono-block quality certification modules, and IO was responsible for the high heat load testing of the modules, and all the modules successfully completed the test project. Each module consisted of seven tungsten-copper blocks and one CuCrZr pipeline, with a 0.4 mm gap between the tungsten-copper blocks maintained by molybdenum spacer. A copper spoiler was inserted into the CuCrZr pipeline. To prevent the presence of impurities in the cooling water from affecting heat transfer, the cooling water used in the experiment is deionized water. The six pieces of small-scale W/Cu mono-blocks were irradiated with high heat loads on the IDTF equipment of JSC NIEFA in St. Petersburg, as shown in Figure 7.26. Four tungsten-copper blocks (from the second to the fifth) on each module were used for high heat load testing, with 5,000 heat load irradiation sessions of 10 MW/m^2, 300 heat load irradiation sessions of 20 MW/m^2, and an additional 700 heat load irradiation of 20 MW/m^2, and a thermal scan of 10 MW/m^2 of the module was performed in the test.

Figure 7.26: Test module mounted on sample table [49].

Although there are preliminary results in the manufacture and testing of tungsten divertor modules, there is still a long way to go before the service conditions of future fusion reactors are required. Hefei University of Technology, together with the Institute of Plasma Physics of the Chinese Academy of Sciences, the SWIP of China National Nuclear Corporation, and the Institute of Solid State Physics of the Chinese Academy of Sciences, are undertaking major national R&D projects to carry out research on the optimization and screening of tungsten-based materials, module design, manufacturing, evaluation and testing of deflectors, and strive to make new breakthroughs in the divertor module, the core component of fusion reactors.

7.4 Summary

The divertor is the core component of the magnetic confinement fusion reactor experimental device, which is in the transition region where the high-temperature plasma is in direct contact with the material. It has developed from the water-cooled type to the helium-cooled type, and the corresponding divertor materials are also constantly changing and developing. With excellent properties such as high melting point, high thermal conductivity, and low tritium retention, tungsten has been selected as the PFM for the divertor of ITER. Copper alloys, with their high thermal conductivity, strength, good thermal stability, and resistance to neutron irradiation activation, are key candidates for heat-sink materials of divertors for modern Tokamak devices and ITER. The heat flux of the divertor can reach 5–20 MW/m^2, requiring forced cooling of the PMF. As a complete divertor component, the connection between the PFC, the PFM, and the heat sink material, as well as the heat sink material and the structural material, presents significant technical challenges. During the process of connection and service, the component will generate high thermal stress. Therefore, choosing an appropriate connection technology and process to reduce the thermal stress at the connection interface is conducive to improving the reliability and stability of the divertor. All kinds of connection technologies are being explored and continuously improved.

The design and manufacturing concepts of W/Cu modules and PFC units have been basically agreed, with substantial progress made by Japan, the European Union, and China. However, there is still ample room for further development, including optimizing materials and properties, and making breakthroughs in manufacturing processes for the integration of PFMs, heat sink materials, and structural components.

References

[1] J. Wenbao, H. Daqian, Q. Zhongyou. Analysis of the temperature field and thermal structure of the first wall of the experimental cladding module of fusion reactor [J]. Atomic Energy Science and Technology, 2010, 4: 340–345.

[2] A.R. Raffray, R. Nygren, D.G. Whyte, S. Abdel-Khalik, R. Doerner, F. Escourbiac, T. Evans, R.J. Goldston, D.T. Hoelzer, S. Konishi, P. Lorenzetto, M. Merola, R. Neu, P. Norajitra, R.A. Pitts, M. Rieth, M. Roedig, T. Rognlien, S. Suzuki, M.S. Tillack, C. Wong. High heat flux components-readiness to proceed from near term fusion systems to power plants [J]. Fusion Engineering and Design, 2010, 85(1): 93–108.

[3] M.S. Tillack, A.R. Raffray, X.R. Wang, S. Malang, S. Abdel-Khalik, M. Yoda, D. Youchison. Recent US activities on advanced He-cooled W-alloy divertor concepts for fusion power plants [J]. Fusion Engineering and Design, 2011, 86(1): 71–98.

[4] P. Norajitra, L.V. Boccaccini, A. Gervash, R. Giniyatulin, N. Holstein, T. Ihli, G. Janeschitz, W. Krauss, R. Kruessmann, V. Kuznetsov, A. Makhankov, I. Mazul, A. Moeslang, I. Ovchinnikov, M. Rieth, B. Zeep. Development of a helium-cooled divertor: Material choice and technological studies [J]. Journal of Nuclear Materials, 2007, 367–370: 1416–1421.

[5] Institute of Plasma Physics, Chinese Academy of Sciences. EAST Big Science Project [Z]. http://east.ipp.ac.cn.

[6] G. Pintsuk. 4.18-Tungsten as a fusion plasma facing material [J]. Comprehensive Nuclear Materials, 2012, 4: 551–581.

[7] T. Hirai, S. Panayotis, V. Barabash, C. Amzallag, F. Escourbiac, A. Durocher, M. Merola, J. Linke, T. Loewenhoff, G. Pintsuk, M. Wirtz, I. Uytdenhouwen. Use of tungsten material for the ITER divertor[J]. Nuclear Materials and Energy, 2016, 9: 616–622.

[8] E. Visca, E. Cacciotti, S. Libera, A. Mancini, A. Pizzuto, S. Roccella, B. Riccardi, F. Escourbiac, G.P. Sanguinetti. Manufacturing and testing of reference samples for the definition of acceptance criteria for the ITER divertor [J]. Fusion Engineering and Design, 2010, 85(10–12): 1986–1991.

[9] N. Litunovsky, E. Alekseenko, A. Makhankov, I. Mazul. Development of the armoring technique for ITER Divertor Dome [J]. Fusion Engineering and Design, 2011, 86(9–11): 1749–1752.

[10] I.V. Mazul, V.A. Belyakov, R.N. Giniatulin, A.A. Gervash, V.E. Kuznetsov, A.N. Makhankov, V.S. Sizenev. Preparation to manufacturing of ITER plasma facing components in Russia [J]. Fusion Engineering and Design, 2011, 86(6–8): 576–579.

[11] T. Hirai, V. Barabash, F. Escourbiac, A. Durocher, L. Ferrand, V. Komarov, M. Merola. ITER divertor materials and manufacturing challenges [J]. Fusion Engineering and Design, 2017, 125: 250–255.

[12] J.H. Song, K.M. Kim, S.H. Hong, H.T. Kim, S.H. Park, H.K. Park, H.J. Ahn, S.K. Kim, D.W. Lee. High heat flux test of tungsten brazed mock-ups developed for KSTA divertor [J]. Fusion Engineering and Design, 2016, 109–111: 576–579.

[13] Q. Li, S. Qin, W. Wang, P. Qi, S. Roccella, E. Visca, G. Liu, G. Luo. Manufacturing and testing of W/Cu mono-block small scale mock-up for EAST by HIP and HRP technologies [J]. Fusion Engineering and Design, 2013, 88: 1808–1812.

[14] K. Sugiyama, M. Mayer, V. Rohde, M. Balden, T. Dürbeck, A. Herrmann, S. Lindig, A. Wiltner, H.W. Müller, R. Neu, The ASDEX Upgrade Team. Deuterium inventory in the full tungsten divertor of ASDEX Upgrade [J]. Nuclear Fusion, 2010, 50(3): 035001.

[15] T. Hirai, F. Escourbiac, S. Carpentier-Chouchana, A. Fedosov, L. Ferrand, T. Jokinen, V. Komarov, A. Kukushkin, M. Merola, R. Mitteau, R.A. Pitts, W. Shu, M. Sugihara, B. Riccardi, S. Suzuki, R. Villari. ITER tungsten divertor design development and qualification program [J]. Fusion Engineering and Design, 2013, 88: 1798–1801.

[16] T. Hirai, F. Escourbiac, V. Barabash, A. Durocher, A. Fedosov, L. Ferrand, T. Jokinen, V. Komarov, M. Merola, S. Carpentier-Chouchana, N. Arkhipov, V. Kuznetcov, A. Volodin, S. Suzuki, K. Ezato, Y. Seki, B. Riccardi, M. Bednarek, P. Gavila. Status of technology R&D for the ITER tungsten divertor monoblock [J]. Journal of Nuclear Materials, 2015, 463: 1248–1251.

[17] V.R. Barabash, G.M. Kalinin, S.A. Fabritsiev, S.J. Zinkle. Specification of CuCrZr alloy properties after various thermo-mechanical treatments and design allowables including neutron irradiation effects [J]. Journal of Nuclear Materials, 2011, 417(1–3): 904–907.

[18] P. Norajitra, S.I. Abdel-Khalik, L.M. Giancarli, T. Ihli, G. Janeschitz, S. Malang, I.V. Mazul, P. Sardain. Divertor conceptual designs for a fusion power plant [J]. Fusion Engineering and Design, 2008, 83 (7–9): 893–902.

[19] P. Norajitra, L.V. Boccaccini, A. Gervash, R. Giniyatulin, N. Holstein, T. Ihli, G. Janeschitz, W. Krauss, R. Kruessmann, V. Kuznetsov, A. Makhankov, I. Mazul, A. Moeslang, I. Ovchinnikov, M. Rieth, B. Zeep. Development of a helium-cooled divertor: Material choice and technological studies [J]. Journal of Nuclear Materials, 2007, 367–370: 1416–1421.

[20] T. Chehtov, J. Aktaa, O. Kraft. Mechanical characterization and modeling of brazed EUROFER-tungsten-joints [J]. Journal of Nuclear Materials, 2007, 367–370: 1228–1232.

[21] Q. Huang, C. Li, Y. Li, M. Chen, M. Zhang, L. Peng, Z. Zhu, Y. Song, S. Gao. Progress in development of China Low Activation Martensitic steel for fusion application [J]. Journal of Nuclear Materials, 2007, 367–370: 142–146.

[22] S. Shuxiang. Preparation and performance evaluation of tungsten-based coatings for components with high heat load in fusion reactors [D]. Ph.D. degree. Beijing: University of Science and Technology Beijing, 2007.

[23] M. Roedig, W. Kuehnlein, J. Linke, M. Merola, E. Rigal, B. Schedler, E. Visca. Investigation of tungsten alloys as plasma facing materials for the ITER divertor [J]. Fusion Engineering and Design, 2002, 61–62: 135–140.

[24] I. Smid, M. Akiba, G. Vieider, L. Plöchl. Development of tungsten armor and bonding to copper for plasma-interactive components [J]. Journal of Nuclear Materials, 1998, 258–263: 160–172.
[25] M. Onozuka, S. Hirai, K. Kikuchi, Y. Oda, K. Shimizu. Manufacturing study of Be, W and CFC bonded structures for plasma-facing components [J]. Journal of Nuclear Materials, 2004, 329–333: 1553–1557.
[26] V. Barabash, M. Akiba, A. Cardella, I. Mazul, B.C. Odegard Jr, L. Plöchl, R. Tivey, G. Vieider. Armor and heat sink materials joining technologies development for ITER plasma facing components [J]. Journal of Nuclear Materials, 2000, 283–287: 1248–1252.
[27] S.V. Maksimova, D.V.F. Khorunov, V.R. Barabash. Problems in fabricating sections of a divertor of a thermonuclear synthesis reactor by brazing [J]. Welding International, 1995, 9(1): 47–50.
[28] G. Shuangquan, F. Yunbiao, Y. Qingzhi, L. Jian. Research progress on connection technology between tungsten and dissimilar materials in deflector [J]. Welding Technology, 2010, 39(9): 3–7.
[29] S. Saito, K. Fukaya, S. Ishiyama, K. Sato. Mechanical properties of HIP bonded W and Cu-alloys joint for plasma facing components [J]. Journal of Nuclear Materials, 2002, 307–311: 1542–1546.
[30] J.-E. Döring, R. Vaßen, G. Pintsuk, D. Stöver. The processing of vacuum plasma-sprayed tungsten–copper composite coatings for high heat flux components [J]. Fusion Engineering and Design, 2003, 66–68: 259–263.
[31] J.C. Mana, J. Li. Study on the connection and interfacial stress of W/Cu material on the first wall of Tokamak[J]. Rare Metals & Cemented Carbide, 2005, 33(4): 38–42.
[32] A.R. Raffray, L. El-Guebaly, T. Ihli, S. Malang, X. Wang. Engineering design and analysis of the ARIES-CS power plant [J]. Fusion Science and Technology, 2008, 54: 725–746.
[33] P. Norajitra, R. Giniyatulin, T. Ihli, W. Krauss, V. Kuznetsov, I. Mazul, I. Ovchinnikov, J. Reiser, G. Ritz, H.-J. Ritzhaupt-Kleissl, V. Widak. Current status of He-cooled divertor development for DEMO [J]. Fusion Engineering and Design, 2009, 84(7–11): 1429–1433.
[34] T. Chehtov, J. Aktaa, O. Kraft. Mechanical characterization and modeling of brazed EUROFER-tungsten-joints [J]. Journal of Nuclear Materials, 2007, 367–370: 1228–1232.
[35] L. Wensheng, L. Shuhua, M. Yunzhu, C. Qingshan, L. Haoyang, Y. Qiang, W. Lei. Microstructure and properties of tungsten/steel vacuum-welded joints based on nickel-based microcrystalline brazing metal [J]. Chinese Journal of Nonferrous Metals, 2014, 24(12): 3051–3058.
[36] J. de Prado, M. Sánchez, A. Ureña. Evaluation of mechanically alloyed Cu-based powders as filler alloy for brazing tungsten to a reduced activation ferritic-martensitic steel [J]. Journal of Nuclear Materials, 2017, 490: 188–196.
[37] B.A. Kalin, V.T. Fedotov, O.N. Sevrjukov, A. Moeslang, M. Rohde. Development of rapidly quenched brazing foils to join tungsten alloys with ferritic steel [J]. Journal of Nuclear Materials, 2004, 329–333: 1544–1548.
[38] B.A. Kalin, V.T. Fedotov, O.N. Sevrjukov, A.N. Kalashnikov, A.N. Suchkov, A. Moeslang, M. Rohde. Development of brazing foils to join monocrystalline tungsten alloys with ODS-EUROFER steel [J]. Journal of Nuclear Materials, 2007, 367–370: 1218–1222.
[39] N. Oono, S. Noh, N. Iwata, T. Nagasaka, R. Kasada, A. Kimura. Microstructures of brazed and solid-state diffusion bonded joints of tungsten with oxide dispersion strengthened steel [J]. Journal of Nuclear Materials, 2011, 417(1–3): 253–256.
[40] W.W. Basuki, J. Aktaa. Investigation on the diffusion bonding of tungsten and EUROFER97 [J]. Journal of Nuclear Materials, 2011, 417(1–3): 524–527.
[41] D. Ye, D. Minghui, N. Ben, C. Dajun, L. Zhongsheng. Research progress on steel/tungsten diffusion bonding technology [J]. Journal of Ordnance Equipment Engineering, 2019, 40(5): 205–209.
[42] S. Zhaoxuan. High thermal load test and finite element simulation of ITER tungsten copper tube feedthrough module [D]. Master's Degree, Hefei: University of Science and Technology of China, 2017.

[43] T. Hirai, F. Escourbiac, S. Carpentier-Chouchana, A. Durocher, A. Fedosov, L. Ferrand, T. Jokinen, V. Komarov, M. Merola, R. Mitteau, R.A. Pitts, W. Shu, M. Sugihara, V. Barabash, V. Kuznetsov, B. Riccardi, S. Suzuki. ITER full tungsten divertor qualification program and progress [J]. Physica Scripta, 2014, T159: 014006.

[44] K. Ezato, S. Suzuki, Y. Seki, K. Mohri, K. Yokoyama, F. Escourbiac, T. Hirai, V. Kuznetcov. Progress of ITER full tungsten divertor technology qualification in Japan [J]. Fusion Engineering and Design, 2015, 98–99: 1281–1284.

[45] K. Ezato, S. Suzuki, Y. Seki, H. Yamada, T. Hirayama, K. Yokoyama, F. Escourbiac, T. Hirai. Progress of ITER full tungsten divertor technology qualification in Japan: Manufacturing full-scale plasma-facing unit prototypes [J]. Fusion Engineering and Design, 2016, 109–111: 1256–1260.

[46] P. Gavila, B. Riccardi, G. Pintsuk, G. Ritz, V. Kuznetsov, A. Durocher. High heat flux testing of EU tungsten monoblock mock-ups for the ITER divertor [J]. Fusion Engineering and Design, 2015, 98–99: 1305–1309.

[47] G.-N. Luo, Q. Li, J.M. Chen, X. Liu, Z.J. Zhou, D.M. Yao. Overview of R & D on plasma-facing materials and components in China [J]. Fusion Science and Technology, 2012, 62: 9–15.

[48] L. Cao, Z. Zhou, D. Yao. EAST full tungsten divertor design [J]. Journal of Fusion Energy, 2015, 34(6): 1451–1456.

[49] G.-N. Luo, G.H. Liu, Q. Li, S.G. Qin, W.J. Wang, Y.L. Shi, C.Y. Xie, Z.M. Chen, M. Missirlian, D. Guilhem, M. Richou, T. Hirai, F. Escourbiac, D.M. Yao, J.L. Chen, T.J. Wang, J. Bucalossi, M. Merola, J.G. Li, EAST Team. Overview of decade-long development of plasma-facing components at ASIPP [J]. Nuclear Fusion, 2017, 57: 065001.

[50] Q. Li. Research on the material design and component preparation of the EAST tungsten-copper divertor element [D]. Doctor's Degree, Beijing: University of Chinese Academy of Sciences, 2012.

[51] G.H. Liu, Q. Li, R. Wei. The manufacture of W/Cu mono-block PFCs for fusion device and its irradiation behavior [J]. Nuclear f usion a nd p lasma physics, 2015, 35(1): 75–78.

Index

alloyed tungsten
– W-K alloy 50
– W-Re alloy 31
– W-Ta alloy 39
– W-Ti alloy 46
– W-V alloy 44
– W-Y alloy 35
– W-Zr alloy 48
application of tungsten in China's EAST
– EAST device 27
– PFM 26
– W/Cu divertor 27
application of tungsten in international fusion devices
– ASDEX 25
– CMSII 25
– ITER reactors 25
– VPS 25
applications and challenges of tungsten materials
– neutron irradiation 28
– two divertor design options 28

brittleness of tungsten
– low-temperature brittleness 123
– neutron irradiation effects 124
– recrystallization brittleness 123

connection between the tungsten divertor and heat sink
– brazing connection 217
– casting connection 217
– electron beam welding 216
– hot isostatic pressing connection 217
– vacuum brazing 219
– vacuum diffusion welding 220
– vacuum plasma spraying 218

damage forms of PFM materials
– irradiation damage 11
– sputtering corrosion and hot corrosion 11
design of the divertor
– cassette bodies 210
– CuCrZr alloy heat sink 211
– flat-type 211
– ITER divertor 209
– mono-block 211
– plasma-facing units 210
development of fusion energy
– DEMO 5
– ITER program 4
– ITER structural device 5
development of fusion energy in China
– ASIPP 9
– ITER construction 10
– Southwestern Institute of Physics 9

forming and processing of tungsten materials
– equal-channel angular pressing 114
– high-pressure torsion 115
– hydrostatic extrusion 113
– multidirectional forging 115
– rolling processing 111
– rotary forging 113
fundamentals of fusion
– fusion reaction 1

high heat load testing of mono-block tungsten-copper module at home and abroad
– China W/Cu-mono-block test evaluation 230
– evaluation of W/Cu-monoblock test in Europe 226
– Japan W/Cu-mono-block test evaluation 222

international fusion reactor device
– European JET device 8
– JT-60 device in Japan 8
– TFTR and DIII-D installations in the United States 8
– tokamak device in Russia 9
intrinsic cause of tungsten brittleness
– chemical bonds 125
– cleavage fracture 128
– dislocation mobility 125
– grain boundary characteristics 127
– internal defects in actual materials 126
irradiation damage behavior of tungsten materials
– heavy ion 157
– helium ion 144
– hydrogen isotope 135
– neutron 151
ITER full-tungsten divertor certification test
– high heat flux 221

 | https://doi.org/10.1515/9783111556987-008

oxide dispersion-strengthened tungsten
- W-La_2O_3 56
- W-ThO_2 65
- W-Y_2O_3 59

plasma-facing materials (PFMs)
- damage forms of PFM materials 11
- plasma-material interactions 10
- selection of PFMs 13
properties comparison of candidate PFMs
- beryllium 16
- carbon-based materials and composite materials 19
- tungsten and tungsten-matrix materials 19

selection of materials for divertor components
- heat sink materials 212
- plasma-facing materials 211
- structural materials 214
sintering densification of tungsten alloy
- conventional sintering 104
- powder activation pretreatment 108
- special sintering technologies 105

transient thermal load damage behavior of commercial tungsten
- DBTT 177
- electron beam thermal shock 177
- heat flux-induced cracking 175
- laser thermal shock 179
- laser/high magnetic field synergistic thermal shock 187
- plasma thermal shock 183
- plasma/electron beam synergistic thermal shock 186
transient thermal load damage behavior of dispersion-strengthened tungsten composites
- carbide dispersion-strengthened tungsten-based materials 201
- oxide dispersion-strengthened tungsten-based materials 198
transient thermal load damage behavior of tungsten alloy
- W-K alloys 192
- W-Ta alloy 195
- W-V alloys 197
transient thermal load damage behavior of ultrafine-grained tungsten
- different grain sizes 190
- impurities 190
transient thermal load experimental simulation device
- elm-like transient thermal shocks 172
- JUDITH-1 and JUDITH-2 electron beam devices 174
tungsten composite powders
- bottom-up technology 100
- condensing drying method 98
- mechanical alloying 98
- top-down technology 99
- vacuum plasma spray deposition 99
tungsten-carbide composites
- W-HfC 83
- W-TaC 80
- W-TiC 66
- W-ZrC 74
tungsten-matrix materials
- brittleness 30
- CFETR 30
- high heat load 30
types of nuclear fusion
- inertial confinement fusion 3
- magnetic confinement fusion 4

ultrafine-grained/nanocrystalline tungsten
- ductility and irradiation resistance 85
- ECAP 85
- HPT 85
- severe plastic deformation method 85

ways to solve tungsten's brittleness
- addition of alloying elements 128
- second phase doping 129
- severe plastic deformation 130